# *Spin Orbitronics*
*and*
# *Topological Properties*
*of Nanostructures*

*ecture Notes of*
*e 12th*
*ternational*
*chool on*
*heoretical*
*hysics*

# Spin Orbitronics and Topological Properties of Nanostructures

***Symmetry and Structural Properties of Condensed Matter***

***Rzeszów, Poland***
***5 – 10 September 2016***


*Editors*

**Vitalii Dugaev**

**gor Tralle**

**Andrzej Wal**

**Józef Barnaś**

W JERSEY • LONDON • SINGAPORE • BEIJING • SHANGHAI • HONG KONG • TAIPEI • CHENNAI • TOKYO

*Published by*

World Scientific Publishing Co. Pte. Ltd.
5 Toh Tuck Link, Singapore 596224
*USA office:* 27 Warren Street, Suite 401-402, Hackensack, NJ 07601
*UK office:* 57 Shelton Street, Covent Garden, London WC2H 9HE

**British Library Cataloguing-in-Publication Data**
A catalogue record for this book is available from the British Library.

**SPIN ORBITRONICS AND TOPOLOGICAL PROPERTIES OF NANOSTRUCTURES**
**Lecture Notes of the Twelfth International School on Theoretical Physics**



ISBN 978-981-3234-33-8

For any available supplementary material, please visit
http://www.worldscientific.com/worldscibooks/10.1142/10830#t=suppl

Printed in Singapore

# Preface

This book includes some of the lectures presented at the 12th International School on Theoretical Physics "Symmetry and Structural Properties of Condensed Matter SSPCM 2016", which took place in Rzeszów (Poland) in September 2016. The SSPCM school is a traditional summer school, which is mostly focused on new ideas and modern methods in theoretical condensed matter physics. This is also a school for young scientists starting their carrier in physics as well as for graduate students. At the SSPCM schools they can meet well-known physicists from the United States, Japan, Russia, and European countries. The young researches can also present and discuss their own results during the School. Although the School is devoted mainly to theoretical physics, we traditionally invite lecturers working in computer physics and experiment as well. This demonstrates an important connection between the theory and experiment, and also gives young scientists a good chance to choose their own place in the modern physics.

Obviously, the main topics of the presented lecture notes reflect the recent interest of physicists. This includes the so-called spin-related phenomena in low-dimensional systems and nanostructures, spin dynamics, new materials and new states of matter, as well as some new methods based on topology and/or symmetry. All these topics appeared in the condensed matter physics in recent several years, and also paved the way to very interesting practical applications. The importance of spin-orbit interaction, which was the main element in many lectures, justifies the title of this book "Spin orbitronics", which became recently very popular in the condensed matter community, similarly like spintronics or spin caloritronics. It should be also emphasized that there are some expectations that these areas of activity will lead to a substantial advancement of modern nanotechnology as well as to the development of new quantum devices.

It is worth mentioning that the idea of the school, which is not focused on a single topic, makes the school in Rzeszów different from the other conferences and workshops. The main reason to organize such meetings is that good mathematical methods and novel physical ideas can be applied in different areas of physics. This is especially important for theoretical physics, because very often a breakthrough in the theory opens a way to solve physical problems in the areas which are far away or completely different from the initial ones and for which the theory was initially developed.

The first School on theoretical physics was held in Zajączkowo (Poland) in 1990. The first organizer and the leader of first ten schools was professor Tadeusz Lulek. His lecture on the Bethe ansatz method and integrability, delivered at the school in 2016, is also included in this book.

We hope that the book will be helpful for those who start their own research in modern condensed matter physics and also want to be familiar with new mathematical methods and physical ideas.

*Vitalii Dugaev, Igor Tralle, Andrzej Wal and Józef Barnaś*

# Contents

Preface v

Quantum Heat Engines with Multiferroic Working Substance 1
*L. Chotorlishvili, M. Azimi, S. Stagraczyński and J. Berakdar*

Proximity Spin-orbit Coupling Physics of Graphene in Transition-metal Dichalcogenides 18
*M. Gmitra, D. Kochan, P. Högl and J. Fabian*

Bethe Ansatz and Integrability 40
*T. Lulek*

Plasmon-photon Coupling in High-electron-mobility Heterostructures: Tutorial on Magnetoplasmon Spectroscopy 56
*M. Białek, I. Grigelionis, K. Nogajewski and J. Łusakowski*

Spin-charge Coupling Effects in a Two-dimensional Electron Gas 80
*R. Raimondi*

Dynamics of Quenched Topological Edge Modes 110
*P. D. Sacramento*

Effect of Confinement on Melting in Nanopores 140
*M. Sliwinska-Bartkowiak, M. Jazdzewska, K. Domin and K. E. Gubbins*

3D Topological Dirac Semimetal Based on the HgCdTe 156
*G. Tomaka, J. Grendysa, M. Marchewka, P. Śliż, C. R. Becker, D. Żak, A. Stadler and E.M. Sheregii*

Effects of Anomalous Velocity in Spin-orbit Coupled Systems 178
*Sh. Mardonov, M. Modugno and E. Ya. Sherman*

Fermion Condensation in Strongly Interacting Fermi Liquids 200
*E. V. Kirichenko and V. A. Stephanovich*

Theory of Electron Transport and Magnetization Dynamics in Metallic Ferromagnets 227
*G. Tatara*

Introduction to the Topic of Jack Polynomials in the Context of Fractional Quantum Hall Effect 258
*B. Kuśmierz and A. Wójs*

Zitterbewegung (Trembling Motion) of Electrons in Narrow Gap Semiconductors 286
*W. Zawadzki and T. M. Rusin*

# Quantum Heat Engines with Multiferroic Working Substance*

L. Chotorlishvili[†], M. Azimi, S. Stagraczyński and J. Berakdar
*Institut für Physik, Martin-Luther-Universität Halle-Wittenberg*
*06099 Halle, Germany*
[†]*E-mail: levan.chotorlishvili@physik.uni-halle.de*

The work provides an overview on some recent advances in the area of quantum thermodynamics and quantum heat engines. A particular emphasis is put on the possibility of constructing finite time quantum cycles and adiabatic shortcuts. We discuss in details the particular quantum heat engines operating with a multiferroic working substance.

*Keywords*: Quantum thermodynamics, quantum heat engines, multiferroics, adiabatic shortcuts, quantum dissipative systems, Bochkov-Kuzovlev and Jarzynski equalities.

## 1. Introduction

For any thermal heat engines and thermodynamic cycles the three main criteria are central: The produced maximal work, the efficiency, and the output power of the engine. The high efficiency of the heat engine is important for performing operations with low energy consumption, while the amount of the produced work and the power of the heat machine are crucial for swift performance. Is it realistic to meet all those three criteria simultaneously? The present work addresses this question and provides a brief overview on the current status of knowledge.

To introduce the problem and notations we start by recalling the basics of the classical thermodynamics.[1] For a thermally isolated system consisting of two parts which are not in thermal equilibrium, a certain work is performed on the surrounding bodies during the transition to the equilibrium state. We exclude the work associated with the general expansion, assuming that the total volume of the system is conserved. Then the produced work is a function of the internal energy of the system $|W| = U_0 - U(S)$ where $U_0$ is the initial energy. Since the transition to the equilibrium state

*This work is supported by the DFG through SFB 762.

might occur in a variety of ways the final energy and the entropy $U(S)$ might be different. Considering the produced work as a functional of the system's entropy we write $\delta_S|W| = -\big(\partial U(S)/\partial S\big)_{V=const} = -T$. As $\delta_S|W|$ is always negative (we use the absolute temperature scale $T > 0$), any increment in the entropy while producing the work lowers work. Hence, the maximal work $|W|_{max}$ is produced during the process when the entropy of the system stays constant $S = const$. More rigorous deliberation leads to the following formula for the maximal work $|W|_{max} = -\delta\big(U - T_0 S + P_0 V\big)$. Here $T_0, P_0$ stand for the temperature and the pressure of the environment, while $U, S, V$ define the internal energy, the entropy, and the volume of the working body. If the volume and the temperature are constant during the process, the produced maximal work is equal to the change in the free energy $|W|_{max} = -\delta F$. Note, the adiabaticity of the process excludes any direct energy exchange between the bodies with the different temperatures.

The ideal heat engine envisioned by Carnot has four strokes: The working substance at temperature $T_h$ absorbs isothermally energy from the hot heat bath, and then is cooled down adiabatically to the temperature $T_l$. Thereafter, the working substance releases isothermally energy to a cold heat bath at $T_l$, and eventually returns adiabatically to the initial state. Two thermal baths with temperatures $T_h > T_c$ are needed to perform a reversible cycle. The existence of two heat baths allows excluding a direct irreversible energy exchange between the systems with different temperatures. The efficiency and the produced work of the ideal heat engine read: $\eta = \frac{T_h - T_l}{T_h}$, $W_{max} = \frac{T_h - T_l}{T_h} Q_{in}$, where $Q_{in}$ is the heat absorbed from the hot bath. The adiabaticity imposes certain restrictions on the cycle's swiftness. An ideal cycle takes an infinite time $\tau \mapsto \infty$, and therefore the output power of the ideal Carnot cycle vanishes $P = W_{max}/\tau = 0$. So it is of relevance to find ways yielding a good cycle efficiency with a reasonable power. This issue is not only a technical but also it is a conceptual one, even for classical systems. For quantum heat engines additional aspects are important. When the size of the working medium is scaled down to the mesoscopic scale purely quantum effects such as quantum fluctuations and interlevel transitions become important. The problem of the thermally assisted interlevel transitions can be solved relatively easily by detaching the working body from the heat bath. Quantum adiabaticity is more subtle. To be precise we specify the concept of adiabaticity separately for quantum and classical systems. The stroke of the cycle, which is adiabatic in the sense of classical thermodynamics may be nonadiabatic for a quantum working substance. The reason lies in quantum interlevel transitions that naturally occur in the case of fast driving and a finite time thermodynamic cycles. Thus, quantum adiabaticity implies not only a decoupling of the system from the thermal source, but also requires an elimination of interlevel transitions that are of a pure quantum nature. Shortcuts to quantum adiabaticity is a recent theoretical concept that allows eliminating the effect of those interlevel transitions.[2–7,9,10] In what follows we provide a brief introduction to shortcuts to quantum adiabaticity. Before that we discuss the efficiency of the Carnot engine at maximum output power for a classical system.[11]

## 2. Efficiency of the Carnot engine at a maximum output power

An ideal Carnot engine assumes that during isothermal strokes the working substance is in equilibrium with the thermal reservoirs, meaning that the isothermal strokes are performed infinitely slowly. Therefore, the power output of the engine is zero, since the finite work is produced in an infinite time. Ref. 11 assumes that during the isothermal expansion the heat flux through the vessel enfolding the working medium is proportional to the temperature gradient across the vessel. Therefore, the expressions for the absorbed heat and the heat rejected to the heat sink read $Q_{in} = \alpha t_1(T_h - T_{hw}), Q_{out} = \beta t_1(T_{lw} - T_l)$. Here $\alpha, \beta$ are constants, $t_1, t_2$ are the durations of the isothermal strokes, and $T_{hw}, T_{lw}$ are the temperatures of the working medium during the isothermal strokes. The key issue in this assumption is that the duration of the isothermal strokes are finite. However, the cycle is reversible and the total entropy production is zero $Q_{in}/T_{hw} = Q_{out}/T_{lw}$. The output power of the cycle is $P = (Q_{in} - Q_{out})/(t_1 + t_2)$ and the total time spent for the two adiabatic strokes is $(\gamma - 1)(t_1 + t_2)$. With this expressions one can maximize the power output $\partial P/\partial x = 0, \partial P/\partial y = 0$, where $x = T_h - T_{hw}, y = T_{lw} - T_l$. After a little algebra for the cycle efficiency and the maximum output power one obtains: $\acute{\eta} = 1 - (T_l/T_h)^{1/2}$. Note a finite time cycle needed for the maximal power output comes at the cost of a lower efficiency $\acute{\eta} < \eta$. Unfortunately, the results obtained in Ref. 11 are not directly applicable to quantum heat engines since in the quantum case pure quantum interlevel transitions may lift the adiabaticity. Adiabatic shortcuts are thus needed.

## 3. First law of thermodynamics for quantum systems

The first law of thermodynamics states that the change in internal energy of a system is equal to the heat added to the system minus the work done by the system. This is a very general formulation applicable to quantum systems as well. However, for quantum systems the definitions of a quantum heat and a quantum work need to be revisited.[12–31]

At nonzero temperature the energy of a system can be evaluated as follows $U = Tr(\hat{\varrho}\hat{H})$. Here $\hat{H}$ is the Hamiltonian of the system and $\hat{\varrho}$ is the density matrix. For the change in the internal energy we deduce: $dU = \sum_{n=1}^{N}\big(E_n d\varrho_{nn} + \varrho_{nn} dE_n\big)$. The first term $\delta Q = E_n d\varrho_{nn}$ corresponds to the heat exchange and is related to the change of the level populations $\varrho_{nn}(E_n, T)$ occurring due to a change in the temperature $T$ for $E_n = const$. The second term $\delta W = \varrho_{nn} dE_n$ corresponds to the produced work. The working substance produces work due to a change in the energy spectrum $dE_n$ of the system. The relation of heat exchange and the quantum level populations is clear. The concept of a quantum work needs however further specification.[32–40]

## 4. Bochkov-Kuzovlev and Jarzynski equalities

Let us consider a thermally isolated classical system $H\big(p, q, \lambda(t)\big)$ driven by an external parameter $\lambda$. A change in the parameter $\lambda(0) = \lambda_0, \lambda(t_f) = \lambda_f$ produces work delivered to the system. This work is assumed to be small compared to the energy of the system. At $t = 0$ the system is thermalized to a temperature $T = 1/\beta$. The work done on the system reads[32]

$$W = \int_0^{t_f} \frac{\partial H}{\partial \lambda}\frac{d\lambda}{dt} dt = H\big(p_f, q_f, \lambda(t_f)\big) - H\big(p_0, q_0, \lambda(0)\big).$$

The work averaged over the statistical ensemble is

$$\big\langle W \big\rangle = \big\langle H\big(p_f, q_f, \lambda(t_f)\big)\big\rangle - \big\langle H\big(p_0, q_0, \lambda(0)\big)\big\rangle.$$

According to the Bochkov-Kuzovlev equalities, not the work $\big\langle W\big\rangle$ itself, but the exponential of the work $\big\langle \exp(-\beta W)\big\rangle$ is the key point:

$$\big\langle \exp(-\beta W)\big\rangle = \int \exp\big[\beta(F_0 - H_0)\big] \exp\big[-\beta(H_f - H_0)\big] d\Gamma_0.$$

Here $\exp\big[\beta(F_0 - H_0)\big]$ is the distribution function of the equilibrated system at $t = 0$, $F_0$ is the free energy, and $\Gamma_0 = p_0 q_0$ is the phase volume of the system. Due to the Liouville theorem the phase volume of the system is invariant $d\Gamma_0 = d\Gamma_f$. Therefore, one writes

$$\big\langle \exp(-\beta W)\big\rangle = \exp(\beta F_0)\int \exp(-\beta H_f)d\Gamma_f.$$

Finally we obtain the expression $\big\langle \exp(-\beta W)\big\rangle = \exp(-\beta\Delta F)$, where $\Delta F = F_f - F_0$. This equality for cyclic process $\lambda_f = \lambda$ was obtained by Bochkov and Kuzovlev in 1977 and by Jarzynski in 1997 for a more general setting $\lambda_f \neq \lambda$.

According to statistical physics, for an adiabatic process when the parameter $\lambda$ changes slowly compared to the system's relaxation time, the produced work is equal to the change in free energy $W = \Delta F$. For the non-adiabatic case $W > \Delta F$ part of the work is wasted on the entropy production. The nonequilibrium entropy associated with the nonadiabatic process is defined as $\Delta S_{ir} = \beta\big\langle W_{ir}\big\rangle$. Here $\big\langle W_{ir}\big\rangle = \big\langle W\big\rangle - \Delta F$ is the difference between the delivered work in the nonadiabatic process and the change in free energy. We can rewrite Jarzynski equality in the following form (cf. Ref. 39): $\Delta F = -T\log\big\langle \exp\big(-\beta W\big)\big\rangle$. Using the ansatz for the work fluctuations $W = \big\langle W\big\rangle + \delta W$ we obtain $\Delta F = \big\langle W\big\rangle - \beta\big\langle(\delta W)^2\big\rangle/2$. Notably, the work for a quantum system is not an observable. This means that the average of the total work performed on the system doesn't corresponds to the expectation values of an operator representing the work.[32] In particular the work delivered to a quantum system is related to the time ordered correlation functions of the exponentiated Hamiltonian. The nonequilibrium entropy associated with a nonadiabatic process can be calculated straightforwardly.[40] First we define the probability distribution of the quantum work

$$p(W) = \sum_{n,m}\delta\big(W - (E^n_m - E^0_n)\big)p^\tau_{n,m}p^0_n.$$

Here

$$p^0_n = \exp\big(-\beta E^0_n\big)/Z_0$$

is the initial thermal Gibbs distribution, $Z_0$ is the partition function, $E^0_n$ are the initial energies, and $p^\tau_{n,m}$ are the transition probabilities. Using $p(W)$ one can calculate the non-equilibrium quantum work:

$$\big\langle W\big\rangle = 1/\beta\sum_n p^0_n \ln p^0_n - 1/\beta\sum_{n,m} p^0_n p^\tau_{n,m}\ln p^\tau_n - \Delta F$$

$p^\tau_n$ is the final equilibrium distribution function. In the absence of purely quantum inter-level transitions $p^\tau_{n,m} = \delta_{nm}$ the first two terms disappear and the quantum work becomes equal to the change in free energy.

## 5. Adiabatic shortcuts and finite time quantum cycles

For a general discussion of shortcuts to adiabaticity and an overview of the interrelation between the various approaches as well as their historical developments we refer to the review article[4] and references therein. Here we will basically follow Berry's transitionless driving formulation[5] which is equivalent to the counterdiabatic approach of Demirplak and Rice.[2,3]

Let us suppose that the Hamiltonian of the system has the form $\hat{H}(p,q,\lambda)$. Here $p,q$ are canonical coordinates and $\lambda$ is a parameter in the sense discussed above. For the solution of Schrödinger equation

$$\imath\frac{\partial\Psi}{\partial t}=\hat{H}\Psi$$

we implement the following ansatz:

$$\Psi=\sum_n a_n(t)\varphi_n(p,q,\lambda)\exp\{-\imath\int_{-\infty}^{t}E_n(\lambda)dt\},$$

where $E_n(\lambda)$ are the instantaneous quasi-energies that depend adiabatically on the parameter $\lambda$. After standard derivations for the time dependent coefficients $a_n(t)$ we obtain the iterative solution

$$a_n^{(1)}(t)=-\int_{-\infty}^{t}d\tau\sum_{m\neq n}\frac{\langle\varphi_n|\frac{\partial H}{\partial\lambda}|\varphi_m\rangle\dot{\lambda}}{E_m-E_n}\times a_m(-\infty)\exp\{-\imath\int_{-\infty}^{\tau}(E_m-E_n)d\acute{\tau}\}.$$

The adiabatic approximation is valid when the following criteria hold

$$\frac{a_n^{(2)}}{a_n^{(1)}}\sim\frac{\partial H}{\partial t}\frac{1}{(E_m-E_n)^2}.$$

Here $a_n^{(2)}$ is a second order correction to $a_n(t)$. If the characteristic time scale of the parameter $\lambda$ is $\dot{\lambda}\sim 1/\tau$ and $\frac{\tau}{(E_m-E_n)^2}\gg 1$ then the dynamic of the system is adiabatic and the effect of the non-adiabaticity is exponentially small.

In the case of an adiabatic evolution the general state $|\Psi_n(t)\rangle$ driven by $\hat{H}_0(t)$ is cast as

$$|\Psi_n(t)\rangle=\exp\left[-\frac{i}{\hbar}\int_0^t dt'E(t') - \int_0^t dt'\langle\Phi_n(t')|\partial_{t'}\Phi_n(t')\rangle\right]|\Phi_n(t)\rangle. \quad (1)$$

In essence the method of adiabatic shortcuts is an inverse engineering problem with the aim of finding of a new Hamiltonian for which the states (1) behave as $\imath\partial_t\Psi_n(t) = \hat{H(t)}\Psi_n(t)$. Note that the time dependence of the new Hamiltonian can be arbitrary fast. With the aid of the unitary time-evolution operator

$$\hat{U}(t) = \sum_n \exp\left[-\frac{i}{\hbar}\int_0^t dt' E(t') - \int_0^t dt' \langle\Phi_n(t')|\partial_{t'}\Phi_n(t')\rangle\right]|\Phi_n(t)\rangle\langle\Phi_n(0)|, \tag{2}$$

we construct the auxiliary (counter-diabatic) Hamiltonian

$$\hat{H}_{CD}(t) = i\hbar(\partial_t\hat{U}(t))\hat{U}^\dagger(t). \tag{3}$$

The reverse state engineering relies on the requirement that the states (1) solve for the Hamiltonian (3), meaning that

$$i\hbar\partial_t|\Psi_n(t)\rangle = \hat{H}_{CD}(t)|\Psi_n(t)\rangle. \tag{4}$$

In this way even for a fast driving the transitions between the eigenstates $|\Phi_n(t)\rangle$ are prevented. After a relatively simple algebra the counter-diabatic (CD) Hamiltonian $\hat{H}_{CD}(t)$ takes the form

$$\hat{H}_{CD}(t) = \hat{H}_0(t) + \hat{H}_1(t), \tag{5}$$

where

$$\hat{H}_1(t) = i\hbar\sum_{m\neq n}\frac{|\Phi_m\rangle\langle\Phi_m|\partial_t\hat{H}_0(t)|\Phi_n\rangle\langle\Phi_n|}{E_n - E_m}. \tag{6}$$

We adopt the initial conditions for the driving protocol as

$$\hat{H}_{CD}(0) = \hat{H}_0(0), \quad \hat{H}_{CD}(\tau) = \hat{H}_0(\tau).$$

Thus, on the time interval $t \in [0, \tau]$ we achieve a fast adiabatic dynamic by means of the counter-diabatic Hamiltonian $\hat{H}_{CD}(t)$.

This result can be straightforwardly generalized to systems with a degenerated spectrum.[41] In this case we have

$$\hat{H}_1(t) = i\hbar\sum_{m\neq n}\sum_{q=1}^{\lambda_n}\sum_{k=1}^{\lambda_m}\frac{|\Phi_m^k\rangle\langle\Phi_m^k|\partial_t\hat{H}_0(t)|\Phi_n^q\rangle\langle\Phi_n^q|}{E_n - E_m}. \tag{7}$$

Here we assumed that the eigenvalue $E_m$ is $\lambda_m$ times degenerated and $|\Phi_m^k\rangle$ are the corresponding degenerated eigenfunctions $k = 1, ...\lambda_m$.

## 6. Superadiabatic quantum heat engine with a multiferroic working medium

A central point for any quantum heat engine is the choice of the appropriate working substance. We identified multiferroics (MF) and in particular magnetoelectrics nanostructures as promising candidates.[10,31] MFs are materials of herostructures with coupled order parameters such as elastic, magnetic, and ferroelectric ordering[42–50,52–54] and can be well integrated in solid-state electronic circuits (in particular in oxide-based electronics). Hence, an engine based on a MF substance performs magnetic, electric and possibly (via piezoelectricity) mechanical works, at the same time. Particularly relevant are quantum spiral magnetoelectric substances.[55] A prototypical one dimensional chiral MF system can be modeled by a frustrated spin $= 1/2$ chain of $N$ sites aligned along the $x$ axis. Spin frustrations is due to competing ferromagnetic nearest neighbor $J_1 > 0$ and antiferromagnetic next-nearest neighbor $J_2 < 0$ interactions. We apply a time dependent electric field $\wp(t)$ which is linearly polarized along the $y$ axis, and an external magnetic field $B$ is applied along the $z$ axis. The corresponding Hamiltonian reads

$$\hat{H}_0(t) = \hat{H}_S + \hat{H}_{SF}(t), \tag{8}$$
$$\hat{H}_S = -J_1 \sum_i \vec{\sigma}_i \cdot \vec{\sigma}_{i+1} - J_2 \sum_i \vec{\sigma}_i \cdot \vec{\sigma}_{i+2} - \gamma_e \hbar B \sum_i \sigma_i^z .$$

Here $\hat{H}_S$ is time independent, while $\hat{H}_{SF}$ is time dependent and contains the coupling of the external electric field to the electric polarization of the chain. The electric polarization $\vec{P}_i$ tagged to spin non-collinearity reads

$$\vec{P}_i = g_{ME} \vec{e}_{i,\,i+1} \times (\vec{\sigma}_i \times \vec{\sigma}_{i+1}),$$

where $\vec{e}_{i,\,i+1}$ is the unit vector connecting the sites $i$ and $i+1$, $g_{ME}$ is a magneto-electric coupling constant. The spatially homogeneous, time dependent electric field $\wp(t)$ couples to the chain electric polarization $\vec{P}$ such that

$$\vec{\wp}(t) \cdot \vec{P} = d(t) \sum_i (\vec{\sigma}_i \times \vec{\sigma}_{i+1})^z,$$

with $d(t) = \wp(t) g_{ME}$. The quantity $(\vec{\sigma}_i \times \vec{\sigma}_{i+1})^z$ is known as the $z$ component of the vector chirality. With this notation $\hat{H}_{SF}(t)$ reads

$$\hat{H}_{SF}(t) = -\vec{\wp}(t) \cdot \vec{P} = d(t) \sum_i (\sigma_i^x \sigma_{i+1}^y - \sigma_i^y \sigma_{i+1}^x). \tag{9}$$

The quantum Otto cycle consists of two quantum isochoric and two adiabatic strokes. The quantum isochoric strokes correspond to a heat exchange between the working substance and the cold and the hot heat baths. During the quantum isochoric strokes the level populations are altered, see Fig. 1.

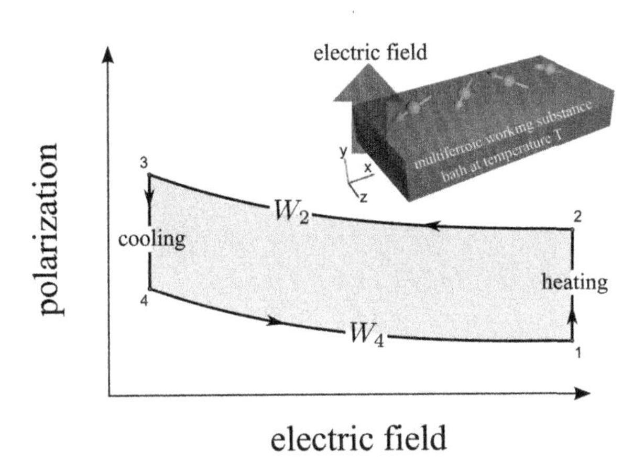


Fig. 1. Scheme of the quantum Otto cycle based on a chiral multiferroic chain. Adapted from Ref. 10

The MF working substance produces work during the adiabatic process. Changing the amplitude of the applied external electric field modifies the energy spectrum of the system. This is the mechanism behind producing work. The quantum Otto cycle and the MF-based engine is detailed in recent works.[10,31] A particular type of the time dependence for the external electric field is given by

$$d(t) = \epsilon\left(\frac{t^3}{3\tau} - \frac{t^2}{2}\right) + d_0, \tag{10}$$

which ensures that the requirement for the shortcuts of adiabaticity

$$\hat{H}_{CD}(0) = \hat{H}_0(0),\ \hat{H}_{CD}(\tau) = \hat{H}_0(\tau),$$

is fulfilled. The power output of the quantum Otto cycle is given by

$$\Re = \frac{-\left(\langle W_2\rangle_{\mathrm{ad}} + \langle W_4\rangle_{\mathrm{ad}}\right)}{\tau_1(T_H) + \tau_2 + \tau_3(T_L) + \tau_4}. \tag{11}$$

Here $\tau_1(T_H)$, $\tau_3(T_L)$ are the relaxation times of the MF working substance in contact with the hot and the cold thermal baths (strokes $1 \to 2$, and $3 \to 4$), $\tau_2$ and $\tau_4$ correspond to the duration of the adiabatic strokes,

$\langle W_2\rangle_{\rm ad}$ and $\langle W_4\rangle_{\rm ad}$ correspond to the work produced during the quantum adiabatic strokes. The condition

$$\langle W_2\rangle_{\rm ad} + \langle W_4\rangle_{\rm ad} + Q_{\rm in} + Q_{\rm out} = 0,$$

during the whole cycle should be satisfied. The corresponding absorbed heat $Q_{\rm in}$ and the released heat $Q_{\rm out}$ by the working substance are defined as follows

$$\begin{aligned} Q_{\rm in} &= \sum_n E_n(0)\left(\frac{e^{-\beta_H E_n(0)}}{\sum_n e^{-\beta_H E_n(0)}} - \frac{e^{-\beta_L E_n(\tau)}}{\sum_n e^{-\beta_L E_n(\tau)}}\right), \\ Q_{\rm out} &= \sum_n E_n(\tau)\left(\frac{e^{-\beta_L E_n(\tau)}}{\sum_n e^{-\beta_L E_n(\tau)}} - \frac{e^{-\beta_H E_n(0)}}{\sum_n e^{-\beta_H E_n(0)}}\right). \end{aligned} \tag{12}$$

We adopt the dimensionless parameters

$$J_1 = 1,\ J_2 = -1,\ B = 0.1,\ d_0 = 2.5,\ \epsilon = 1.$$

In explicite units these parameters correspond to the one phase MF material[56] $LiCu_2O_2$, $J_1 = -J_2 = 44[\mathrm{K}]$. The external driving fields strengths are $B = 3[\mathrm{T}]$, $\wp = 5\times 10^3[\mathrm{kV/cm}]$. We assume that the duration of the adiabatic strokes of the cycle are equal to $\tau_2 = \tau_4 = \tau$. The time unit in our calculations corresponds to the $\hbar/J_1 \approx 0.1[\mathrm{ps}]$. CD driving allows reducing the driving time. Implementing a short driving protocol is supposed to maximize the output power of the cycle. Duration of the isochoric strokes can be estimated via the Lindblad master equation.[57]

We supplement the CD Hamiltonian $\hat{H}_{CD}(t)$ by the Hamiltonian of the heat bath $\hat{H}_{\rm bath}$ and the system-bath interaction $\hat{H}_{\rm int}$. In addition, we assume that the phononic heat bath is coupled to the $z$ component of the vector chirality $K_n^z = (\sigma_n^x\sigma_{n+1}^y - \sigma_n^y\sigma_{n+1}^x)$. The argument behind this doing is that the vector chirality is a characteristic measure for the non-collinearity in the spin order and is directly influenced by the lattice distortion and the phononic modes

$$\begin{aligned} \hat{H} &= \hat{H}_{CD}(t) + \hat{H}_{\rm int} + \hat{H}_{\rm bath}, \\ \hat{H}_{\rm bath} &= \int dk\omega_k \hat{b}_k^\dagger \hat{b}_k, \\ \hat{H}_{\rm int} &= \sum_{n=1}^{4} K_n^z \int dk g_k(\hat{b}_k^\dagger + \hat{b}_k). \end{aligned} \tag{13}$$

Here $\hat{b}_k^\dagger$, $\hat{b}_k$ are the phonon creation and annihilation operators, and $g_k$ is the coupling constant between the system and the bath. After a straightforward derivations we obtain

$$\frac{d\rho_S(t)}{dt} = \sum_{\omega,\omega'}\sum_{\alpha,\gamma} e^{i(\omega-\omega')t}\Gamma(\omega)\big(K_\beta^z(\omega)\rho_S(t)K_\alpha^{z^\dagger}(\omega')$$
$$- K_\alpha^{z^\dagger}(\omega')\big(K_\beta^z(\omega)\rho_S(t)\big) + h.c.,$$
$$\Gamma(\omega) = \int_0^\infty ds e^{i\omega s}\langle B^\dagger(t)B(t-s)\rangle. \tag{14}$$

Here

$$B(t) = \int dk g_k(\hat{b}_k^\dagger e^{i\omega_k t} + \hat{b}_k e^{-i\omega_k t}),\ K_\alpha^z(\omega) = \sum_{\omega=E_m-E_n} \pi(E_n)K_\alpha^z\pi(E_n)$$

and $\pi(E_n) = |\Psi_n\rangle\langle\Psi_n|$ is the projection operator onto the eigenstates $|\Psi_n\rangle$ of the CD Hamiltonian. For the bath correlation functions $\Gamma(\omega)$ we deduce

$$\gamma(\omega) = \Gamma(\omega) + \Gamma^*(\omega),$$
$$\gamma(\omega) = \pi J(\omega)\begin{cases}\frac{1}{\exp[\beta\omega]-1}, & \omega < 0\\ \frac{1}{\exp[\beta\omega]-1} + 1, & \omega > 0\end{cases}. \tag{15}$$

Here $J(\omega) = \frac{\pi}{\omega}\sum_j g_j^2\delta(\omega-\omega_j) = \pi\gamma$ is the spectral density of the thermal bath.[57] The efficiency of the cycle reads $\eta = \frac{\delta Q_H + \delta Q_c}{\delta Q_H}$.

## 7. Summary

Shortcuts to adiabaticity allows realizing transitionless fast quantum adiabatic dynamics to achieve a finite power output from a quantum engine. We derived analytical results for the mean square fluctuation for the work, the irreversible work and output power of the cycle. We observed that the work mean square fluctuations are increasing with the duration of the adiabatic strokes $\tau$ (cf. Fig. 2). The irreversible work exhibits a non-monotonic behavior (see Fig. 3) and has a maximum for $\tau = 0.26$(ps). At the end of the adiabatic stroke the irreversible work becomes zero confirming so that the cycle is reversible. We found that the quantum heat engine with a MF working medium has an optimal duration corresponding to the largest power output (see Fig. 4).

By implementing a Lindblad master equation we studied the thermal relaxation of the system. We evaluated the transferred heat $\delta Q_H$ to the

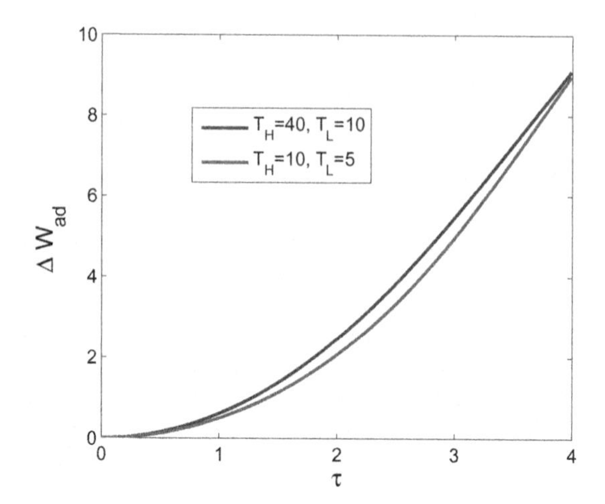


Fig. 2. Standard deviation of the work $\Delta W_{\rm ad}$ in scaled units for two different hot and cold bath temperatures. The other parameters are: $\varepsilon = 1$, $J_1 = 1, J_2 = -1, B = 0.1, d_0 = 2.5$. Unscaled unit of $\Delta W_{ad}$ amounts to $6 \times 10^{-22}[J]$. Adapted from Ref. 10.

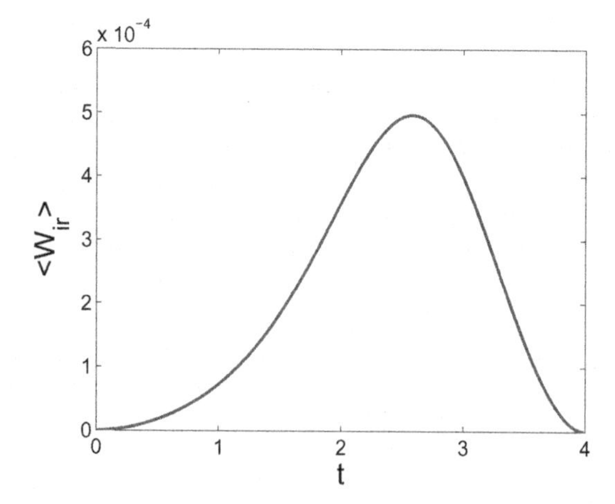


Fig. 3. $\langle W_{\rm ir} \rangle$ for the values of the parameters $J_1 = 1, J_2 = -1, B = 0.1, d_0 = 2.5$. Adapted from Ref. 10.

working substance and the heat released by system to the cold bath $\delta Q_c$. We inferred a cycle efficiency of $\eta = 1 + \delta Q_c/\delta Q_H \approx 47\%$. If the system thermalizes to the Gibbs ensemble the efficiency is lower at $\eta \approx 23\%$ (see Fig. 5).

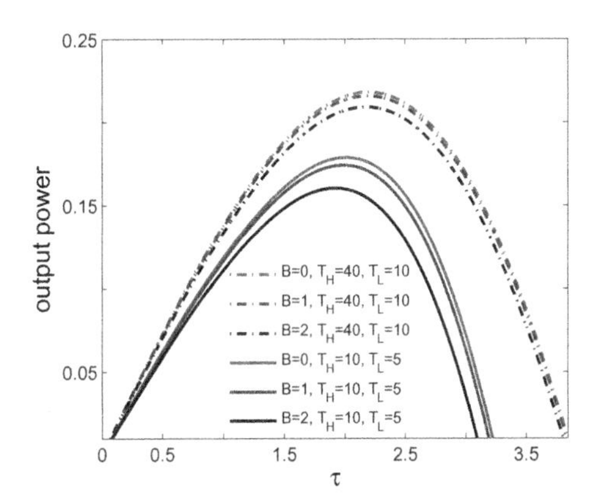


Fig. 4. Output power for $J_1 = 1, J_2 = -1, d_0 = 2.5$, $\varepsilon = 1$ and for the parameters depicted on the figure. Adapted from Ref. 10.

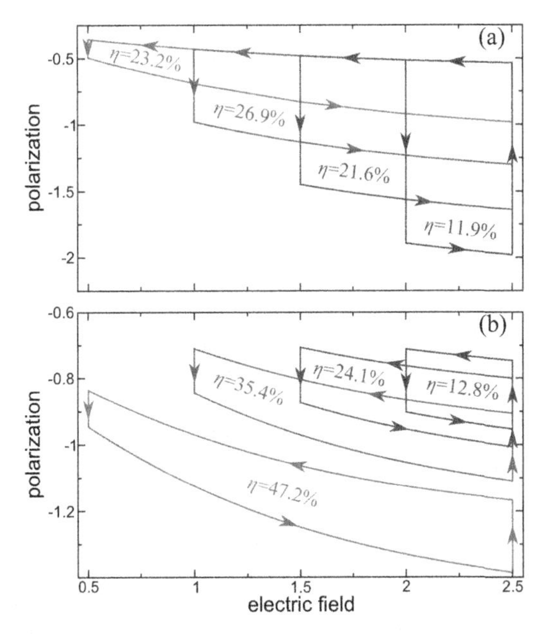


Fig. 5. A complete quantum Otto cycle (a) using the level population corresponding to Gibbs distribution and (b) the level population obtained from the Lindblad master equation for the parameters $\gamma = 0.1, T_H = 40, T_L = 10, d_0 = 2.5$ and $d_1$ as in the figures. Adapted from Ref. 10.

## Acknowledgment

This work is supported by the DFG through the SFB 762.


## References


1. L.D. Landau, E.M. Lifshitz (1980). Statistical Physics. Vol. 5, Butterworth-Heinemann 3rd ed., (1980).
2. M. Demirplak and S. A. Rice, Adiabatic Population Transfer with Control Fields, *J. Phys. Chem. A* **107**, 9937 (2003).
3. M. Demirplak and S. A. Rice, Assisted Adiabatic Passage Revisited, *J. Phys. Chem. B* **109**, 6838 (2005).
4. E. Torrontegui, S. Ibanez, S. Martinez-Garaot, M. Modugno, A. del Campo, D. Guery-Odelin, A. Ruschhaupt, X. Chen, and J. G. Muga, Shortcuts to Adiabaticity *Adv. Atom. Mol. Opt. Phys.* **62**, 117 (2013).
5. M. V. Berry, Transitionless quantum driving, *J. Phys. A: Math. Theor.* **42**, 365303 (2009).
6. A. del Campo, M. M. Rams, W. H. Zurek, Assisted Finite-Rate Adiabatic Passage Across a Quantum Critical Point: Exact Solution for the Quantum Ising Model, *Phys. Rev. Lett.* **109**, 115703 (2012).
7. A. del Campo, J. Goold, and M. Paternostro, More bang for your buck: Super-adiabatic quantum engines, *Sci. Rep.* **4**, 6208 (2014).
8. H. Saberi, T. Opatrný, K. Mølmer, and A. del Campo, Adiabatic tracking of quantum many-body dynamics, *Phys. Rev. A* **90**, 060301(R) (2014).
9. M. Beau, J. Jaramillo, and A. del Campo, Scaling-Up Quantum Heat Engines Efficiently via Shortcuts to Adiabaticity, *Entropy* **18**, 168 (2016).
10. L. Chotorlishvili, M. Azimi, S. Stagraczyński, Z. Toklikishvili, M. Schüler, and J. Berakdar, Superadiabatic quantum heat engine with a multiferroic working medium, *Phys. Rev. E* **94**, 032116 (2016).
11. F.L. Curzon, B. Ahlborn, Efficiency of a Carnot engine at maximum power output, Amer. J. Phys. **43** 22 (1975).
12. H. T. Quan, Y. D. Wang, Y.-X Liu, C. P. Sun, and Franco Nori, *Phys. Rev. Lett.* **97**, 180402 (2006).
13. H. T. Quan, Y.-X Liu, C. P. Sun, and Franco Nori, Quantum thermodynamic cycles and quantum heat engines, *Phys. Rev. E* **76**, 031105 (2007).
14. H. T. Quan, Quantum thermodynamic cycles and quantum heat engines. II., *Phys. Rev. E* **79**, 031101 (2009).

15. V. Mukherjee, W. Niedenzu, A. G. Kofman, and G. Kurizki *Phys. Rev. E* **94**, 062109 (2016).
16. R. Kosloff, Y. Rezek The quantum harmonic Otto cycle *arXiv:1612.03582*, (2016).
17. O Abah, E Lutz - Performance of superadiabatic quantum machines *arXiv:1611.09045*, (2016).
18. M. Esposito, N. Kumar, K. Lindenberg, and C. Van den Broeck, Stochastically driven single-level quantum dot: A nanoscale finite-time thermodynamic machine and its various operational modes, *Phys. Rev. E* **85**, 031117 (2012).
19. N. Kumar, C. Van den Broeck, M. Esposito, and K. Lindenberg, Thermodynamics of a stochastic twin elevator, *Phys. Rev. E* **84**, 051134 (2011).
20. A. Alecce, F. Galve, N. Lo. Gullo, L. Dell'Anna, F. Plastina, and R. Zambrini, Quantum Otto cycle with inner friction: finite-time and disorder effects, *New J. Phys.* **17**, 075007 (2015).
21. O. Abah, J. Roßnagel, G. Jacob, S. Deffner, F. Schmidt-Kaler, K. Singer, and E. Lutz, Single-Ion Heat Engine at Maximum Power, *Phys. Rev. Lett.* **109**, 203006 (2012).
22. O. Abah and E. Lutz, Efficiency of heat engines coupled to nonequilibrium reservoirs, *Europhys. Lett.* **106**, 20001 (2014).
23. J. Roßnagel, O. Abah, F. Schmidt-Kaler, K. Singer, and E. Lutz, Nanoscale Heat Engine Beyond the Carnot Limit, *Phys. Rev. Lett.* **112**, 03602 (2014).
24. R. Wang, J.Wang, J. He, and Y. Ma, Performance of a multilevel quantum heat engine of an ideal N-particle Fermi system, *Phys. Rev. E* **86**, 021133 (2012).
25. S. Çakmak, F. Altintas, Ö. E. Müstecaplioglu, Lipkin-Meshkov-Glick model in a quantum Otto cycle, *Eur. Phys. J. Plus* **131**, 197 (2016).
26. A. M. Zagoskin, S. Savel'ev, Franco Nori, and F. V. Kusmartsev, Squeezing as the source of inefficiency in the quantum Otto cycle, *Phys. Rev. B* **86**, 014501 (2012).
27. F. Altintas and Ö. E. Müstecaplioglu, General formalism of local thermodynamics with an example: Quantum Otto engine with a spin-1/2 coupled to an arbitrary spin. *Phys. Rev. E* **92**, 022142 (2015).
28. E. A. Ivanchenko, Quantum Otto cycle efficiency on coupled qudits, *Phys. Rev. E* **92**, 032124 (2015).
29. N. Linden, S. Popescu, and P. Skrzypczyk, How Small Can Thermal Machines Be? The Smallest Possible Refrigerator. *Phys. Rev. Lett.* **105**, 130401 (2010).

30. C. Van den Broeck, N. Kumar, and K. Lindenberg, Efficiency of Isothermal Molecular Machines at Maximum Power, *Phys. Rev. Lett.* **108**, 210602 (2012).
31. M. Azimi, L. Chotorlishvili, S. K. Mishra, T. Vekua, W. Hübner, and J. Berakdar, Quantum Otto heat engine based on a multiferroic chain working substance, *New J. of Phys.* **16**, 063018 (2014).
32. P. Talkner, E. Lutz, and P. Hänggi Fluctuation theorems: Work is not an observable *Phys. Rev. E* **75**, 050102(R) (2007).
33. G. N. Bochkov, Yu. E. Kuzovlev General theory of thermal fluctuations in nonlinear systems *Sov. Phys. JETP* **45**, 125 (1977).
34. G. N. Bochkov, Yu. E. Kuzovlev Fluctuation-dissipation relations for nonequilibrium processes in open systems *Sov. Phys. JETP* **49** 543 (1979).
35. G. N. Bochkov, Yu. E. Kuzovlev Generalized fluctuation-dissipation theorem *Physica A* **106** 443 (1981).
36. G. N. Bochkov, Yu. E. Kuzovlev Kinetic potential and variational principles for nonlinear irreversible processes *Physica A* **106** 480 (1981).
37. C. Jarzynski Nonequilibrium Equality for Free Energy Differences *Phys. Rev. Lett.* **78** 2690 (1997).
38. L. P. Pitaevskii, Rigorous results of nonequilibrium statistical physics and their experimental verification *Phys. Usp.* **54**, 625 (2011); *Usp. Fiz. Nauk.* **181**, 647 (2011).
39. J. Hermans Simple analysis of noise and hysteresis in (slow-growth) free energy simulations *J. Phys. Chem.* **95** 9029 (1991).
40. S. Deffner and E. Lutz Generalized Clausius Inequality for Nonequilibrium Quantum Processes *Phys. Rev. Lett.* **105**, 170402 (2010).
41. Xue-Ke Song, Hao Zhang, Qing Ai, Jing Qiu and Fu-Guo Deng Shortcuts to adiabatic holonomic quantum computation in decoherence-free subspace with transitionless quantum driving algorithm *New J. Phys.* **18** 023001 (2016).
42. W. Eerenstein, N. D. Mathur, and J. F. Scott, Multiferroic and magnetoelectric materials, *Nature* **442**, 759 (2006).
43. J. Wang, J. B. Neaton, H. Zheng, V. Nagarajan, S. B. Ogale, B. Liu, D. Viehland, V. Vaithyanathan, D. G. Schlom, U. V. Waghmare, N. A. Spaldin. K. M. Rabe, M. Wuttig, and R. Ramesh, Epitaxial $BiFeO_3$ Multiferroic Thin Film Heterostructures, *Science* **299**, 1719 (2003).
44. V. Garcia, M. Bibes, L. Bocher, S. Valencia, F. Kronast, A. Crassous, X. Moya, S. Enouz-Vedrenne,A.Gloter, D. Imhoff, C. Deranlot, N. D.

Mathur, S. Fusil, K. Bouzehouane, and A. Barthélémy, Ferroelectric Control of Spin Polarization, *Science* **327**, 1106 (2010).
45. M. Bibes and A. Barthélémy, Multiferroics: Towards a magnetoelectric memory, *Nat. Mater.* **7**, 425 (2008).
46. N. A. Spaldin and M. Fiebig, The Renaissance of Magnetoelectric Multiferroics, *Science* **309**, 391 (2005).
47. S. Valencia, A. Crassous, L. Bocher, V. Garcia, X. Moya, R. O. Cherifi, C. Deranlot, K. Bouzehouane, S. Fusil, A. Zobelli, A. Gloter, N. D. Mathur, A. Gaupp, R. Abrudan, F. Radu, A. Barthélémy, and M. Bibes, Interface-induced room-temperature multiferroicity in $BaTiO_3$, *Nat. Mater.* **10**, 753 (2011).
48. D. I. Khomskii, Spin chirality and nontrivial charge dynamics in frustrated Mott insulators: spontaneouscurrents and charge redistribution, *J. Phys.: Condens. Matter* **22**, 164209 (2010).
49. H. L. Meyerheim, F. Klimenta, A. Ernst, K. Mohseni, S. Ostanin, M. Fechner, S. Parihar, I. V. Maznichenko, I. Mertig, and J. Kirschner, Structural Secrets of Multiferroic Interfaces, *Phys. Rev. Lett.* **106**, 087203 (2011).
50. S. W. Cheong and M. Mostovoy, Multiferroics: a magnetic twist for ferroelectricity, *Nat. Mater.* **6**, 13 (2007).
51. Y. Tokura, Multiferroics—toward strong coupling between magnetization and polarization in a solid, *J. Magn. Magn. Mat.* **310**, 1145 (2007).
52. P. P. Horley, A. Sukhov, C. Jia, E. Martínez, and Jamal Berakdar, Influence of magnetoelectric coupling on electric field induced magnetization reversal in a composite unstrained multiferroic chain, *Phys. Rev. B* **84**, 064425 (2011).
53. M. Menzel, Y. Mokrousov, R. Wieser, J. E. Bickel, E. Vedmedenko, S. Blugel, S. Heinze, K. vonBergmann, A. Kubetzka, and R. Wiesendanger, Information Transfer by Vector Spin Chirality in Finite Magnetic Chains, *Phys. Rev. Lett.* **108**, 197204 (2012).
54. K. F. Wang, J.-M. Liu, and Z. F. Ren, Multiferroicity: the coupling between magnetic and polarization orders, *Adv. Phys.* **58**, 321–448 (2009).
55. M. Azimi, L. Chotorlishvili, S. K. Mishra, S. Greschner, T. Vekua, and J. Berakdar, Helical multiferroics for electric field controlled quantum information processing, *Phys. Rev. B* **89**, 024424 (2014).
56. S. Park, Y. J. Choi, C. L. Zhang, and S. W. Cheong, Ferroelectricity in an S=1/2 Chain Cuprate *Phys. Rev. Lett.* **98**, 057601 (2007).
57. H. P. Breuer and F. Petruccione, *The Theory of Open Quantum Systems*, Oxford University Press, Oxford, (2002).

# Proximity Spin-orbit Coupling Physics of Graphene in Transition-metal Dichalcogenides

M. Gmitra*, D. Kochan, P. Högl and J. Fabian

*Institute for Theoretical Physics, University of Regensburg*
*Universitätsstrasse 31, D-93040 Regensburg, Germany*
**E-mail: martin.gmitra@ur.de*
*http://www.physik.uni-regensburg.de*

Proximity induced spin-orbit coupling effects in graphene on transition-metal dichalcogenides (TMD) have potential to bring graphene spintronics to the next level. Here we discuss electronic structure of graphene on monolayer $MoS_2$, $MoSe_2$, $WS_2$ and $WSe_2$. The Dirac electrons of graphene lie within the semiconducting gap and exhibit a giant global proximity spin-orbit coupling, without compromising the semimetallic character. We found that graphene on $MoS_2$, $MoSe_2$, and $WS_2$ has a topologically trivial band structure, while graphene on $WSe_2$ exhibits inverted bands. Within the inverted regime the graphene zigzag nanoribbon hosts protected helical edge states demonstrating the quantum spin Hall effect, which opens a new route to access topological states of matter in graphene-based systems. Additionally the use of graphene/TMD vertical heterostructures serve as a promising platform for optospintronics, in particular for optical spin injection into graphene.

*Keywords*: Heterostructures, graphene, transition-metal dichalcogenides, spin-orbit coupling, spintronics, optospintronics.

## 1. Introduction

Designing heterostructures of graphene and other two-dimensional atomic crystals in a precisely defined sequence is one of the leading topics in material science and condensed matter physics.[1] Building of such van der Waals heterostructures opens a venue to create new materials with novel properties. Recently there has been a strong push to enhance spin-orbit coupling in graphene to enable spintronics applications.[2,3] Graphene spintronics[4] has relied exclusively on electrical spin injection.[5–7] Decorating graphene with adatoms[8,9] has proven particularly promising, as demonstrated experimentally by the giant spin Hall effect signals.[10,11] In parallel, there have been intensive efforts to predict realistic graphene structures that would exhibit the quantum spin (and anomalous) Hall effect,[12–15] introduced by Kane and Mele[16] as a precursor of topological insulators.[17–19]

Combining graphene with semiconducting two-dimensional transition-metal dichalcogenides (TMD)[20,21] can open new routes for graphene spintronics applications. In this chapter we discuss fundamental electronic properties and the spin-orbit fine structure of the graphene Dirac bands for graphene on monolayer $MoS_2$, $MoSe_2$, $WS_2$, and $WSe_2$. The proximity effect induces strong spin-orbit coupling in graphene, while simultaneously preserving the Dirac band structure at the Fermi level. We introduce an effective spin-orbit Hamiltonian which explains the proximity induced spin splittings of the Dirac states. We show that the induced spin-orbit coupling is giant, and the metal atoms of TMD are responsible for its enhancement. This overview is based on original references 20 and 21.

## 2. Heterostructure description

The heterostructure of our interest contains a vertical stack of graphene which physically adsorbs on TMD. Typical interlayer distance is about 3 Å governed by the van der Waals forces. TMD has the chemical symbol $MX_2$, where M stands for the metal atom and X for the chalcogen atom. Here we consider as metal atoms Mo and W, and as chalcogen S and Se. A schematic view of the graphene/TMD heterostructure is shown in Fig. 1.

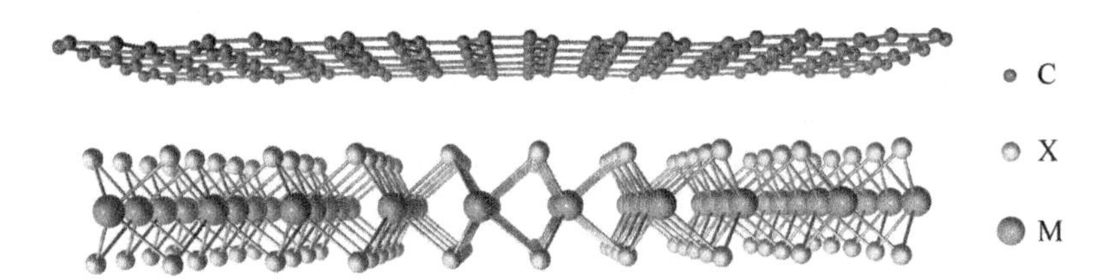


Fig. 1. Atomic structure of graphene on a monolayer transition-metal dichalcogenide $MX_2$.

TMDs are becoming increasingly popular in optoelectronics as sensitive photodetectors[22] or, forming lateral heterostructures,[23,24] as two-dimensional solar cells.[25] Important, TMDs have a sizeable spin-orbit coupling and lack space inversion symmetry. As a result, their band structure[26] allows for a valley resolved optical spin excitation by circularly polarized light.[27–29] TMDs can thus facilitate optical spin injection into graphene in hybrid structures.

The possibility to build such a heterostructure has been demonstrated by an efficient growth of graphene on $MoS_2$.[30–32] Graphene layers are reported to be ultraflat, having large mean free paths.[33] Angle-resolved photoemission found an intact Dirac point but a strong hybridization elsewhere in the $\pi$ system.[34] The potential devices of technological applications of these hybrid structures have already been discussed,[35] mainly as a basis for nonvolatile memory,[36] sensitive photodetection,[37] and gate-tunable persistent photoconductivity.[38]

Recently it was predicted that monolayer $MoS_2$ induces strong spin-orbit coupling in graphene, of about 1 meV[20] (compared to 10 $\mu$eV in pristine graphene[39]) which grows with the atomic number of the metal atom.[21] A recent experiment[11] on the room temperature spin Hall effect in graphene on few layers of $WS_2$ found a large spin-orbit coupling, about 17 meV, attributing it to defects in the thin $WS_2$, rather than to the genuine proximity effect. Enhanced spin-orbit coupling in graphene on TMD manifests in a pronounced low-temperature weak anti-localization effect and in a spin-relaxation time two to three orders of magnitude smaller than in graphene on conventional substrates.[40,41]

The physical adsorption preserves the low energy electronic band structure of graphene. The linear Dirac dispersion retains for larger momenta with respect the K point. In the vicinity of the K point the electronic band structure is locally distorted due to proximity effects. We found that the proximity spin-orbit coupling in graphene on TMD grows with the increasing atomic number of the transition-metals. In most cases we find conventional Dirac cones, affected by the proximity effects. But for graphene on $WSe_2$ we see a robust band inversion and emergent spin Hall effect in the corresponding zigzag nanoribbons.

## 3. Effective Hamiltonian

The $\pi$ conjugated bands of graphene extend over several hundreds of meV with linear dispersion around the Fermi energy. The model describing those orbital effects in pristine graphene close to the K (K′) points reads

$$\mathcal{H}_0 = \hbar v_{\mathrm{F}}(\kappa\sigma_x k_x + \sigma_y k_y). \tag{1}$$

Here $v_{\mathrm{F}}$ is the Fermi velocity of Dirac electrons, $k_x$ and $k_y$ are the Cartesian components of the electron wave vector measured from K (K′), parameter $\kappa = 1\,(-1)$ for K (K′), and $\sigma_x$ and $\sigma_y$ are the pseudospin Pauli matrices acting on the two-dimensional vector space formed by the two triangular sublattices of graphene. The Hamiltonian $\mathcal{H}_0$ describes gapless Dirac states with the conical dispersion $\varepsilon_0 = \nu\hbar v_{\mathrm{F}}|\boldsymbol{k}|$ near the Dirac points; $\nu = 1(-1)$ for the conduction (valence) band, see Fig. 2(a).

To describe the effective orbital energy difference on A and B sublattices of graphene, we assume staggered potential in the form,

$$\mathcal{H}_{\Delta} = \Delta\, \sigma_z, \tag{2}$$

where $\sigma_z$ is the pseudospin Pauli matrix and parameter $\Delta$ controls proximity induced energy gap that opens in the Dirac spectrum. The dispersion for the orbital Hamiltonian $\mathcal{H}_{\mathrm{orb}} = \mathcal{H}_0 + \mathcal{H}_{\Delta}$ equals $\varepsilon_{\mathrm{orb}} = \nu\sqrt{\Delta^2 + (\hbar v_{\mathrm{F}}|\boldsymbol{k}|)^2}$ and is shown in Fig. 2(b), with the energy gap at K (K′) of $2\Delta$.

The minimal spin-orbit coupling Hamiltonian $\mathcal{H}_{\mathrm{so}} = \mathcal{H}_{\mathrm{I}}(\lambda_{\mathrm{I}}^{\mathrm{A}}, \lambda_{\mathrm{I}}^{\mathrm{B}}) + \mathcal{H}_{\mathrm{R}}(\lambda_{\mathrm{R}}) + \mathcal{H}_{\mathrm{PIA}}(\lambda_{\mathrm{PIA}}^{\mathrm{A}}, \lambda_{\mathrm{PIA}}^{\mathrm{B}})$ we consider has three contributions conventionally called as intrinsic (I)—parameterized by spin-conserving couplings $\lambda_{\mathrm{I}}^{\mathrm{A}}$ and $\lambda_{\mathrm{I}}^{\mathrm{B}}$, Rashba (R)—parameter $\lambda_{\mathrm{R}}$, and the pseudospin (sublattice) inversion asymmetry (PIA) term[9] that is characterized by spin-flipping couplings $\lambda_{\mathrm{PIA}}^{\mathrm{A}}$ and $\lambda_{\mathrm{PIA}}^{\mathrm{B}}$. Broken sublattice symmetry in graphene due to TMD results in sublattice-resolved second-nearest neighbor spin-orbit coupling parameters, in our case the intrinsic spin-orbit couplings $\lambda_{\mathrm{I}}^{\mathrm{A}}$ and $\lambda_{\mathrm{I}}^{\mathrm{B}}$ and PIA terms $\lambda_{\mathrm{PIA}}^{\mathrm{A}}$ and $\lambda_{\mathrm{PIA}}^{\mathrm{B}}$. The corresponding proximity induced intrinsic spin-orbit coupling Hamiltonian close to K (K′), reads

$$\mathcal{H}_{\mathrm{I}} = \lambda_{\mathrm{I}}^{\mathrm{A}}[(\sigma_z + \sigma_0)/2]\kappa s_z + \lambda_{\mathrm{I}}^{\mathrm{B}}[(\sigma_z - \sigma_0)/2]\kappa s_z. \tag{3}$$

It represents a generalization of the McClure-Yafet Hamiltonian[42] to situations with the broken pseudospin (sublattice) symmetry. For example in hydrogenated graphene the sublattice symmetry is broken explicitly,[9] while in graphene on TMD it is broken implicitly.[4,21] In Eq. (3) the $s_z$ is the spin Pauli matrix and $\sigma_0$ is the unit matrix acting on the pseudospin (sublattice) space. If $\lambda_{\mathrm{I}}^{\mathrm{A}} = \lambda_{\mathrm{I}}^{\mathrm{B}}$, the intrinsic contribution reduces to that in bare graphene and its main effect is the spin-orbit induced anticrossing at the Dirac cones,[39] that leaves the spin degeneracy intact. However, in the presence of a staggered potential the spin degeneracy would be lifted since the staggering breaks the space inversion symmetry. In the general case with non-zero stagger $\Delta$ and $\lambda_{\mathrm{I}}^{\mathrm{A}} \neq \lambda_{\mathrm{I}}^{\mathrm{B}}$, describing graphene on TMD, the energy dispersion equals to $\mu(\lambda_{\mathrm{I}}^{\mathrm{A}} - \lambda_{\mathrm{I}}^{\mathrm{B}})/2 + \nu\sqrt{(\hbar v_{\mathrm{F}}|\boldsymbol{k}|)^2 + (\Delta + \mu(\lambda_{\mathrm{I}}^{\mathrm{A}} + \lambda_{\mathrm{I}}^{\mathrm{B}})/2)^2}$, where $\mu = 1(-1)$ for spin up (down) branches. The spectrum is schematically shown in Fig. 2(c).

Like the intrinsic spin-orbit coupling, PIA coupling is also represented as the next-nearest-neighbor (same sublattice) hopping, that is accompanied with a spin flip.[9] The PIA terms near the Dirac points are $k$ dependent. PIA

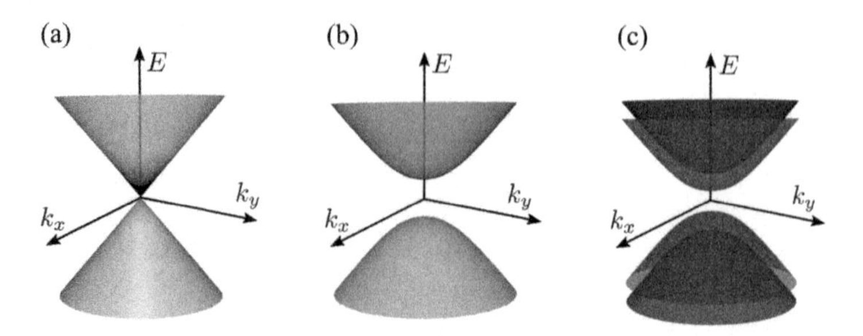


Fig. 2. Schematic plot of low energy electronic band structure topologies in graphene. (a) pristine graphene, (b) effect of staggered potential, (c) band spin splitting due to presence of staggered potential and intrinsic spin-orbit coupling, or Rashba or PIA term.

spin-orbit couplings turn the spin quantization axes of the electronic states towards the graphene plane. The sublattice resolved PIA Hamiltonian in first order in $k$ reads

$$\mathcal{H}_{\mathrm{PIA}} = \frac{a}{2}\left[(\lambda_{\mathrm{PIA}}^{\mathrm{A}}(\sigma_z + \sigma_0) + \lambda_{\mathrm{PIA}}^{\mathrm{B}}(\sigma_z - \sigma_0)\right](k_x s_y - k_y s_x), \tag{4}$$

where $\lambda_{\mathrm{PIA}}^{\mathrm{A}}$ and $\lambda_{\mathrm{PIA}}^{\mathrm{B}}$ are the PIA spin-orbit parameters and $a = 2.46$ Å is the pristine graphene lattice constant. The PIA spin-orbit terms become relevant for larger momenta away from the K point.

Placing graphene on TMD also breaks the lateral mirror symmetry, giving rise to the Rashba type spin-orbit coupling,[43]

$$\mathcal{H}_{\mathrm{R}} = \lambda_{\mathrm{R}}(\kappa\sigma_x s_y - \sigma_y s_x), \tag{5}$$

where $\lambda_{\mathrm{R}}$ is the Rashba parameter and $s_x$, $s_y$ are the spin Pauli matrices. In the tight-binding language, the Rashba coupling is the nearest-neighbor spin-flip hopping. It contributes to the spin splitting of the bands, and defines the spin quantization axis for each Bloch state, away from the time reversal points Γ and M. The spin expectation values for $k$s around the K points have in-plane components and form circular spin-orbit field texture around the K points.

The Hamiltonian $\mathcal{H} = \mathcal{H}_0 + \mathcal{H}_\Delta + \mathcal{H}_{\mathrm{so}}$ fully describes graphene's bands around K (K′). Its eigenenergies at the K (K′) are

$$\begin{aligned}\varepsilon_{\nu\mu} =& \frac{1+\nu\mu}{2}\left[\nu\Delta + \frac{1+\nu}{2}\lambda_{\mathrm{I}}^{\mathrm{A}} + \frac{1-\nu}{2}\lambda_{\mathrm{I}}^{\mathrm{B}}\right] - \\ &\frac{1-\nu\mu}{4}\left[\lambda_{\mathrm{I}}^{\mathrm{A}} + \lambda_{\mathrm{I}}^{\mathrm{B}} - \nu\sqrt{(2\Delta - \lambda_{\mathrm{I}}^{\mathrm{A}} + \lambda_{\mathrm{I}}^{\mathrm{B}})^2 + 16\lambda_{\mathrm{R}}^2}\right],\end{aligned} \tag{6}$$

where $\nu = 1(-1)$ stands for the conduction (valence) and $\mu = 1(-1)$ for the spin up (down) branches, respectively. The expectation values of the spin along $z$ for the corresponding states are given by

$$\langle s_z \rangle_{\nu\mu} = \frac{\mu\kappa\hbar}{2}\left[\frac{1+\nu\mu}{2} + \frac{1-\nu\mu}{2}\frac{2\Delta - \lambda_{\mathrm{I}}^{\mathrm{A}} - \lambda_{\mathrm{I}}^{\mathrm{B}}}{\sqrt{(2\Delta - \lambda_{\mathrm{I}}^{\mathrm{A}} + \lambda_{\mathrm{I}}^{\mathrm{B}})^2 + 16\lambda_{\mathrm{R}}^2}}\right]. \tag{7}$$

The normalized eigenstates $\psi_{\nu\mu}$ at the K point in the basis $\sigma \otimes s = |\mathrm{A}\uparrow, \mathrm{A}\downarrow, \mathrm{B}\uparrow, \mathrm{B}\downarrow\rangle$ read

$$\begin{aligned}
\psi_{+-} &= |0,\, i\alpha_+ Q_+,\, 4\lambda_{\mathrm{R}}\alpha_+,\, 0\rangle \simeq |\mathrm{A}\downarrow\rangle, \\
\psi_{++} &= |1, \quad 0, \qquad 0, \quad 0\rangle = |\mathrm{A}\uparrow\rangle, \\
\psi_{--} &= |0, \quad 0, \qquad 0, \quad 1\rangle = |\mathrm{B}\downarrow\rangle, \\
\psi_{-+} &= |0,\, i\alpha_- Q_-,\, 4\lambda_{\mathrm{R}}\alpha_-,\, 0\rangle \simeq |\mathrm{B}\uparrow\rangle,
\end{aligned} \tag{8}$$

where $\alpha_\pm = 1/\sqrt{16\lambda_{\mathrm{R}}^2 + Q_\pm^2}$ and $Q_\pm = \pm\sqrt{(2\Delta + \lambda_{\mathrm{I}}^{\mathrm{A}} + \lambda_{\mathrm{I}}^{\mathrm{B}})^2 + 16\lambda_{\mathrm{R}}^2} + 2\Delta + \lambda_{\mathrm{I}}^{\mathrm{A}} + \lambda_{\mathrm{I}}^{\mathrm{B}}$. The eigenvectors are ordered assuming the following increase in energy: $\varepsilon_{+-} > \varepsilon_{++} > \varepsilon_{--} > \varepsilon_{-+}$. Analyzing the eigenvectors we see that the valence bands are localized on the sublattice B, while the conduction bands on the sublattice A, in the so called normal order. The $z$ component of the spin alternates from band to band. The top valence $\psi_{--}$ and bottom conduction $\psi_{++}$ states are pure pseudospin and spin states. On the other hand, Rashba spin-orbit coupling mixes spin and pseudospin of the outermost states. The eigenstates at K$'$ have the same form, but opposite spins.

Using the formulas for the eigenenergies, Eq.(6), and for the spin expectation values, Eq.(7), we can algebraically extract the orbital band gap $\Delta$ and the three spin-orbit parameters $\lambda_{\mathrm{I}}^{\mathrm{A}}$, $\lambda_{\mathrm{I}}^{\mathrm{B}}$, and $\lambda_{\mathrm{R}}$ by comparing to first-principles data for the fine structure at the K point.

## 4. Survey of first-principles band structures

Electronic band structure of graphene on TMD was calculated within density functional theory[44] (DFT) as implemented in plane wave suite QUANTUM ESPRESSO.[45] To properly describe the low energy electronic states of graphene a set of appropriate parameters and extensions were used.[46] Both graphene and TMDs have incommensurate lattice constants. To reduce structural strain we constructed a large supercell comprising $3 \times 3$ supercell of TMD and $4 \times 4$ supercell of graphene. In this way we deal with residual lattice strains up to 1.5%. Structural relaxation provides graphene to TMD distance of van der Waals order $3.3 - 3.4$ Å. Graphene surface, in addition, is slightly corrugated with the averaged value of few pm.

In Fig. 3 we show calculated band structures of graphene on $MoS_2$, $MoSe_2$, $WS_2$ and $WSe_2$. Physical adsorption of graphene preserves linear dispersion of graphene $\pi$ states which appear within band gap of TMDs. We found that with increasing atomic number of chalcogen the Dirac point shifts towards valence edge of TMD.

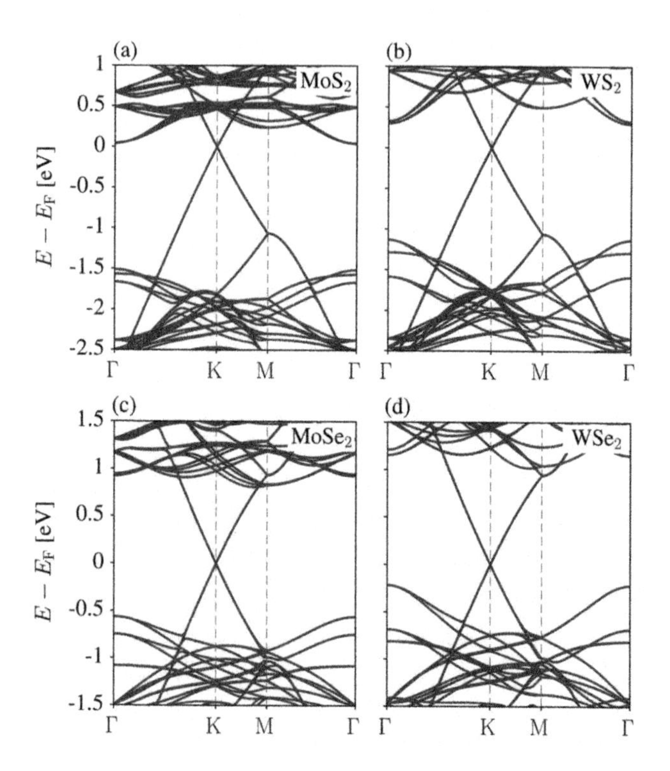


Fig. 3. Calculated electronic band structures for heterostructure of graphene on TMD along high symmetry lines: (a) $MoS_2$, (b) $WS_2$, (c) $MoSe_2$ and (d) $WSe_2$.

When graphene and TMD form van der Waals bilayer structures, the dipole moment pointing towards graphene is induced. Its amplitude is about 0.62 to 0.66 Debye for Mo based and W based TMDs, respectively. A gating voltage in experiment should effectively manipulate the induced dipole moment. To mimic the gating experiment we investigate the influence of an applied transverse electric field. The electric field is included within self-consistent DFT calculations. As positive electric fields we denote such a field that points from TMD towards graphene.

In Fig. 4 we show energy offsets for $MoS_2$ and $WSe_2$ as a function of the transverse electric field. For positive fields the Dirac cone is shifted within the TMD band gap towards conduction band edge of TMD, while negative transverse fields move the cone towards valence band edge. In

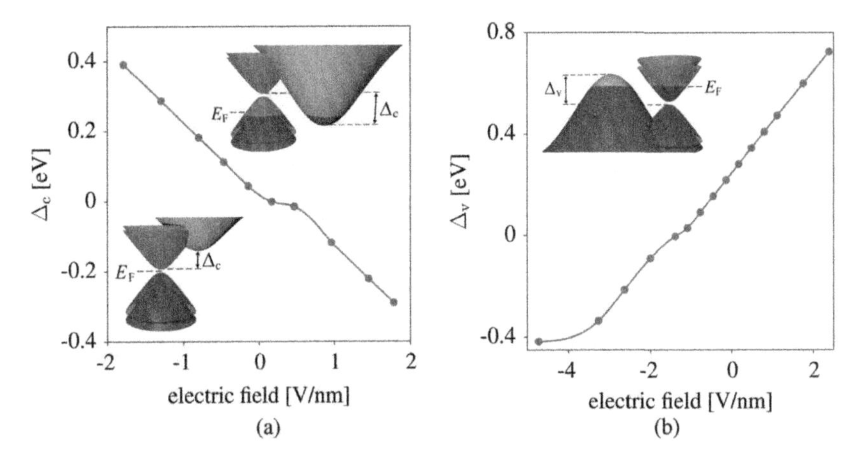


Fig. 4. Calculated band offset (a) $\Delta_c$ from the conduction band minimum of $MoS_2$, (b) $\Delta_v$ from valence band maximum of $WSe_2$ to graphene valence band maximum as a function of the applied transverse electric field.

case of graphene on $MoS_2$ we found that for negative transverse fields and positive fields up to 0.2 V/nm the band offset $\Delta_c$ stays positive, see Fig. 4. The $\Delta_c$ is the energy offset from conduction band minimum of $MoS_2$ to graphene conduction band minimum. For fields larger than 0.2 V/nm the Dirac cone is shifted beyond the conduction band edge. Graphene gets $p$ doped while the $MoS_2$ is $n$ doped. Fermi surface is a mixture of massless-massive electron-hole system. For graphene on $WSe_2$ the Dirac cone is much closer to the valence band edge of $WSe_2$. Therefore for negative transverse fields we may observe the Fermi surface of a mixture of graphene Dirac electrons and $WSe_2$ massive holes. Indeed, for fields below $-1.4$ V/nm graphene gets $n$ doped and $WSe_2$ gets $p$ doped. In Fig. 4(b) we show band offset $\Delta_v$, which is the difference between the valence band maximum of graphene and $WSe_2$, as a function of the electric field.

## 5. Topology of Dirac band structure

As the physisorption in graphene/TMD heterostructures preserves orbital character of electronic band structure, the spin-orbit proximity effect would be apparent at smaller energy scales. The proximity-induced fine topological features depend on the TMD material and should be apparent in vicinity of the Dirac point. In Fig. 5 we show zoom into the Dirac point within few meV energy range.

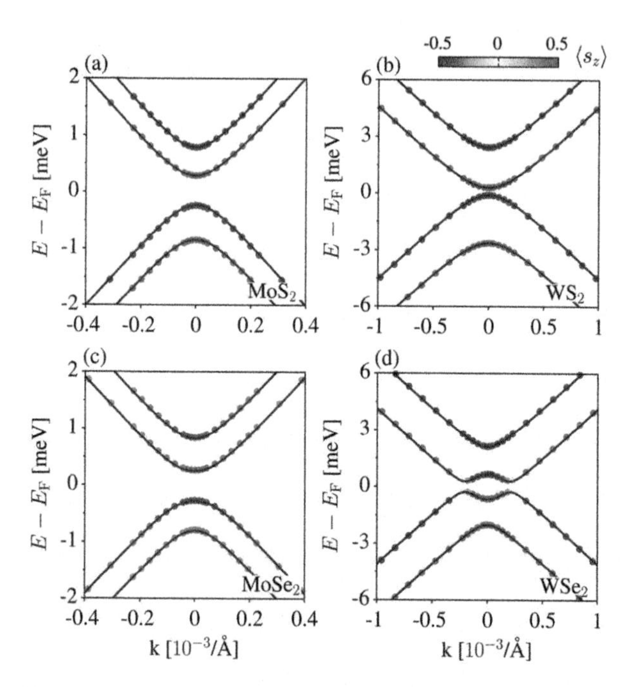


Fig. 5. Calculated electronic band structures in the vicinity of the Dirac point for graphene/TMD heterostructures: (a) $MoS_2$, (b) $WS_2$, (c) $MoSe_2$, and (d) $WSe_2$. The solid lines are model fits, while the circles are first-principles data. Colors code the $z$-component of the spin expectation value. Adapted from Ref. 21.

For graphene on $MoS_2$, $WS_2$, and $MoSe_2$ the band structure topology in the vicinity of the Dirac point share the same topology, see Fig. 5(a-c). The essential features are (i) opening of an orbital gap due to the effective staggered potential (on average, atoms A and B in the graphene supercell see a different environment coming from the TMD layer), (ii) anticrossing of the bands due to the intrinsic spin-orbit coupling, and (iii) spin splittings of the bands due to spin-orbit coupling and breaking of the space inversion symmetry. Both the orbital gap and spin-orbit couplings are on the meV scales, which are giant compared to the 10 $\mu$eV spin-orbit splitting in pristine graphene.[39] In Fig. 5 we also show the spin character of the bands around K. We find that the valence states are formed at the B sublattice while the conduction states live on A. The same orbital ordering is at K′. The spin alternates as we go through the bands. At K′ the spin orientation is opposite.

The case of graphene on $WSe_2$ is exceptional. Figure 5(d) shows an *inverted band structure*, which is an indication for a nontrivial topological ordering. While far from K the band ordering in the Dirac band structure of graphene on $WSe_2$ looks the same as in the other three cases, close to K the two lowest energy bands anticross. The anticrossing gap is proportional to Rashba coupling, in this case it is about 0.56 meV. The top of the valence band and the bottom of the conduction band have opposite spins to the rest of the states of the same bands.

## 6. Dirac bands captured by the model

The solid lines in Fig. 5 correspond to the model described in Sec. 3. The extracted orbital and spin-orbit coupling parameters are listed in Tab. 1. Fermi velocity $v_F$ and the orbital hopping $t$ are related via $\hbar v_F = \sqrt{3}\,ta/2$. Their renormalized values, listed in Tab. 1, for studied TMDs are a direct consequence of the implemented lateral strain in order to build the commensurate supercells for DFT calculations. It has been shown that other commensurate structural models (different supercell sizes) with different lateral strain as well as vertical misalignment influence spin-orbit coupling parameters up to 10%.[21]

The proximity spin-orbit parameters $\lambda_I^A$ and $\lambda_I^B$ are in the range $0.2-1.2$ meV which is about 20 to 100 times more than in pristine graphene.[39] Similar giant values of spin-orbit coupling in graphene are induced by hydrogen adatoms,[8–10] and even more by fluorine[47] or copper,[48] though the mechanisms are different. Unlike in the adatom cases in which the induced spin-orbit coupling is only local, in our case the giant coupling is global. While the intrinsic parameters $\lambda_I^A$ and $\lambda_I^B$ change rather moderately with applying the transverse electric field, the Rashba parameter $\lambda_R$, is rather sensitive to the field.[21]

Table 1. Calculated orbital and spin-orbit parameters for graphene/TMD heterostructures. Labels: $v_F$ is the Fermi velocity of the Dirac states in graphene on TMDs, $t$ is the hopping energy of graphene's $\pi$ electrons, $\Delta$ is the induced orbital gap of graphene, $\lambda_I^A$ and $\lambda_I^B$ are the intrinsic spin-orbit couplings for A and B graphene sublattices, $\lambda_R$ is the Rashba spin-orbit coupling, and $\lambda_{PIA}^A$ and $\lambda_{PIA}^B$ are the pseudospin-inversion-asymmetry (PIA) spin-orbit terms for the two sublattices.

| TMD | $v_F/10^5$ [m/s] | $t$ [eV] | $\Delta$ [meV] | $\lambda_I^A$ [meV] | $\lambda_I^B$ [meV] | $\lambda_R$ [meV] | $\lambda_{PIA}^A$ [meV] | $\lambda_{PIA}^B$ [meV] |
|---|---|---|---|---|---|---|---|---|
| $MoS_2$ | 8.506 | 2.668 | 0.52 | -0.23 | 0.28 | 0.13 | -1.22 | -2.23 |
| $MoSe_2$ | 8.223 | 2.526 | 0.44 | -0.19 | 0.16 | 0.26 | 2.46 | 3.52 |
| $WS_2$ | 8.463 | 2.657 | 1.31 | -1.02 | 1.21 | 0.36 | -0.98 | -3.81 |
| $WSe_2$ | 8.156 | 2.507 | 0.54 | -1.22 | 1.16 | 0.56 | -2.69 | -2.54 |

In the case of graphene on $MoS_2$, $WS_2$, and $MoSe_2$ the similarity of the band structure topology near Dirac point, see Fig. 5(a-c), is also captured by the extracted values of parameters listed in Tab. 1. There the staggered potential $\Delta$ is larger than the absolute values of the induced intrinsic spin-orbit coupling parameters. For graphene on $WSe_2$ the $\Delta$ is smaller than the absolute values of the intrinsic spin-orbit coupling reflecting thus band inversion. In all the cases we see that the $\lambda_I^B \simeq -\lambda_I^A$. An interplay between the intrinsic spin-orbit coupling and staggered potential drives band inversion as illustrated in Fig. 6 where the spectrum at the K point is shown as a function of $\lambda_I$ ($\lambda_I \equiv \lambda_I^B = -\lambda_I^A$). The four levels are doubly spin degenerated at $\lambda_I = 0$ but separated by the staggered potential energy of $2\Delta$. With increasing $\lambda_I$ the levels further spin split. The outermost states are a combination of both the pseudospin and spin states, $\psi_{+-} = a_+|A\downarrow\rangle + b_+|B\uparrow\rangle$ and $\psi_{-+} = a_-|B\uparrow\rangle + b_-|A\downarrow\rangle$. For both branches, electrons ($\nu = +1$) and holes ($\nu = -1$), the $|b_\nu| \ll |a_\nu|$, and the eigenvalues equal $\nu\sqrt{(\Delta + \lambda_I)^2 + 4\lambda_R^2}$. For explicit expressions of $a_\nu$ and $b_\nu$ compare to Eqs. (8). The inner states are pure spin and pseudospin resolved. Their order determines the regime of the system. For $\lambda_I < \Delta$ the spin is alternating, and we call it as normal regime. At $\lambda_I = \Delta$ the levels cross and the system enters band inversion regime, where for $\lambda_I > \Delta$ the pseudospin is becoming alternating for the consecutive levels. At the critical point, $\lambda_I = \Delta$, the global gap of the system closes at K point and the band inversion regime suggests for existence of non-trivial, topological regime of the system.

Let us briefly discuss what is the origin of the induced giant spin-orbit coupling in graphene on TMD. We trace the enhancement to the hybridiza-

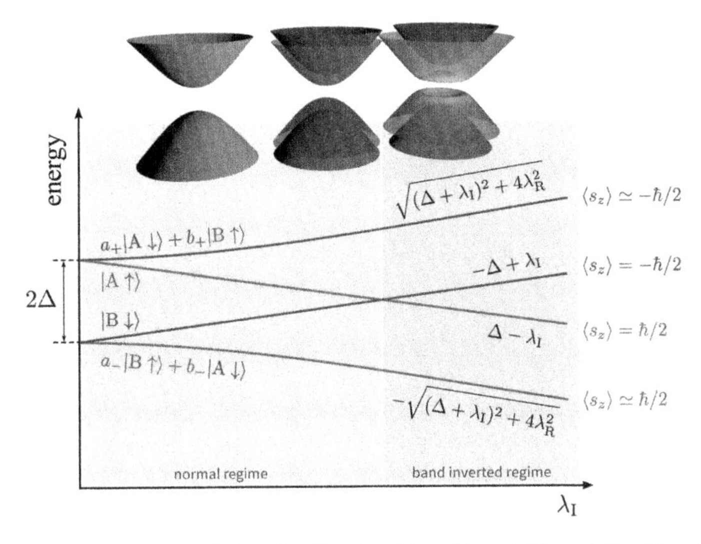


Fig. 6. Energy spectrum of the Hamiltonian $\mathcal{H} = \mathcal{H}_{\text{orb}} + \mathcal{H}_{\text{so}}$ at the K point as a function of the intrinsic spin orbit-coupling parameter $\lambda_{\text{I}}$ ($\lambda_{\text{I}} \equiv \lambda_{\text{I}}^{\text{B}} = -\lambda_{\text{I}}^{\text{A}}$). At the top the evolution of the graphene band structure topologies near the Dirac point for $\lambda_{\text{I}} = 0$ and $\lambda_{\text{I}} < \Delta$ within normal regime as well as for $\lambda_{\text{I}} > \Delta$ in band inverted non-trivial topological regime are sketched. Here $\lambda_{\text{PIA}}^{\text{A}} = \lambda_{\text{PIA}}^{\text{B}} = 0$.

tion of the carbon orbitals with the $d$-orbitals of metal atom of TMD. We find only 0.3% of $d$-orbitals at the K point by analyzing the calculated density of states for graphene on $\text{MoS}_2$. But when we turn the spin-orbit coupling on Mo atoms off, the orbital gap in zero field remains almost unchanged ($\Delta = 0.506$ meV), while the spin-orbit parameters drop to their pristine graphene values $\lambda_{\text{I}}^{\text{A}} \simeq \lambda_{\text{I}}^{\text{B}} = 24$ $\mu$eV and $\lambda_{\text{R}} = 10$ $\mu$eV, which are, curiously, also determined by $d$ orbitals, but from carbon atoms.[39]

## 7. Origin of effective parameters

To understand proximity-induced changes in graphene's band structure we proposed in the previous sections an effective model describing its low energy states. As graphene and TMD have incommensurate lattices, not only graphene sublattices A and B differ, but even the sites that belong to the same sublattice experience different local environments. This would imply that in reality all the orbital and spin-orbit hoppings are positions dependent, i.e the full Hamiltonian in the local atomic basis would read

$$\begin{aligned}\mathcal{H}_{\text{full}} = &-\sum_{\sigma}\sum_{\langle m,n\rangle} t(m,n)\,|X_m\,\sigma\rangle\,\langle X_n\,\sigma| + \sum_{\sigma,m}\Delta^{\mathrm{X}}(m)|X_m\,\sigma\rangle\,\langle X_m\,\sigma| \\ &+\sum_{\sigma}\sum_{\langle\langle m,n\rangle\rangle}\frac{\nu_{m,n}}{\sqrt{3}}\left[i\hat{s}_z\right]_{\sigma\sigma}\frac{1}{3}\lambda_{\mathrm{I}}^{\mathrm{X}}(m,n)\,|X_m\,\sigma\rangle\,\langle X_n\,\sigma| \\ &+\sum_{\sigma\neq\sigma'}\sum_{\langle\langle m,n\rangle\rangle}\left[i\hat{\boldsymbol{s}}\times\mathbf{d}_{m,n}\right]_{\sigma\sigma'}\frac{2}{3}\lambda_{\mathrm{PIA}}^{\mathrm{X}}(m,n)\,|X_m\,\sigma\rangle\,\langle X_n\,\sigma'| \\ &+\sum_{\sigma\neq\sigma'}\sum_{\langle m,n\rangle}\left[i\hat{\boldsymbol{s}}\times\mathbf{d}_{m,n}\right]_{\sigma\sigma'}\frac{2}{3}\lambda_{\mathrm{R}}(m,n)\,|X_m\,\sigma\rangle\,\langle X_n\,\sigma'|\,.\end{aligned} \tag{9}$$

Here $|X_m\,\sigma\rangle$ is a carbon $2p_z$ orbital with spin $\sigma$ located on a lattice site $m$ that belongs to sublattice $X = \{\mathrm{A},\mathrm{B}\}$, $\hat{\boldsymbol{s}}$ stands for a vector of Pauli spin matrices, $\mathbf{d}_{m,n}$ is a unit vector pointing from the site $n$ to $m$ and $\nu_{m,n} = \mathrm{sign}[(\vec{m}-\vec{j})\times(\vec{j}-\vec{n})]_z$, where $j$ is a lattice site that is a common nearest neighbor of $m$ and $n$. Summations over $m$, $\langle m,n\rangle$, and $\langle\langle m,n\rangle\rangle$ run over the lattice sites (staggered potential), direct nearest neighbor pairs (orbital and Rashba terms) and the next nearest neighbor pairs (intrinsic and PIA spin-orbit terms), respectively. The superscript $X$ on a particular coupling constant accompanying a hopping $|X_m\rangle\langle X_n|$ between two sites on the same sublattice emphasizes this sublattice explicitly.

To extract the full set of model parameters: $t(m,n)$, $\Delta^{\mathrm{X}}(m)$, $\lambda_{\mathrm{I}}^{\mathrm{X}}(m,n)$, $\lambda_{\mathrm{PIA}}^{\mathrm{X}}(m,n)$, $\lambda_{\mathrm{R}}(m,n)$, would be a formidable task. They will depend on the specific supercell arrangement of graphene on TMD. For example, taking a larger supercell that could further reduce strain, or considering twist or glide between the two layers, as could happen in experiments, would lead to a different and enlarged set of model parameters. Such an approach is in essence not practical, and, therefore, we look for an effective alternative. It is clear, that for each supercell arrangement, the particular hopping, say $\lambda(m,n)$, would be a periodic function of the supercell translations, i.e. $\lambda(m,n) = \lambda(m+\mathbf{R}_a, n+\mathbf{R}_a)$ for any superlattice vector $\mathbf{R}_a$. Therefore, $\lambda(m,n)$ can be expanded into the series with respect to some suitable system of functions $f_i(m,n)$ having the supercell periodicity, i.e. $\lambda(m,n) = \lambda_0 + \sum_i \lambda_i f_i(m,n)$. Our effective model aims to capture the basic physics of graphene on TMD near the K points, i.e. for electrons having long wave lengths. Therefore, they would rather weakly experience fluctuations governed by $\sum_i \lambda_i f_i(m,n)$. So in the long wave length limit it is reasonable to approximate $\lambda(m,n)$ just by a constant term $\lambda_0$. Doing so, we reduce the full model Hamiltonian $\mathcal{H}_{\text{full}}$, given by Eq. (9), to its effective long wave length form:

$$\begin{aligned}\mathcal{H} = &-t\sum_{\sigma}\sum_{\langle m,n\rangle}|X_m\,\sigma\rangle\langle X_n\,\sigma| + \sum_{\sigma,m}\Delta^{\mathrm{X}}|X_m\,\sigma\rangle\langle X_m\,\sigma| \\ &+\sum_{\sigma}\sum_{\langle\langle m,n\rangle\rangle}\frac{\nu_{m,n}}{\sqrt{3}}\left[i\hat{s}_z\right]_{\sigma\sigma}\frac{1}{3}\lambda_{\mathrm{I}}^{\mathrm{X}}\,|X_m\,\sigma\rangle\langle X_n\,\sigma| \\ &+\sum_{\sigma\neq\sigma'}\sum_{\langle\langle m,n\rangle\rangle}\left[i\hat{\mathbf{s}}\times\mathbf{d}_{m,n}\right]_{\sigma\sigma'}\frac{2}{3}\lambda_{\mathrm{PIA}}^{\mathrm{X}}\,|X_m\,\sigma\rangle\langle X_n\,\sigma'| \\ &+\sum_{\sigma\neq\sigma'}\sum_{\langle m,n\rangle}\left[i\hat{\mathbf{s}}\times\mathbf{d}_{m,n}\right]_{\sigma\sigma'}\frac{2}{3}\lambda_{\mathrm{R}}\,|X_m\,\sigma\rangle\langle X_n\,\sigma'|\,,\end{aligned} \tag{10}$$

where we also assume $\Delta^{\mathrm{A}}+\Delta^{\mathrm{B}}=0$. For more details and symmetry related discussion see Ref. 49. Transforming $\mathcal{H}$ to Bloch basis and expanding it around the Dirac points would lead to the Hamiltonian $\mathcal{H}_0+\mathcal{H}_\Delta+\mathcal{H}_{\mathrm{so}}$ that was discussed in the previous sections. As a comment, it is not generally excluded, that for some supercell arrangements, e.g. long wave length Moire patterns, the averaging over the supercell for some model parameters (such as $\Delta$ and the difference between $\lambda_{\mathrm{I}}^{\mathrm{A}}$ and $\lambda_{\mathrm{I}}^{\mathrm{B}}$) can give zero.

## 8. Topological states in graphene nanoribbon on $WSe_2$

The inverted band structure is a precursor of the quantum spin Hall effect. Although zigzag graphene nanoribbons were predicted to host helical edge states,[16] intrinsic spin-orbit coupling in graphene is too weak[39] for such states to be experimentally realized. Instead, 2d (Hg,Cd)Te quantum wells have emerged as a prototypical quantum spin Hall system.[17–19]

The results presented in Sec. 5 and 6 strongly suggest that band inversion in graphene on monolayer $WSe_2$, due to the strong *proximity* spin-orbit coupling, acts as a quantum spin Hall insulator. Rashba spin-orbit coupling is responsible for the anticrossing

$$\Delta_{\mathrm{R}} = 4\lambda_{\mathrm{R}}\sqrt{\frac{\lambda_{\mathrm{I}}^{\mathrm{B}}-\Delta}{\lambda_{\mathrm{I}}^{\mathrm{B}}-\lambda_{\mathrm{I}}^{\mathrm{A}}}}\left[1+\frac{(\lambda_{\mathrm{I}}^{\mathrm{B}}+\Delta)^2}{(\lambda_{\mathrm{I}}^{\mathrm{B}}-\Delta)(\Delta-\lambda_{\mathrm{I}}^{\mathrm{A}})}\right]^{-1/2}, \tag{11}$$

that emerges along the circle centered at the K points, see the Mexican hat features at Fig. 5(d). We call $\Delta_{\mathrm{R}}$ as the Rashba gap,[21] while the strong Rashba spin-orbit coupling makes graphene on $WSe_2$ to be an insulator. This behavior is robust against an applied transverse electric field and vertical strain.[21]

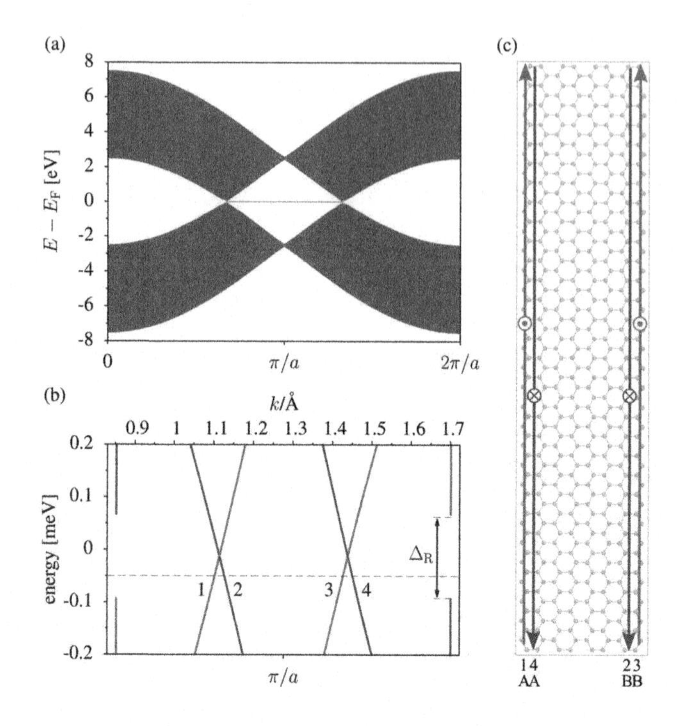


Fig. 7. Helical edge states in a zigzag graphene nanoribbon on $WSe_2$, with the width of 200 nm. (a) Calculated band structure for a zigzag graphene nanoribbon on $WSe_2$. Tight-binding model described in Ref. 21 and parameters used in the calculation are from Tab. 1. (b) Electronic states with spin up (red) and spin down (blue) within the Rashba anticrossing gap $\Delta_R$. Labels 1&4 and 2&3 denote time-reversed pair states. (c) Sketch of the helical states for energy -0.05 meV with the labels for spatial and sublattice localization, as well spin up $\odot$ and spin down $\otimes$ character. In-plane vertical up (down) arrows indicate positive (negative) group velocities.

To demonstrate the presence of helical edge states we have converted our effective Hamiltonian $\mathcal{H}$ into a tight-binding model[21], see Eq. (10) in the previous section, or an earlier work on hydrogenated graphene,[9] and analyzed the energy spectra and states of zigzag nanoribbons of graphene on TMDs. The results for graphene on $WSe_2$ are shown in Fig. 7(a), for a nanoribbon of 200 nm width. A closer look at the Fermi level reveals four bands within the Rashba gap, see Fig. 7(b). The four bands appear to be paired forming time-reversed Kramers doublets that cross at two Dirac cones, these are the helical edge states defining the quantum spin Hall regime. For example, doublet states 1&4 (2&3) in Fig. 7(c) are counter propagating spin-polarized edge states localized on a sublattice A (B) as indicated at the figure. These states are topologically protected by time-reversal against the backscattering on non-magnetic impurities. Employing model parameters from Table 1 for graphene on $MoS_2$, $MoSe_2$, and $WS_2$, which have trivial Dirac bulk bands, see Fig. 5(a-c), zigzag nanoribbons remain insulating as well, featuring no helical edge states.

## 9. Platform for optospintronics

Graphene on TMDs may serve as an ideal platform for optospintronics. In Fig. 8 we show a proposal of a device that utilizes an optical spin injection into graphene. A circularly polarized light, tuned to the band gap of TMD, excites electron spins by optical orientation.[2,50] In effect, the light produces spin-polarized excitons which dissociate into spin-polarized electrons and holes. As in the recent optical experiment,[38,51] we expect that electrons will be transferred to graphene, leaving holes behind in TMD, although in which way electrons and holes split may depend on the TMD material as well as on gating. The spin-polarized electrons (or holes) diffuse in graphene. One can detect this spin accumulation either optically, by observing a circular polarization of the photoluminescence[50] elsewhere in graphene on TMD, or electrically, or using Kerr spectroscopy.[52] The latter is illustrated in Fig. 8 where a ferromagnetic electrode on top of graphene detects the presence of the spin accumulation in graphene.[2,3] Spin precession in graphene can be observed as the Hanle signal (which is not possible to see in the spin-valley coupled TMD[53]) by applying an external magnetic field transverse to the injected spin, providing Larmor precession.[3]

It has been already demonstrated that TMD[54] and graphene[55] have a range of unique and complementary optoelectronic properties. Assembling them in a vertical heterostructures enables to combine separate properties in one device, thus creating multifunctional optoelectronic systems with superior performance. It has been demonstrated that such vertical devices allow to study electron-hole separation and recombination and set quantum efficiency of these van der Waals heterostructures[51] Spin transport *per se* in graphene/TMD bilayers should be fascinating too. The presence of the giant, effectively uniform spin-orbit fields gives large spin Hall signals, even greater than in hydrogenated graphene.[10] Weak-antilocalization experiments[40,41] confirm the proximity enhanced spin-orbit coupling in graphene and bilayer graphene on TMD. Most important, the spin, like charge, properties of these structures are expected to be highly field

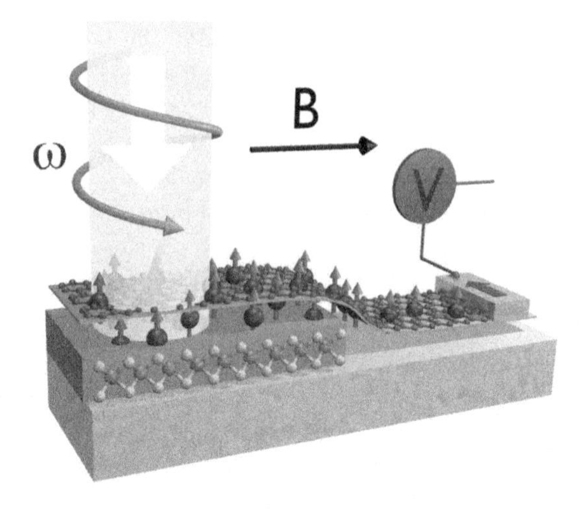


Fig. 8. Optospintronic device based on graphene/TMD hybrids. Optical spin injection into graphene, facilitated by the semiconducting TMD. A circularly polarized light excites spin-polarized electrons in the semiconductor. The spin is transferred to graphene where it can be detected as a Hanle signal by the ferromagnetic electrode.

tunable. The fascinating prospect of realizing the massive-massless electron gas coupling of the two electron gases, if the Fermi level is positioned in both band structures, calls for new theories of spin transport and spin relaxation in such hybrid systems.

## 10. Conclusions

Proximity enhanced uniform spin-orbit coupling acting on Dirac electrons in graphene on TMD opens new routes for graphene-based spintronics applications. There is an apparent transition between two topological regimes – normal and the band inverted one – that can be controlled by tuning the strength of the intrinsic spin-orbit coupling and the staggered potential. This can be achieved, for instance, by application of a vertical strain. Combining unique optical properties of TMD one can utilize the vertical heterostructures in new optospintronics devices. In addition, the giant enhancement of spin-orbit coupling allows to realize quantum spin Hall states in graphene nanoribbons. There are robust helical spin polarized edge states, forming non zero spin current distinct from Kane-Mele model,[16] along the zig-zag direction of the graphene layer on TMD that are preserved against the backscattering. This indicates that the helical edge states deserve further theoretical and experimental studies.

## Acknowledgments

This work was supported by DFG SFB 689, GRK 1570, International Doctorate Program Topological Insulators of the Elite Network of Bavaria, and by the EU Seventh Framework Programme under Grant Agreement No. 604391 Graphene Flagship.


## References


1. A. K. Geim and I. V. Grigorieva, Van der Waals heterostructures, *Nature* **499**, 419 (2013).
2. I. Žutić, J. Fabian and S. Das Sarma, Spintronics: Fundamentals and applications, *Rev. Mod. Phys.* **76**, 323 (2004).
3. J. Fabian, A. Matos-Abiague, C. Ertler, P. Stano and I. Žutić, Semiconductor spintronics, *Acta Phys. Slovaca* **57**, 565 (2007).
4. W. Han, R. K. Kawakami, M. Gmitra and J. Fabian, Graphene spintronics, *Nat. Nanotechnol.* **9**, 794 (2014).
5. N. Tombros, C. Józsa, M. Popinciuc, H. T. Jonkman and B. J. van Wees, Electronic spin transport and spin precession in single graphene layers at room temperature, *Nature (London)* **448**, 571 (2007).
6. K. Pi, W. Han, K. M. McCreary, A. G. Swartz, Y. Li and R. K. Kawakami, Manipulation of spin transport in graphene by surface chemical doping, *Phys. Rev. Lett.* **104**, 187201 (2010).
7. T.-Y. Yang, J. Balakrishnan, F. Volmer, A. Avsar, M. Jaiswal, J. Samm, S. R. Ali, A. Pachoud, M. Zeng, M. Popinciuc, G. Güntherodt, B. Beschoten and B. Özyilmaz, Observation of long spin-relaxation times in bilayer graphene at room temperature, *Phys. Rev. Lett.* **107**, 047206 (2011).
8. A. H. Castro Neto and F. Guinea, Impurity-induced spin-orbit coupling in graphene, *Phys. Rev. Lett.* **103**, 026804 (2009).
9. M. Gmitra, D. Kochan and J. Fabian, Spin-orbit coupling in hydrogenated graphene, *Phys. Rev. Lett.* **110**, 246602 (Jun 2013).
10. J. Balakrishnan, G. Kok, W. Koon, M. Jaiswal and A. H. C. Neto, Colossal enhancement of spin-orbit coupling in weakly hydrogenated graphene, *Nat. Phys.* **9**, 1 (2013).
11. A. Avsar, J. Y. Tan, T. Taychatanapat, J. Balakrishnan, G. K. W. Koon, Y. Yeo, J. Lahiri, A. Carvalho, A. S. Rodin, E. C. T. O'Farrell, G. Eda, A. H. Castro Neto and B. Özyilmaz, Spin-orbit proximity effect in graphene, *Nat. Commun.* **5**, 4875 (2014).

12. Z. Qiao, S. A. Yang, W. Feng, W.-K. Tse, J. Ding, Y. Yao, J. Wang and Q. Niu, Quantum anomalous hall effect in graphene from rashba and exchange effects, *Phys. Rev. B* **82**, 161414 (2010).
13. C. Weeks, J. Hu, J. Alicea, M. Franz and R. Wu, Electrically tunable quantum anomalous hall effect in graphene decorated by 5d transition-metal adatoms, *Phys. Rev. X* **1**, 021001 (2011).
14. H. Zhang, C. Lazo, S. Blügel, S. Heinze and Y. Mokrousov, Electrically tunable quantum anomalous hall effect in graphene decorated by 5d transition-metal adatoms, *Phys. Rev. Lett.* **108**, 056802 (2012).
15. Z. Qiao, W. Ren, H. Chen, L. Bellaiche, Z. Zhang, A. MacDonald and Q. Niu, Quantum Anomalous Hall Effect in Graphene Proximity Coupled to an Antiferromagnetic Insulator, *Phys. Rev. Lett.* **112**, 116404 (2014).
16. C. L. Kane and E. J. Mele, $Z_2$ topological order and the quantum spin hall effect, *Phys. Rev. Lett.* **95**, 146802 (2005).
17. B. A. Bernevig, T. L. Hughes and S.-C. Zhang, Quantum Spin Hall Effect and Topological Phase Transition in HgTe Quantum Wells, *Science* **314**, 1757 (2006).
18. M. König, S. Wiedmann, C. Brüne, A. Roth, H. Buhmann, L. W. Molenkamp, X.-L. Qi and S.-C. Zhang, Quantum Spin Hall Insulator State in HgTe Quantum Wells, *Science* **318**, 766 (2007).
19. H. Zhang, C.-X. Liu, X.-L. Qi, X. Dai, Z. Fang and S.-C. Zhang, Topological insulators in $Bi_2Se_3$, $Bi_2Te_3$ and $Sb_2Te_3$ with a single Dirac cone on the surface, *Nat. Phys.* **5**, 438 (2009).
20. M. Gmitra and J. Fabian, Graphene on transition-metal dichalcogenides: A platform for proximity spin-orbit physics and optospintronics, *Phys. Rev. B* **92**, 155403 (2015).
21. M. Gmitra, D. Kochan, P. Högl and J. Fabian, Trivial and inverted dirac bands and the emergence of quantum spin hall states in graphene on transition-metal dichalcogenides, *Phys. Rev. B* **93**, 155104 (2016).
22. O. Lopez-Sanchez, D. Lembke, M. Kayci, A. Radenovic and A. Kis, Ultrasensitive photodetectors based on monolayer $MoS_2$, *Nat. Nanotechnol.* **8**, 497 (2013).
23. C. Huang, S. Wu, A. M. Sanchez, J. J. P. Peters, R. Beanland, J. S. Ross, P. Rivera, W. Yao, D. H. Cobden and X. Xu, Lateral heterojunctions within monolayer $MoSe_2$-$WSe_2$ semiconductors, *Nat. Mater.* **13**, 1096 (2014).

24. C.-H. Lee, G.-H. Lee, A. M. van der Zande, W. Chen, Y. Li, M. Han, X. Cui, G. Arefe, C. Nuckolls, T. F. Heinz, J. Guo, J. Hone and P. Kim, Atomically thin p-n junctions with van der Waals heterointerfaces, *Nat. Nanotechnol.* **9**, 676 (2014).
25. A. Pospischil, M. M. Furchi and T. Mueller, Solar-energy conversion and light emission in an atomic monolayer p-n diode, *Nat. Nanotechnol.* **9**, 257 (2014).
26. A. Kormányos, G. Burkard, M. Gmitra, J. Fabian, V. Zólyomi, N. D. Drummond and V. Fal'ko, **k.p** theory for two-dimensional transition metal dichalcogenide semiconductors, *2D Mater.* **2**, 022001 (2015).
27. D. Xiao, G.-B. Liu, W. Feng, X. Xu and W. Yao, Coupled Spin and Valley Physics in Monolayers of $MoS_2$ and Other Group-VI Dichalcogenides, *Phys. Rev. Lett.* **108**, 196802 (2012).
28. K. F. Mak, K. He, J. Shan and T. F. Heinz, Control of valley polarization in monolayer $MoS_2$ by optical helicity, *Nat. Nanotechnol.* **7**, 494 (2012).
29. H. Zeng, J. Dai, W. Yao, D. Xiao and X. Cui, Valley polarization in $MoS_2$ monolayers by optical pumping, *Nat. Nanotechnol.* **7**, 490 (2012).
30. Y.-C. Lin, N. Lu, N. Perea-Lopez, J. Li, Z. Lin, X. Peng, C. H. Lee, C. Sun, L. Calderin, P. N. Browning, M. S. Bresnehan, M. J. Kim, T. S. Mayer, M. Terrones and J. A. Robinson, Direct synthesis of van der Waals solids, *ACS Nano* **8**, 3715 (2014).
31. M.-Y. Lin, C.-E. Chang, C.-H. Wang, C.-F. Su, C. Chen, S.-C. Lee and S.-Y. Lin, Toward epitaxially grown two-dimensional crystal hetero-structures: Single and double $MoS_2$/graphene hetero-structures by chemical vapor depositions, *Appl. Phys. Lett.* **105**, 073501 (2014).
32. A. Azizi, S. Eichfeld, G. Geschwind, K. Zhang, B. Jiang, D. Mukherjee, L. Hossain, A. F. Piasecki, B. Kabius, J. A. Robinson and N. Alem, Freestanding van der Waals Heterostructures of Graphene and Transition Metal Dichalcogenides, *ACS Nano* **9**, 4882 (2015).
33. C.-P. Lu, G. Li, K. Watanabe, T. Taniguchi and E. Y. Andrei, $MoS_2$: Choice Substrate for Accessing and Tuning the Electronic Properties of Graphene, *Phys. Rev. Lett.* **113**, 156804 (2014).
34. H. Coy Diaz, J. Avila, C. Chen, R. Addou, M. C. Asensio and M. Batzill, Direct Observation of Interlayer Hybridization and Dirac Relativistic Carriers in Graphene/$MoS_2$ van der Waals Heterostructures, *Nano Lett.* **15**, 1135 (2015).

35. N. A. Kumar, M. A. Dar, R. Gul and J. Baek, Graphene and molybdenum disulfide hybrids: synthesis and applications, *Mater. Today* **18**, 286 (2015).
36. S. Bertolazzi, D. Krasnozhon and A. Kis, Nonvolatile Memory Cells Based on $MoS_2$/Graphene Heterostructures, *ACS Nano* **7**, 3246 (2013).
37. W. Zhang, C.-P. Chuu, J.-K. Huang, C.-H. Chen, M.-L. Tsai, Y.-H. Chang, C.-T. Liang, Y.-Z. Chen, Y.-L. Chueh, J.-H. He, M.-Y. Chou and L.-J. Li, Ultrahigh-Gain Photodetectors Based on Atomically Thin Graphene-$MoS_2$ Heterostructures, *Sci. Rep.* **4**, 3826 (2014).
38. K. Roy, M. Padmanabhan, S. Goswami, T. P. Sai, G. Ramalingam, S. Raghavan and A. Ghosh, Graphene-$MoS_2$ hybrid structures for multifunctional photoresponsive memory devices, *Nat. Nanotechnol.* **8**, 826 (2013).
39. M. Gmitra, S. Konschuh, C. Ertler, C. Ambrosch-Draxl and J. Fabian, Band-structure topologies of graphene: Spin-orbit coupling effects from first principles, *Phys. Rev. B* **80**, 235431 (2009).
40. Z. Wang, D.-K. Ki, H. Chen, H. Berger, A. H. MacDonald and A. F. Morpurgo, Strong interface-induced spin-orbit interaction in graphene on WS2, *Nat. Commun.* **6**, 8339 (2015).
41. Z. Wang, D.-K. Ki, J. Y. Khoo, D. Mauro, H. Berger, L. S. Levitov and A. F. Morpurgo, Origin and magnitude of 'designer' spin-orbit interaction in graphene on semiconducting transition metal dichalcogenides, *Phys. Rev. X* **6**, 041020 (2016).
42. J. W. McClure and Y. Yafet, Theory of the g-factor of the current carriers in graphite single crystals, in *Proceedings of the 5th Conference on Carbon*, (Pergamon New York, 1962).
43. S. Konschuh, M. Gmitra and J. Fabian, Tight binding theory of spin-orbit coupling in graphene, *Phys. Rev. B* **82**, 245412 (2010).
44. J. Hohenberg and W. Kohn, Inhomogeneous electron gas, *Phys. Rev.* **136**, B864 (1964).
45. P. Giannozzi and et al., Quantum espresso: a modular and open-source software project for quantum simulations of materials, *J.Phys.: Condens. Matter* **21**, 395502 (2009).
46. We used norm conserving pseudopotentials with a kinetic cutoff of 60 Ry for wave functions and 240 Ry for charge density. For exchange-correlation functional we considered generalized gradient approximation.[56] The reduced Brillouin zone was sampled by $12 \times 12$ $k$ points. Atomic positions were relaxed using quasi-Newton algorithm based on trust radius procedure until forces were smaller as $10^{-4}$ Ry/bohr

including the van der Waals interactions within semiempirical approach.[57,58] The supercell calculations were embeded in a slab geometry with vacuum of about 13 Å.
47. S. Irmer, T. Frank, S. Putz, M. Gmitra, D. Kochan and J. Fabian, Spin-orbit coupling in fluorinated graphene, *Phys. Rev. B* **91**, 115141 (2015).
48. T. Frank, S. Irmer, M. Gmitra, D. Kochan and J. Fabian, Copper adatoms on graphene: theory of orbital and spin-orbital effects, *arXiv:1610.01798* (2016).
49. D. Kochan, S. Irmer and J. Fabian, Model spin-orbit coupling hamiltonians for graphene systems, *arXiv: 1610.08794* (2016).
50. F. Meier and B. P. Zakharchenya, *Optical Orientation* (North-Holand, New York, 1984).
51. M. Massicotte, P. Schmidt, F. Vialla, K. G. Schädler, A. Reserbat-Plantey, K. Watanabe, T. Taniguchi, K. J. Tielrooij and F. H. L. Koppens, Picosecond photoresponse in van der Waals heterostructures, *Nat Nano* **11**, 42 (2016).
52. L. Yang, N. A. Sinitsyn, W. Chen, J. Yuan, J. Zhang, J. Lou and S. A. Crooker, Long-lived nanosecond spin relaxation and spin coherence of electrons in monolayer MoS2 and WS2, *Nat. Phys.* **11**, 830 (2015).
53. G. Sallen, L. Bouet, X. Marie, G. Wang, C. R. Zhu, W. P. Han, Y. Lu, P. H. Tan, T. Amand, B. L. Liu and B. Urbaszek, Robust optical emission polarization in $MoS_2$ monolayers through selective valley excitation, *Phys. Rev. B* **86**, 081301 (2012).
54. Q. H. Wang, K. Kalantar-Zadeh, A. Kis, J. N. Coleman and M. S. Strano, Electronics and optoelectronics of two-dimensional transition metal dichalcogenides, *Nat. Nano.* **7**, 699 (2012).
55. F. Bonaccorso, Z. Sun, T. Hasan and A. C. Ferrari, Graphene photonics and optoelectronics, *Nat. Photon.* **4**, 611 (2010).
56. J. P. Perdew, K. Burke and M. Ernzerhof, Generalized gradient approximation made simple, *Phys. Rev. Lett.* **77**, 3865 (1996).
57. S. Grimme, Semiempirical GGA-type density functional constructed with a long-range dispersion correction, *J. Comput. Chem.* **27**, 1787 (2006).
58. V. Barone, M. Casarin, D. Forrer, M. Pavone, M. Sambi and Vittadini, Role and effective treatment of dispersive forces in materials: Polyethylene and graphite crystals as test cases, *J. Comput. Chem.* **30**, 934 (2009).

# Bethe Ansatz and Integrability

T. Lulek

*Faculty of Physics, Adam Mickiewicz University*
*Umultowska 85, 61-614 Poznań, Poland*
*Faculty of Mathematics and Natural Sciences, University of Rzeszow*
*Pigonia 1, 35-310 Rzeszów, Poland*
*E-mail: tadlulek@amu.edu.pl*

Bethe Ansatz (BA) is known as a genious mathematical guess which admits an exact solution to some non-trivial many-body problems (integrable systems) which include the eigenproblem of the Heisenberg Hamiltonian for the ring of $N$ spins 1/2 with isotropic nearest-neighbour interaction (the XXX model) as the seminal example. Algebraic Bethe Ansatz (ABA) of Faddeev and Tachtajan is a formalism which reproduces this guess in an elegant algebraic form, in terms of monodromy, Lax operators and transfer matrix. Both approaches, the original BA (known nowadays as the coordinate BA) and ABA, imply a need to solve the system of BA equations for spectral Bethe parameters - the system of $r$ highly nonlinear algebraic equations, with $r$ being the number of spin deviations from the ferromagnetic saturation state $|\,0\rangle$. Then, each regular solution $(\lambda_1, ..., \lambda_r)$ yields the corresponding exact eigenstate of the system in a form $B(\lambda_1)...B(\lambda_r)\ |0\rangle$, where $B(\lambda)$ is the creation operator for an excitation characterized by the spectral parameter $\lambda$. The aim of our report is to point out another way for determination of exact eigenstates within ABA formalism, namely - by application of the set of $N$ commuting observables $T^{(l)}$ (constants of motion) from the transfer matrix $T(\lambda)$. Knowledge of spectra of these constants of motion allows us to determine exact eigenstates of the Heisenberg Hamiltonian, along the standard quantum-mechanical prescription of Dirac. It is worth to mention that the numerical problem of solving non-linear BA equations is replaced by elementary algebraic manipulations with matrices of finite dimension, and the results are given in the form of density matrices - the one-dimensional projectors onto exact eigenstates. Bethe parameters can be then recovered from already known eigenstates by famous Baxter $TQ$ equation.

*Keywords*: Commuting observables, rigged string configurations.

## 1. Introduction

It is well known that Bethe Ansatz (BA), proposed in a famous paper[1] of Bethe in 1931, provides a unique solution of a non-trivial many-body problem of quantum mechanics, namely the eigenproblem of the Heisenberg Hamiltonian for the one-dimensional ring of $N$ nodes, each with the spin 1/2, with isotropic nearest-neighbour interaction (the XXX model). This result has stimulated a variety of developments and achievements in mathematics and theoretical physics, in particular within quantum integrable systems.[2,3] An important notion associated with integrability of a quantum system is existence of a number of mutually commuting observables which provide a classification of basis states in the relevant Hilbert space. Such a set of observables has arised from the transfer matrix proposed by Baxter[2,4] within the context of more general models (XXZ, six- and eight-vertex models, and some their variants) which can be solved using BA form of eigenfunctions. A similar proposal, specified to the XXX model, is given by Faddeev and Takhtajan[5] (cf. also Refs. 6,7), known as the algebraic Bethe Ansatz (ABA). The latter reproduces BA in an algebraic form, providing thus a convenient tool for quantum information processing,[8] with the space of quantum states interpreted as a prototype of memory of a quantum computer consisting of $N$ qubits, and Lax operators, monodromy matrix and other ingredients of ABA as natural realizations of various quantum gates.

The strategy of reaching the desired solution of the relevant eigenproblem in both BA (referred recently to as the coordinate BA) and ABA involves need to solve the following system of equations

$$\left(\frac{\lambda_\alpha + i/2}{\lambda_\alpha - i/2}\right)^N = \prod_{\beta \in \tilde{r} \setminus \{\alpha\}} \frac{\lambda_\alpha - \lambda_\beta + i}{\lambda_\alpha - \lambda_\beta - i}, \quad \alpha \in \tilde{r}, \tag{1}$$

where $\tilde{r} = \{1, ..., r\}$, $r$ is the number of reversed spins ($r \leq N/2$), and $\lambda_\alpha$ is the spectral parameter associated with the $\alpha$'th Bethe pseudoparticle. Clearly, it is the system of $r$ algebraic equations for $r$ unknown parameters $\lambda_\alpha$, $\alpha \in \tilde{r}$. This system is internally coupled and so highly non-linear that its explicit solutions are available only in some special cases. Thus BA provides indeed the exact solution, but this solution is, in general, implicit: its explicit form for given integers $N$ and $r$ is hidden in non-linear BA equations (1). In this note we aim to propose another strategy of solving the eigenproblem of the Heisenberg Hamiltonian of the XXX model, which consists in exploiting the integrability of this model. Namely, we construct the transfer matrix $T(\lambda)$ along the prescription of ABA as the $N-$th degree polynomial with respect to the spectral parameter $\lambda$, derive mutually commuting observables from (operator) coefficients of this polynomial, and then determine spectra and common eigenstates of these observables along the standard quantum-mechanical prescription of Dirac.[9]

We point out that such an approach yields essential computational simplification since it replaces the need to solve BA equations (1) by some spectral problems for hermitian operators, each of them essentially reduced to determination of all roots of polynomials of a single variable - the eigenvalue of an observable, and next to routine matrix operations with appropriate projectors. These computations can be further simplified by explicit use of the translational symmetry of the XXX model within the so called basis of wavelets.[10–12] The results for a particular fixed value of $N$ can be unified due to Galois symmetry of a finite extension of the cyclotomic number field $\mathbb{Q}(\omega)$, $\omega = \exp(2\pi i/N)$, associated with appropriate characteristic polynomials of constants of motion ($\mathbb{Q}$ is the prime number field of rationals, and $\mathbb{Q}(\omega)$ is its extension by the primitive $N-$th root of unity, associated with the translational symmetry of the model).

## 2. The coordinate Bethe Ansatz

Let $\tilde{N} = \{j = 1, 2, ..., N\}$ be the set of all nodes of the magnetic ring, and $\tilde{2} = \{0, 1\} \equiv \{+, -\}$ be the standard basis of the space $\mathbb{C}^2$ of a 1/2-spin. Then

$$\tilde{2}^{\tilde{N}} = \{f : \tilde{N} \to \tilde{2}\} \tag{2}$$

is the set of all magnetic configurations, and

$$\mathcal{H} = \left(\mathbb{C}^2\right)^{\otimes N} \equiv \ell c_{\mathbb{C}} \left(\tilde{2}^{\tilde{N}}\right) \tag{3}$$

is the space of all quantum states of the XXX model, spanned over the orthonormal basis $\tilde{2}^{\tilde{N}}$ ($\ell c_{\mathbb{C}} X$ stands for the linear closure of a set $X$). Each node can be seen as occupied by a qubit $\mathbb{C}^2$, and the space $\mathcal{H}-$ as the memory of a quantum computer consisting of $N$ qubits. The Heisenberg Hamiltonian $\hat{H} \in \text{End}\, \mathcal{H}$ has the form

$$\hat{H} = \frac{1}{2} \sum_{j \in \tilde{N}} \left(\sigma_j \cdot \sigma_{(j+1) \bmod N} - \hat{I}_{\mathcal{H}}\right), \tag{4}$$

where $\sigma_j = (\sigma_j^x, \sigma_j^y, \sigma_j^z)$, $j \in \tilde{N}$, is the vector of standard Pauli matrices at the $j-$th node, and $\hat{I}_{\mathcal{H}} \in \text{End}\, \mathcal{H}$ is the unit operator. The main problem consists in determination of all eigenvalues and eigenstates of the Hamiltonian $\hat{H}$. Let

$$\tilde{2}^{\tilde{N}} = \bigcup_{r=0}^{N} Q^{(r)}, \tag{5}$$

with

$$Q^{(r)} = \{\mathbf{j} = (j_1, j_2, ..., j_r)\, |1 \leq j_1 < j_2 < ... < j_r \leq N\} \tag{6}$$

be the decomposition of the set of all magnetic configurations into subsets $Q^{(r)}$ of all configurations with exactly $r$ spin deviations from the ferromagnetic saturation state $|++...+\rangle$, $r = 0, 1, ..., N$. These spin deviations are located on the ring $\tilde{N}$ in positions given by the vector $\mathbf{j}$, with components $j_\alpha \in \tilde{N}$, $\alpha \in \tilde{r} \equiv \{1, 2, ..., r\}$, ordered increasingly. Let

$$\mathcal{H} = \bigoplus_{r=0}^{N} \mathcal{H}^r \tag{7}$$

be the corresponding decomposition of the state space $\mathcal{H}$ into subspaces

$$\mathcal{H}^r = \ell c_{\mathbb{C}} Q^{(r)}. \tag{8}$$

The spherical symmetry of the nodes imposes that the eigenproblem of $\hat{H}$ separates into subproblems, one for each subspace $\mathcal{H}$. Moreover, problems for $\mathcal{H}^r$ and $\mathcal{H}^{N-r}$ are related by a particle-hole symmetry $(+ \rightleftarrows -)$, so it is sufficient to consider only cases $r \leq N/2$ ("above equator"). The space $\mathcal{H}^r$ is the starting point of BA.

A reversed spin can be seen as a kind of a particle (a Bethe pseudoparticle), which in a state $|\mathbf{j}\rangle \in \mathcal{H}^r$ is located at the node $j_\alpha \in \tilde{N}$, $\alpha \in \tilde{r}$. Within the dynamics dictated by the Heisenberg Hamiltonian (4), Bethe pseudoparticle $\alpha \in \tilde{r}$ can jump from the node $j_\alpha$ to $j'_\alpha = (j_\alpha \pm 1) \mod N$, until the node $j'_\alpha$ is empty, otherwise the jump is forbidden More precisely, the action of $\hat{H}$, restricted to the (invariant) space $\mathcal{H}^r$, reads

$$\hat{H}\,|\mathbf{j}\rangle = \sum_{\mathbf{j}' \in Q^{(r)}_{\mathbf{j}}} (|\mathbf{j}'\rangle - |\mathbf{j}\rangle)\,, \tag{9}$$

where $Q^{(r)}_{\mathbf{j}} \subset Q^{(r)}$ is the set of all those magnetic configurations $\mathbf{j}'$ which are admissible by a single jump from the configuration $\mathbf{j}$. Within this picture, $\mathcal{H}^r$ can be interpreted as the space of all quantum states of the system of $r$ Bethe pseudoparticles on the ring $\tilde{N}$. Eq.(8) has the meaning of Schrödinger quantization of this system, that is the set of all its positions, or the classical configuration space. Locally, it is an $r-$dimensional hypercubic lattice, but forbidden jumps $\mathbf{j} \mapsto \mathbf{j}'$, that is, such that

$$j'_{(\alpha+1) \bmod r} = (j_\alpha + 1) \mod N, \tag{10}$$

introduce some $F-$dimensional boundaries, $1 \leq F \leq r-1$. Eventually, the classical configuration space $Q^{(r)}$ constitutes a prism consisting of $\binom{N}{r}$

points of a hypercubic lattice, with boundaries representing all possible islands of adjacent Bethe pseudoparticles.[10–12]

The celebrated BA itself is a guess function $\psi_{BA} : Q^{(r)} \to \mathbb{C}$, of the form

$$\psi_{BA}(j_1, ..., j_r) = \sum_{\pi \in \Sigma_r} A_\pi \prod a_{\pi(\alpha)}^{j_\alpha}, \tag{11}$$

where

$$a_\lambda = e^{ip_\alpha} = \frac{\lambda_\alpha + i/2}{\lambda_\alpha - i/2}, \quad \alpha \in \tilde{r} \tag{12}$$

are formulas which define three kinds of frequently used Bethe parameters: (i) $p_\lambda \in \mathbb{C}$ is the pseudomomentum of the $\alpha$–th Bethe pseudoparticle (ii) $\lambda$ is its spectral parameter, (iii) $a_\lambda$ is its portion of phase when jumping from a node $j$ to $j+1$, $\Sigma_r$ is the symmetric group on the set $\tilde{r}$ of labels of Bethe pseudoparticles, $\pi \in \Sigma_r$ is the scattering channel from $1, 2, ..., r$ to $\pi(1), \pi(2), ..., \pi(r)$, and

$$A_\pi = \exp\left(\frac{i}{2} \sum_{1 \le \alpha < \beta \le r} \phi_{\pi(\alpha)\pi(\beta)}\right), \quad \pi \in \Sigma_r, \tag{13}$$

is the scattering amplitude in the channel $\pi \in \Sigma_r$, with $\psi_{\alpha\beta}$ being antisymmetric two-particle scattering phases, related to Bethe parameters as

$$e^{i\phi_{\alpha\beta}} = -\frac{s_{\alpha\beta}}{s_{\beta\alpha}} = e^{-i\phi_{\beta\alpha}} \tag{14}$$

where

$$s_{\alpha\beta} = a_\alpha a_\beta - 2a_\beta + 1, \quad \alpha \in \tilde{r}, \beta \in \tilde{r} \backslash \{\alpha\}, \tag{15}$$

is the scattering factor for a particle $\alpha$ on another particle $\beta$. It is worth to observe that Bethe pseudoparticles are (i) distinguishable classically, by their location on the ring $\tilde{N}$, (ii) indistinguishable in a quantum BA state. Their identity is kept on the generic part of the classical configuration space $Q^{(r)}$, but is lost at solitonic scatterings at boundaries. Observe that the scattering factor $a_{\alpha\beta}$ of Eq. (15) is asymmetric under exchange $\alpha \rightleftarrows \beta$, but yields antisymmetric scattering phase $\phi_{\alpha\beta} = -\phi_{\beta\alpha} \bmod 2\pi$.

## 3. The Algebraic Bethe Ansatz (ABA)

As mentioned in the introduction, the formalism of ABA is well suited for processing of quantum information. It operates with single-node qubits

$(\mathbb{C}^2)_j$, $j \in \tilde{N}$, and also applies an auxiliary qubit, denoted by $V \cong \mathbb{C}^2$. The principal ingredient is the Lax operator

$$\hat{L}_j(\lambda) = \lambda I_{\mathcal{H}} \otimes I_V + \frac{1}{2} \sum_{i \in \{x,y,z\}} \sigma_j^i \otimes \sigma^i, \tag{16}$$

where $\lambda \in \mathbb{C}$ is the spectral parameter, $I_V$ is the unit operator in $V$, $\sigma^i$ are Pauli matrices in End $V$, and

$$\sigma_j^i = I_2 \otimes ... \otimes \sigma^i \otimes ... \otimes I_2, \quad i \in \{x,y,z\}, \quad j \in \tilde{N}, \tag{17}$$

is the Pauli matrix $\sigma^i$, immersed into the $j-$th factor $\left(\mathbb{C}^2\right)_j$ of the tensor power state space $\mathcal{H}$, with $I_2$ being the unit operator in $\mathbb{C}^2$. Thus $\hat{L}_j(\lambda)$ formally operates in $\mathcal{H} \otimes V$, but effectively only in $\left(\mathbb{C}^2\right)_j \otimes V$. When written in the auxiliary space $V$, it reads

$$\hat{L}_j(\lambda) = \begin{pmatrix} \lambda I_2 + \frac{i}{2} s_j^z & i s_j^- \\ i s_j^+ & \lambda I_2 - \frac{1}{2} s_j^z \end{pmatrix}, \tag{18}$$

where $\mathbf{s} = \sigma/2$ is the spin vector operator, and $s^\pm = s^x \pm i s^y$. Commutation rules for two Lax operators, $\hat{L}_j(\lambda)$ and $\hat{L}_j(\mu)$, each operator associated with its own auxiliary qubit, written in $V \otimes V$ have the form

$$R_{ab}(\lambda - \mu) L_{ja}(\lambda) L_{jb}(\mu) = L_{j_b}(\mu) L_{ja}(\lambda) R_{ab}(\lambda - \mu) \tag{19}$$

where $a, b \in \tilde{2}$, and the matrix $R(\lambda) \in$ End $(V \otimes V)$

$$R(\lambda) = \begin{pmatrix} \lambda + i & 0 & 0 & 0 \\ 0 & \lambda & i & 0 \\ 0 & i & \lambda & 0 \\ 0 & 0 & 0 & \lambda + i \end{pmatrix} \tag{20}$$

defines a quadratic algebra (or a quantum group) with the structure constants adjusted to the XXX model. This algebra satisfies the Yang-Baxter condition,[3,13] and has proved to reproduce the most of original results of Bethe in an elegant algebraic form. Essentially, ABA deals with the representation of this algebra in the space $\mathcal{H}$. Formally, one introduces the monodromy matrix in $\mathcal{H} \otimes V$ as the product of all Lax operators along the ring $\tilde{N}$, i.e.

$$M(\lambda) = L_N(\lambda) L_{N-1}(\lambda) ... L_1(\lambda) = \begin{pmatrix} \hat{A}(\lambda) & \hat{B}(\lambda) \\ \hat{C}(\lambda) & \hat{D}(\lambda) \end{pmatrix}, \tag{21}$$

so that the last formula presents a $2 \times 2$ matrix in the auxiliary space $V$, with entries $\hat{A}(\lambda), \hat{B}(\lambda), \hat{C}(\lambda), \hat{D}(\lambda)$ in End $\mathcal{H}$.

Now, one of the main results of ABA can be stated as follows. Let $(\lambda_1, ..., \lambda_r)$ be a solution of the system (1) of BA equations, such that (i) all $\lambda_\alpha$, $\alpha \in \tilde{r}$, are distinct and finite, (ii) all factors in Eq.(1) are non-zero (so that these equations are non singular). Then

$$|\lambda_1, ..., \lambda_r\rangle = \hat{B}(\lambda_1)...\hat{B}(\lambda_r)|+...+\rangle \tag{22}$$

is an exact eigenstate of the Heisenberg Hamiltonian $\hat{H}$, which coincides with $\psi_{BA}$ of Eq.(11) for this set of spectral parameters (cf. Eq. (12)) up tp normalization, and corresponds to the eigenvalue

$$E(\lambda_1, ..., \lambda_r) = -\sum_{\alpha \in \tilde{r}} \frac{1}{\lambda_\alpha^2 + \frac{1}{4}} = -\sum_{\alpha \in \tilde{r}} 2(\cos p_\alpha - 1). \tag{23}$$

Each exact eigenstate $|\lambda_1, ..., \lambda_r\rangle$ , obtained by means of Eq.(22), requires thus the knowledge of a solution $(\lambda_1, ..., \lambda_r)$ of BA equations (1), together with the non-diagonal operator function $\hat{B}(\lambda) \in \text{End}\, \mathcal{H}$ of the monodromy matrix $\hat{M}(\lambda)$. Accordingly, $\hat{B}(\lambda)$ has the interpretation of a creation operator for a magnon with the spectral parameter (rapidity) $\lambda \in \mathbb{C}$. Moreover, the quadratic algebra resulting from the Yang-Baxter relation (19) implies

$$\left[\hat{B}(\lambda), \hat{B}(\mu)\right] = 0, \tag{24}$$

so that the ordering in Eq.(22) is irrelevant.

Here we discuss another strategy of solving the eigenproblem of the Heisenberg Hamiltonian $\hat{H}$ (Eq.(19)) in the space $\mathcal{H}^r$ (Eq.(8)) with use of the formalism of ABA. We exploit the well known properties of the transfer matrix

$$\hat{T}(\lambda) = \text{trace}_V \hat{M}(\lambda) = \hat{A}(\lambda) + \hat{D}(\lambda). \tag{25}$$

The most important fact reads

$$\left[\hat{T}(\lambda), \hat{T}(\mu)\right] = 0, \quad \lambda \in \mathbb{C}, \mu \in \mathbb{C}, \tag{26}$$

that is, transfer matrices mutually commute for arbitrary values of the spectral parameter. It follows that the transfer matrix $\hat{T}(\lambda)$, generated within ABA by Eqs. (21) and (25), gives rise to a family of mutually commuting operators. In particular, they include the Heisenberg Hamiltonian

$$\hat{H} = i\frac{d}{d\lambda}\ln \hat{T}(\lambda)\bigg|_{\lambda=i/2} - NI_{\mathcal{H}} = iC_N^{-1}\frac{d\hat{T}}{d\lambda}\bigg|_{\lambda=i/2} - NI_{\mathcal{H}} \tag{27}$$

and the shift operator $C_N \in \text{End}\, \mathcal{H}$

$$C_N = i^{-N}\hat{T}\left(\frac{i}{2}\right) \tag{28}$$

which represents the cyclic group $C_N$–the translation symmetry group of the XXX model (we use for brevity the same symbol $C_N$ for the cyclic group and the generator in End $\mathcal{H}$).

We proceed to consider this family in some more detail. It follows from Eqs. (25) and (21) that $\hat{T}(\lambda)$ is a polynomial of degree $N$ in $\lambda$, i.e.

$$\hat{T}(\lambda) = \sum_{l=0}^{N} \hat{T}^{(l)} \lambda^l, \tag{29}$$

so that the commutativity (26) implies that the operator coefficients $\hat{T}^{(l)}$ mutually commute, and thus span an Abelian subalgebra in End $\mathcal{H}$, referred sometimes to as the Bethe aubalgebra.[14] An explicit calculation based on Eq.(18) yields

$$\hat{T}^{(N)} = 2I_{\mathcal{H}}, \quad \hat{T}^{(N-1)} = 0, \tag{30}$$

and the other $N-1$ operators $\hat{T}^{(l)}$, $l = 0, 1, ..., N-2$, are hermitian. When taken together with the $z-$ component $\hat{S}^z$ of the total spin operator

$$\hat{\mathbf{S}} = \frac{1}{2} \sum_{j \in \tilde{N}} \sigma_j, \tag{31}$$

they constitute the sequence $\left(\hat{T}^{(0)}, \hat{T}^{(1)}, ..., \hat{T}^{(N-2)}, \hat{S}^z\right)$ of $N$ mutually commuting observables, which can be used for a classification of states of the model, realizing thus the idea of integrability.

## 4. Dirac method for classification of states

The sequence $(\hat{T}^{(0)}, \hat{T}^{(1)}, ..., \hat{T}^{(N-2)}, \hat{S}^z)$ of mutually commuting observables provides the starting point for classification of exact eigenstates of the Heisenberg Hamiltonian by the corresponding sequences $|t_i^{(0)}, t_i^{(1)}, ..., t_i^{(N-2)}, N/2 - r\rangle$ of eigenvalues. Let

$$\text{spec } \hat{T}^{(l)} = \left\{ t_i^{(l)} \middle| \, i \in \tilde{s}(l) = \{1, 2, ..., s(l)\} \right\} \tag{32}$$

be the spectrum of the observable $\hat{T}_i^{(l)}$, that is, the set of all distinct eigenvalues, the eigenvalue $t^{(l)}$ with the degeneracy $g_i^l = \dim \mathcal{H}(t_i^{(l)})$, $i \in \tilde{s}(l)$, so that

$$\sum_{i=0}^{s(l)} t_i^{(l)} g_i^{(l)} = 2^N, \quad l = 0, 1, ..., N-2 \tag{33}$$

accounts for completeness of each observable $\hat{T}^{(l)}$. The eigenspace $\mathcal{H}(t_i^{(l)})$ corresponding to an eigenvalue $t_i^{(l)}$, $l = 0, 1, ..., N-2$, $i \in \tilde{s}(l)$, is determined by the projector [15] (cf. also Ref. 12)

$$P^{(l)}\left(t_i^{(l)}\right) = \prod_{i' \in \tilde{s}(l) \setminus \{i\}} \frac{\hat{T}^{(l)} - t_i^{(i)} I_{\mathcal{H}}}{t_{i'}^{(l)} - t_i^{(l)}}. \tag{34}$$

Clearly, for fixed $l$ these projectors are mutually orthogonal, i.e.

$$P^{(l)}(t_i^{(l)}) P^{(l)}(t_{i'}^{(l)}) = P^{(l)}(t_i^{(l)}) \delta_{ii'} \tag{35}$$

and complete, i.e.

$$\sum_{i \in \tilde{s}(l)} \hat{P}^{(l)}(t_i^{(l)}) = I_{\mathcal{H}}, \quad l = 0, 1, ..., N-2. \tag{36}$$

Clearly, the corresponding projector $P^r$ onto the eigenspace $\mathcal{H}\left(\frac{N}{2} - r\right)$ of $\hat{S}^z$ satisfies

$$\hat{P}^r \mathcal{H} = \mathcal{H}^r \equiv \mathcal{H}\left(\frac{N}{2} - r\right). \tag{37}$$

Here we mention also that the observable

$$\hat{T}^{(N-2)} = -\left(\hat{\mathbf{S}}\right)^2 + \frac{3}{4} N I_{\mathcal{H}} \tag{38}$$

is related to the total spin $S$ of the system, and thus

$$\hat{P}^r \hat{T}^{(N-2)} \mathcal{H} = \mathcal{H}^{rr} \subset \mathcal{H}^r. \tag{39}$$

In general, eigenspaces $\mathcal{H}\left(t_i^{(l)}\right)$ are multidimensional ($g_i^{(l)} > 1$). The standard method of Dirac consists in looking for common eigenspaces of the whole sequence of mutually commuting (and thus compatible) observables, which corresponds to the product of projectors

$$\hat{P}(i_0, i_1, ..., i_{N-2}, r) = \hat{P}^r \prod_{l=0}^{N-2} P^{(l)}\left(t_{i_l}^{(l)}\right), \quad i_l \in \tilde{s}(l), \ r = 0, 1, ..., N. \tag{40}$$

As a result, one gets either zero (when the sequence of eigenvalues $(i_0, i_1, ..., i_{N-2}, r)$ is incompatible (we simplify notation, with $i_l$ standing for $\hat{t}_{i_l}^{(l)}$), or a common eigenspace of all $N$ observables of the sequence. In the latter case, the fact that the Bethe algebra has a simple spectrum in each highest weight subspace $\mathcal{H}^{rr}$ (cf. Eq.(39)), implies that each non-zero projector (40) is one-dimensional, and thus can be presented as

$$\hat{P}(i_0, i_1, ..., i_{N-2}, r) = |\, i_0 i_1 ... i_{N-2} r \rangle \langle i_0 i_1 ... i_{N-2} r \,|, \tag{41}$$

that is, as the density matrix of an exact eigenstate of the Heisenberg Hamiltonian of the XXX model. This important fact was proved by Mukhin, Tarasov and Varchenko.[14] It moreover implies that

$$\sum_{i_0, i_1, ..., i_{N-2}, r} \hat{P}(i_0, i_1, ..., i_{N-2,r}) = I_{\mathcal{H}}, \tag{42}$$

which means that the eigenstates (40) form a complete set in the space $\mathcal{H}$. In other words, the sequence $\left(\hat{T}^{(0)}, \hat{T}^{(1)}, ..., \hat{T}^{(N-2)}, \hat{S}^z\right)$ constitutes a complete set of compatible observables, consistent with the original definition of Dirac.[9] It is worth to note that the Dirac method yields exact eigenstates in the form (41) of density matrices which involves normalization, whereas the original BA formulas (11)-(15) are unnormalized.

The procedure described above proposes another strategy of reaching solution of the eigenproblem of the Heisenberg Hamiltonian. The original BA, as well as ABA, need to solve the system of $r$ non-linear BA equations (1), for $r = 2, 3, ...$ up to $N/2$ or $(N-1)/2$ for $N$ even and odd, respectively. The procedure proposed above needs to solve $N-1$ spectral problems for observables $\hat{T}^{(l)}$, $l = 0, 1, ..., N-2$, that is, to find all roots of appropriate characteristic polynomials of a single variable, and then to make routine matrix manipulations with projectors along Eqs. (34) and (40). The result of this procedure has the form of density matrices for exact eigenstates expressed in terms of spectra spec$\hat{T}^{(l)}$, rather than spectral parameters $(\lambda_1, ..., \lambda_r)$ which are not necessary for determination of eigenstates. Nevertheless, these spectral parameters can be readily determined from knowledge of the eigenvalue $t(\lambda)$ of the transfer matrix $\hat{T}(\lambda)$ which for the state $|i_0, i_1, ..., i_{N-2}, r\rangle$ reads

$$t(\lambda) = \sum_{l=0}^{N-2} t_{i_l}^{(l)} \lambda^l, \tag{43}$$

by use of known Baxter formula (Refs. 3,5; see also, e.g. Refs. 14,16,17)

$$t(\lambda)Q(\lambda) = \left(\lambda + \frac{i}{2}\right)^N Q(\lambda - i) + \left(\lambda - \frac{i}{2}\right)^N Q(\lambda + i) \tag{44}$$

where

$$Q(\lambda) = \prod_{\alpha \in \tilde{r}} (\lambda - \lambda_\alpha) \equiv \lambda^r + \sum_{\lambda \in \tilde{r}} Q_\alpha \lambda^{\alpha - 1}. \tag{45}$$

The coefficients $Q_\alpha$ in Eq.(45) are determined by comparison of terms with the same power of $\lambda$ in Eq.(44), and the $r$−th degree polynomial $Q(\lambda)$.

Such a strategy was reported by Baxter,[4] and also sketched more recently in paper.[17]

## 5. An example: The hexagonal ring, $N = 6$, $r = 3$

The classical configuration space $Q^{(r)}$ for the three-magnon sector reads

$$Q^{(3)} = \{\mathbf{j} = (j_1, j_2, j_3) | \, 1 \leq j_1 < j_2 < j_3 \leq 6\}, \tag{46}$$

and consists of $\binom{6}{3} = 20$ positions, arranged into four $C_6-$orbits of the cyclic group $C_6-$ the translational symmetry group of the magnetic ring $\tilde{6}$. These orbits are presented in Table 1. Each position $\mathbf{j}$ has there the label $(\mathbf{t}, j)$, with $\mathbf{t} = (t_1, t_2, t_3)$ being the relative position of three Bethe pseudoparticles, and $j-$ the consecutive number within the orbit $\mathbf{t}$. Three orbits are regular, i.e. each of them consists of 6 elements, whereas the fourth, $\mathbf{t} = (2, 2, 2)$, is triply rarefied, and corresponds to two Neel configurations, $-, +, -, +, -, +$ and $+, -, +, -, +, -$. The orbit $(1, 1, 4)$ describes a single island of three adjacent Bethe pseudoparticles, whereas $(1, 2, 3)$ and $(1, 3, 2)$ correspond to $2 + 1$ and $1 + 2$ structure of islands, respectively. In the Neel orbit $(2, 2, 2)$, each Bethe pseudoparticle is separated from others.

Table 1. Orbit structure of the classical configuration space $Q^{(3)}$ for the hexagonal ring $\tilde{6}$.

| $j \backslash \mathbf{t}$ | 132 | 114 | 123 | 222 |
|---|---|---|---|---|
| 1 | 125 | 123 | 124 | 135 |
| 2 | 236 | 234 | 235 | 246 |
| 3 | 134 | 345 | 346 | |
| 4 | 245 | 456 | 145 | |
| 5 | 356 | 156 | 256 | |
| 6 | 146 | 126 | 136 | |

Quantum-mechanical calculations can be conveniently done within the basis of wavelets,[10–12], which consists in performing the Fourier transform on each $C_6-$orbit of Table 1. It corresponds to transition from the initial (positional) basis $|\mathbf{t}, j\rangle$ of the space $\mathcal{H}^3 = lc_{\mathbb{C}} Q^{(3)}$ to the basis $|\mathbf{t}, k\rangle$ of quasimomenta $k \in B$, where

$$B = \{k = 0, \pm 1, \pm 2, 3\} \tag{47}$$

is the Brillouin zone i.e. the set of all labels of irreducible representations

$\Gamma_k$ of the translation group $C_6 \equiv \{j = 1, 2, ..., 6\}$, such that

$$\Gamma_k(j) = \exp(2\pi i k j/6), \quad j \in C_6. \tag{48}$$

Clearly, $k$ is an exact quantum number of the XXX model, so that eigenproblem for each observable $\hat{T}^{(0)}, \hat{T}^{(1)}, \hat{T}^{(2)}, \hat{T}^{(3)}$ and $\hat{T}^{(4)}$, of the size $20\times 20$, decomposes into subproblems, of the sizes $3 \times 3$ and $4 \times 4$ for $k \in \{\pm 1, \pm 2\}$ and $k \in \{0, 3\}$, respectively. Rows and columns of each subproblem are labelled by the index $\mathbf{t}$ of relative position.

It is an easy matter to calculate matrices of each observable $T^{(l)}$ within the basis of wavelets, for each quasimomentum $k \in B$, then to get spectra spec $T^{(l)}$, projectors $P^{(l)}(t_i^{(l)})$, $t_i^{(l)} \in$ spec $T^{(l)}$, and make appropriate intersections. The results are presented in Tables 2 – 4. Table 2 encloses the third descendant $(\hat{S}^-)^3\,|0\rangle$ of the vacuum state and 5 second descendants $(\hat{S}^-)^2\,|k\rangle$ of the one-magnon states, for $k \in \{\pm 1, \pm 2, 3\}$, Table 3 yields 9 first descendants of the two-magnon states, and Table 4 highest weight states in $\mathcal{H}^3$. The eigenstates are labelled by rigged string configurations $(\hat{S}^-)^d\,|\nu\mathcal{L}\rangle$, with $d \in \{0, 1, 2, 3\}$ being the degree of a descendant, $\nu-$ the configuration of strings, i.e. the Young diagram of a partition $\nu \vdash r'$, with $r' = 3 - S$ interpreted as the number of Bethe pseudoparticles coupled into strings, and $\mathcal{L}-$ the collection of riggings, one rigging for each string, represented by a row of the Young diagram $\nu$. Rigged string configurations were proposed by Kerov, Kirillov and Reshetikhin;[18] here we use riggings by quasimomenta, as suggested by Lulek et al.[19]

Table 2. Dirac eigenstates for the case of the third descendant $(\hat{S}^-)^3\,|++++++\rangle$ of the vacuum, and second descendants of the one-magnon states in the space $\mathcal{H}^{(3)}$. $r' = 3-S$ where $S$ is the total spin, $k-$ the quasimomentum, $t^{(l)}$ is the eigenvalue of the observable $T^{(l)}$, and $E$ is the energy of the eigenstate. The last column reports the rigged string configuration of the state.

| $r'$ | $k$ | $T^{(0)}$ | $T^{(1)}$ | $T^{(2)}$ | $T^{(3)}$ | $T^{(4)}$ | E | |
|---|---|---|---|---|---|---|---|---|
| 0 | 0 | $-\frac{1}{32}$ | 0 | $\frac{15}{8}$ | 0 | $\frac{-15}{2}$ | 0 | |
| 1 | 1 | $-\frac{1}{32}$ | $-\frac{\sqrt{3}}{4}$ | $\frac{11}{8}$ | $3\sqrt{3}$ | $-\frac{3}{2}$ | $-1$ | $\boxed{1}$ |
| | 2 | $-\frac{1}{32}$ | $-\frac{3\sqrt{3}}{4}$ | $-\frac{21}{8}$ | $\sqrt{3}$ | | -3 | $\boxed{2}$ |
| | 3 | $\frac{11}{32}$ | 0 | $-\frac{25}{8}$ | 0 | | $-4$ | $\boxed{3}$ |
| | $-2$ | $-\frac{1}{32}$ | $\frac{3\sqrt{3}}{4}$ | $-\frac{21}{8}$ | $-\sqrt{3}$ | | $-3$ | $\boxed{-2}$ |
| | $-1$ | $-\frac{1}{32}$ | $\frac{\sqrt{3}}{4}$ | $\frac{11}{8}$ | $-3\sqrt{3}$ | | $-1$ | $\boxed{-1}$ |

Table 3. Dirac eigenvalues for the first descendants of the two-magnon states. Cf. Table 2. for notation.

| $r'$ | $k$ | $T^{(0)}$ | $T^{(1)}$ | $T^{(2)}$ | $T^{(3)}$ | $T^{(4)}$ | $E$ | |
|---|---|---|---|---|---|---|---|---|
| 2 | 1 | $-\frac{9}{32}$ | $-\frac{5\sqrt{3}}{4}$ | $\frac{11}{8}$ | $-\sqrt{3}$ | $\frac{5}{2}$ | $-5$ | $\begin{matrix}\boxed{3}\\ \boxed{-2}\end{matrix}$ |
| | 2 | $\frac{9}{32}+\frac{\sqrt{17}}{16}$ | $\frac{-7\sqrt{3}-\sqrt{51}}{8}$ | $\frac{-3+2\sqrt{17}}{8}$ | $\frac{\sqrt{3}-\sqrt{51}}{2}$ | | $\frac{-7-\sqrt{17}}{2}$ | $\begin{matrix}\boxed{-2}\\ \boxed{-2}\end{matrix}$ |
| | | $\frac{9}{32}-\frac{\sqrt{17}}{16}$ | $\frac{-7\sqrt{3}+\sqrt{51}}{8}$ | $\frac{-3-2\sqrt{17}}{8}$ | $\frac{\sqrt{3}+\sqrt{51}}{2}$ | | $\frac{-7+\sqrt{17}}{2}$ | $\boxed{2}\boxed{\phantom{0}}$ |
| | 3 | $\frac{3}{32}$ | $0$ | $-\frac{25}{8}$ | $0$ | | $-2$ | $\boxed{3}\boxed{\phantom{0}}$ |
| | $-2$ | $\frac{9}{32}+\frac{\sqrt{17}}{16}$ | $\frac{7\sqrt{3}+\sqrt{51}}{8}$ | $\frac{-3+2\sqrt{17}}{8}$ | $\frac{-\sqrt{3}+\sqrt{51}}{2}$ | | $\frac{-7-\sqrt{17}}{2}$ | $\begin{matrix}\boxed{3}\\ \boxed{-2}\end{matrix}$ |
| | | $\frac{9}{32}-\frac{\sqrt{17}}{16}$ | $\frac{7\sqrt{3}-\sqrt{51}}{8}$ | $\frac{-3-2\sqrt{17}}{8}$ | $\frac{-\sqrt{3}-\sqrt{51}}{2}$ | | $\frac{-7+\sqrt{17}}{2}$ | $\boxed{-2}\boxed{\phantom{0}}$ |
| | $-1$ | $-\frac{9}{32}$ | $\frac{5\sqrt{3}}{4}$ | $\frac{11}{8}$ | $\sqrt{3}$ | | $-5$ | $\begin{matrix}\boxed{2}\\ \boxed{3}\end{matrix}$ |
| | 0 | $-\frac{21}{32}+\frac{\sqrt{5}}{4}$ | $0$ | $\frac{15}{8}+\sqrt{5}$ | $0$ | | $-5+\sqrt{5}$ | $\begin{matrix}\boxed{2}\\ \boxed{-2}\end{matrix}$ |
| | | $-\frac{21}{32}-\frac{\sqrt{5}}{4}$ | $0$ | $\frac{15}{8}-\sqrt{5}$ | $0$ | | $-5-\sqrt{5}$ | $\begin{matrix}\boxed{3}\\ \boxed{3}\end{matrix}$ |

All density matrices corresponding to these 20 eigenstates of tables 2 – 4 are readily computed in a desktop, and presented in an exact form of roots of some quadratic polynomials, with coefficients in the cyclotomic number field $\mathbb{Q}(i\sqrt{3}) \equiv \mathbb{Q}(e^{i\pi/3})$, entering from the Fourier transform for $N = 6$. We quote for example

$$2\rho\left((\hat{S}^-)^2\,|\boxed{3}\rangle\right) = \begin{array}{c|cccc} \mathbf{t}\backslash\mathbf{t} & 114 & 123 & 132 & 222 \\ \hline 114 & 0 & 0 & 0 & 0 \\ 123 & 0 & 1 & 1 & 0 \\ 132 & 0 & 1 & 1 & 0 \\ 222 & 0 & 0 & 0 & 0 \end{array} \tag{49}$$

$$6\rho\left((\hat{S}^-)\,|\boxed{3}\boxed{\phantom{0}}\rangle\right) = \begin{array}{c|cccc} \mathbf{t}\backslash\mathbf{t} & 114 & 123 & 132 & 222 \\ \hline 114 & 1 & -1 & 1 & \sqrt{3} \\ 123 & -1 & 1 & -1 & -\sqrt{3} \\ 132 & 1 & -1 & 1 & \sqrt{3} \\ 222 & \sqrt{3} & -\sqrt{3} & \sqrt{3} & 3 \end{array} \tag{50}$$

Table 4. Dirac eigenstates for the space $\mathcal{H}^{33}$ of the highest weight vectors in $\mathcal{H}^3$.

| $r'$ | $k$ | $T^{(0)}$ | $T^{(1)}$ | $T^{(2)}$ | $T^{(3)}$ | $T^{(4)}$ | $E$ | |
|---|---|---|---|---|---|---|---|---|
| 3 | 1 | $-\frac{1}{32}$ | $-\sqrt{3}$ | $\frac{23}{8}$ | 0 | $\frac{9}{2}$ | $-4$ | [3][ ] / [−2] |
| | 3 | $\frac{31}{32}+\frac{\sqrt{13}}{4}$ | 0 | $\frac{7}{8}+\sqrt{13}$ | 0 | | $-5-\sqrt{13}$ | [3] / [3] / [3] |
| | | $\frac{31}{32}-\frac{\sqrt{13}}{4}$ | 0 | $\frac{7}{8}-\sqrt{13}$ | 0 | | $-5+\sqrt{13}$ | [3][ ][ ] |
| | $-1$ | $-\frac{1}{32}$ | $\sqrt{3}$ | $\frac{23}{8}$ | 0 | | $-4$ | [3][ ] / [2] |
| | 0 | $-\frac{25}{32}$ | 0 | $\frac{15}{8}$ | 0 | | $-6$ | [3][ ] / [3] |

$$4\rho\left((\hat{S}^-)^2\,|\boxed{2}\rangle\right) = \begin{array}{c|ccc} \mathbf{t}\backslash\mathbf{t} & 114 & 123 & 132 \\ \hline 114 & 0 & 0 & 0 \\ 123 & 0 & 2 & -i\sqrt{3} \\ 132 & 0 & i\sqrt{3} & 2 \end{array} \tag{51}$$

We observe from Eq. (49) that the second descendant $(\hat{S}^-)^2\,|\boxed{3}\rangle$ of the one-magnon state at the boundary $k=3$ of the Brillouin zone $B$ (cf. Eq. (47)) occupies only two $C_6-$orbits, $\mathbf{t}=(1,2,3)$ and $(1,3,2)$. Similarly, Eq. (50) says that the first descendant $(\hat{S}^-)\,|\boxed{3}\boxed{\phantom{3}}\rangle$ of the bound state with $k=3$ is spread over all four orbits. Eq. (51) yields that the second descendant $(\hat{S}^-)^2\,|\boxed{2}\rangle$ of the one-magnon state from the interior of the Brillouin zone i.e. $k=2$, cannot reach the orbit $t=(1,1,4)$, for some dynamical reasons. It cannot either reach the Neel orbit $\mathbf{t}=(2,2,2)$, for a kinetical reason: this orbit is triply rarefied, and thus incompatible with this quasimomentum.

As seen, Bethe parameters $(\lambda_1,\ldots,\lambda_r)$ are not necessary for determination of density matrices of exact eigenstates within this method. However, they can be readily calculated from known eigenvalues $t(\lambda)$ of the transfer matrix $\hat{T}(\lambda)$ for a given highest weight eigenstate $|\nu\mathcal{L}\rangle$, by use of the Baxter $TQ$ equation (44). We quote in Table 5 the polynomials $Q(\lambda)$ for each of the eigenstates of Table 4.

Table 5. $Q-$polynomials for eigenstates of Table 4. Bethe parameters $(\lambda_1, \lambda_2, \lambda_3)$ of the state $|\,\nu\mathcal{L}\rangle$ are roots of the corresponding polynomial $Q(\lambda)$.

| $\nu\mathcal{L}$ | $Q(\lambda)$ |
|---|---|
| $\begin{array}{\|c\|c\|}\hline 3 & \\ \hline \mp 2 \\ \cline{1-1}\end{array}$ | $\lambda^3 + \frac{1}{12}\lambda \pm \frac{\sqrt{3}}{12}$ |
| $\begin{array}{\|c\|}\hline 3 \\ \hline 3 \\ \hline 3 \\ \hline\end{array}$ | $\lambda^3 + \left(\frac{5}{12} - \frac{\sqrt{13}}{6}\right)\lambda$ |
| $\begin{array}{\|c\|c\|c\|}\hline 3 & & \\ \hline\end{array}$ | $\lambda^3 + \left(\frac{5}{12} + \frac{\sqrt{13}}{6}\right)\lambda$ |
| $\begin{array}{\|c\|c\|}\hline 3 & \\ \hline 3 \\ \cline{1-1}\end{array}$ | $\lambda^3 + \frac{1}{4}\lambda$ |

## 6. Conclusions

We have presented here a variant of solution of the eigenproblem of the Heisenberg Hamiltonian for XXX model which avoids the need to solve BA system of non-linear equations. The method bases essentially on the transfer matrix $\hat{T}(\lambda)$ which is given by ABA. It allows to determine $N-1$ observables which span the Abelian Bethe subalgebra, and serve as a set of mutually commuting hermitian operators. These operators, taken together with the $z-$component of the total spin of the system, span a maximal Abelian subalgebra in the algebra End $\mathcal{H}$ of all operators in the state space $\mathcal{H}$. Integrability of the model guarantees completeness of the system of common eigenstates of this maximal Abelian subalgebra, along a general recipe of Dirac.

Exact eigenstates are constructed as density matrices which arise as one-dimensional intersections of projection operators onto eigenspaces of each member of the derived set of observables. In this way, solution of the problem is done by determination of spectra for all those observables (which amounts to finding all roots of the corresponding characteristic polynomials, each involving just a single variable), and determination of projectors by routine matrix manipulations. We believe that for moderately low values of $N$ such a program is more easy in performing calculations than BA or ABA based on the system of $r$ nonlinear coupled Bethe equations.

## References

1. H. Bethe, Z. Physik **71**, 205 (1931) (in German; English translation in: D.C. Mattis, The Many-Body Problem, World Sci., Singapore, 689 (1993)).
2. R. J. Baxter, Exactly Solved Models in Statistical Mechanics, Academic Press, New York (1982).
3. L. Faddeev and L. Takhtajan, Hamiltonian Methods in the Theory of Solitons, Springer, Berlin (1987).
4. R. J. Baxter, J. Stat. Phys. **108**, 1(2002).
5. L. D. Faddeev and L.A. Takhtajan, LOMI **109**, 134 (1981) (in Russian; English translation: J. Sov. Math. **24**, 241 (1984)).
6. L.A. Faddeev, arXiv: hep-th/9605187v1.
7. L. A. Faddeev, Acta Appl. Math. **39**, 69 (1995).
8. M.A. Nielsen and I.L. Chuang, Quantum Computation and Quantum Information, Cambridge U. P., Cambridge (2000).
9. P.A.M. Dirac, Principles of Quantum Mechanics, Oxford U. P., Oxford (1967).
10. B. Lulek, T. Lulek, Czechosl. J. Phys. **51**, 357 (2001).
11. B. Lulek, T. Lulek, A. Wal and P. Jakubczyk, Physica B **337**, 375 (2003).
12. T. Lulek, B. Lulek, D. Jakubczyk and P. Jakubczyk, Physica B **382**, 162 (2006).
13. C. N. Yang, Phys. Rev. Lett. **19**, 1312 (1967).
14. E. Mukhin, V. Tarasov, and A. Varchenko, Commun. Math. Phys. **288**, 1 (2009).
15. A.A.Jucys, Lietuvos Fizikos Rinkinys, **6**, 163 (1966).
16. S.E.Derkachov, J. Phys. A: Math. Gen. **32**, 5299 (1999).
17. W. Hao, R. I. Nepomechie, and A.J. Sommerse, Phys. Rev. **E88**, 052113 (2013).
18. S.V. Kerov, A.N. Kirillov, and N.Y. Reshetikhin, LOMI **155**, 50 (1986) (in Russin; English translation: J. Sov. Math. **41**, 916 (1988)).
19. B. Lulek, T. Lulek, M. Labuz, and R. Stagraczynski, Physica B **434**, 14 (2010).

# Plasmon-photon Coupling in High-electron-mobility Heterostructures: Tutorial on Magnetoplasmon Spectroscopy

M. Białek*, I. Grigelionis†, K. Nogajewski‡, and J. Łusakowski*,§

*Faculty of Physics, University of Warsaw
ul. Pasteura 5, 02-093 Warsaw, Poland
§E-mail: jerzy.lusakowski@fuw.edu.pl

†Semiconductor Physics Institute of CPST
Saulėtekio av. 3, LT-10257 Vilnius, Lithuania.

‡LNCMI, CRNS - UGA-UPS-INSA-EMFL
25 rue des Martyrs, 38042 Grenoble, France

A two-dimensional electron plasma in high-electron mobility heterostructures couples to terahertz (THz) radiation which allows to study fundamental properties of this electron system and construct plasma - based devices. The present paper reviews some of recent experimental results obtained on high-electron-mobility heterostructures based on GaAs, CdTe and GaN. Magnetospectroscopy experiments involved measurements of photocurrent, photovoltage and transmission with monochromatic excitation, and photovoltage or transmission with a Fourier spectrometer. Plasmon dispersion relations were determined in each type of samples and showed to follow theoretically predicted dependences. However, a correct interpretation of data required taking into account sample - dependent factors which were related either to the geometry of the sample (this was mainly the case of GaAs-based heterostructures) or to peculiar material properties (mainly in the case of CdTe- and GaN-related heterostructures). The paper gives an elementary description of notions and phenomena related to magnetoplasmon excitations in a two-dimensional plasma as well as a detailed description of procedures of data analysis leading from experimental results to plasmon dispersion relations. This text can be also considered as a tutorial on magnetoplasmon spectroscopy for beginners.

*Keywords*: Magnetoplasmons, THz spectroscopy, 2D electron plasma.

## 1. Introduction

Four decades after the first observation of plasmons in two-dimensional systems in 1977,[1] the physics of these excitations is still vividly developing, both on the theoretical and experimental side. One of the reasons is an increasing quality of the heterostructures resulting in higher and higher electron mobility of a two-dimensional electron plasma (2DEP). In particular, this concerns GaAs/AlGaAs heterostructures which enable to obtain the electron mobility on the level of $10^7$ $cm^2/Vs$ at millikelvin temperatures. These extreme values of mobility allowed to observe, for instance, microwave induced resistance oscillations (MIRO)[2] and zero-resistance states (ZRS)[3] which has been attracting physicists' attention for already almost two decades. For a recent review of theories and experiments devoted to MIRO and ZRS we recommend a paper by Dmitriev *et al.*[4]. Let us note, however, that important contributions have appeared quite recently which shed more light on these intriguing phenomena.[5–7] An extreme mobility of a 2DEP in GaAs/AlGaAs heterostructures allows also to investigate low-energy (microwave) excitations of a quantum electron liquid in the regime of the fractional quantum Hall (FQHE) effect at millikelvin temperatures.[8] An experimental method of all-optical detection of the excitations presented in that paper can also be applied to study plasmons in a 2DEP at higher temperatures and frequencies (see Ref. 25 in Ref. 8).

Although it is the electron mobility in GaAs/AlGaAs heterostructures which attains the record values, technological developments of an epitaxial growth allows to study subtle quantum effects in 2DEP embedded in other materials. We will consider two such cases. The first one are CdTe/CdMgTe quantum wells. There is a number of advanced experiments which show a very high quality of these structures. These include, in particular, observation of the FQHE,[9,10] quantization of resistance in a quantum point contact (QPC)[11] and spin current effects.[12,13] A CdMnTe/CdMgTe quantum well was used to generate THz radiation by optically induced spin waves in a high-electron-mobility 2DEP.[14]

The other case are GaN/AlGaN heterostructures with the electron mobility close to or exceeding $10^5$ $cm^2/Vs$.[15,16] In such heterostructures, integer quantum Hall effect (IQHE) was observed[17] which shows that their quality is high enough to study other quantum effects.

There is a broad literature devoted to far-infrared, or THz, properties of a 2DEP in GaAs/AlGaAs heterostructures. However, the results concerning CdTe/CdMgTe and GaN/AlGaN are substantially less abundant. For this reason, it seems to be interesting to present experimental data on these less frequently studied materials and confront them with a much better known one.

We carried out experiments on three types of samples: single heterostructures of GaAs/AlGaAs and GaN/AlGaN and single quantum wells of CdTe/CdMgTe. Our studies involved measurements at high magnetic fields and low temperatures with a monochromatic THz excitation or a Fourier spectroscopy. The geometry of samples in all cases was the same: they were rectangular mesas with a metallic grid prepared with an optical or electron lithography. Our main goal was to determine plasmon dispersion relations. Generally, we have found that they follow theoretical predictions provided the plasmon wave vector is correctly determined and partial screening by the grid-gate is taken into account

The paper is directed mainly to newcomers to the field of 2D plasma physics. It contains explanations of the basic ideas and notions related to propagation of electromagnetic waves in heterostructures and a detailed description of the procedure which leads from experimentally observed spectra to a plasmon dispersion relation. Let us note that a large, comprehensive review of the physics of 2D plasmons can be found in Ref. 18.

The paper is organized as follows. In Section 2 we describe the mechanism of coupling of the electromagnetic wave to a 2DEP related to the presence of the grid gate. Section 3 is devoted to an analysis of dispersion relations. In Section 4 we describe the samples and experimental techniques which is followed by description of the main experimental results in Section 5. Finally, we conclude the paper pointing out new directions of research which reveal new aspects of the physics of two-dimensional plasmons.

## 2. Grids and mesas

A grid fabricated on the top surface of a heterostructure is a standard "device" used to couple the electromagnetic field to the plasmon (see Fig. 1). It was used in the first experiment on 2D plasmons[1] and still remains a system of choice for a similar study. A grid is characterized by its period $\Lambda$ and the geometrical aspect ratio $\alpha$, i.e., the ratio of the grid line's width to the grid period. The necessity to use a coupler stems from the fact that the wave vector $k$ of a photon which we want to use to excite a plasmon is much smaller than that of the plasmon. If one estimates appropriate values, one finds that the energy (or frequency) of the photon in the THz, or even microwave, band coincides with a typical 2D plasmon energy, while its momentum is only of the order of 1% of that of a plasmon.[a] A grid fabricated on the top surface of the heterostructure periodically modulates, in the

[a]This analysis concerns typically considered situations which can be described within a so-called non-retarded approximation, when the plasmon velocity is much smaller than the speed of light. This is the case of all the experiments described in the present paper.

near field, the amplitude of the incident electromagnetic field. The modulated field carries Fourier components with wave vectors of sufficiently large values to fulfill the momentum conservation law in the photon - plasmon interaction.

One can notice that not only metallic grids but other periodic structures could serve as couplers too. In fact, this was verified by preparing a two-dimensional lattice of etched nanodots[19] or grooves in a dielectric layer covering a heterostructure.[20]

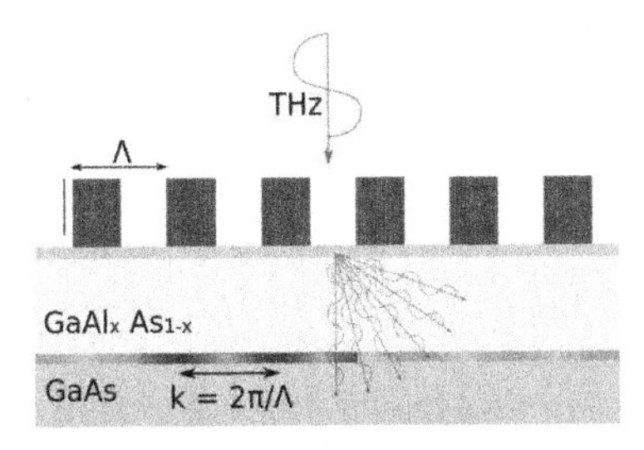


Fig. 1. A scheme of diffraction of a THz wave incident normally on a GaAs/AlGaAs heterostructure covered by a grid with the period $\Lambda$. $k$ is the wave vector of the fundamental plasmon mode.

Let us note that in a typical experiment related to 2D plasmons, the wave length $\lambda$ of an incident THz wave in vacuum is of the order of 100 $\mu$m while the periodicity of the grid (and the plasmon wavelength) is of the order of 1 $\mu$m. A 2DEP is situated typically at about 50 - 100 nm below the top surface of heterostructure. Thus, the effects of modulation we consider are really well in the near-field regime. The grid changes the symmetry of the incident wave which is transverse when falls on the sample (i.e., its electric field is perpendicular to the wave vector) and longitudinal when propagating in the 2DEP layer, with its electric field vector lying also in the 2DEP plane. Only longitudinal waves can couple to plasmons propagating in the 2DEP layer. One can think about a plasmon as an electromagnetic wave propagating in a conductive medium.

The role of the grid is not only to generate a longitudinal field propagating with appropriate $k$-vectors. The grid is a polarizer for the incident wave which means that it transmits only the waves with the polarization plane perpendicular to the grid lines. The efficiency of the transmission depends on geometrical details of the grid, $\Lambda$ and $\alpha$, referred to the wavelength $\lambda$. The grid is also responsible for a partial screening of the plasmon electric (high frequency) field which is schematically shown in Fig. 2.

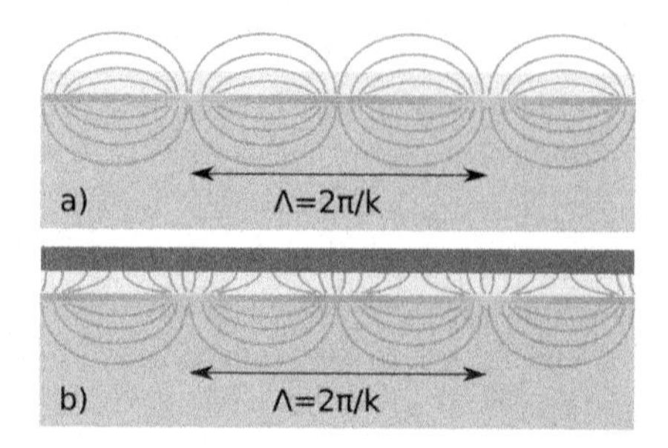


Fig. 2. A schematic picture of a high frequency field distribution in a) ungated and b) gated heterostructure. The plasmon wavelength $\Lambda$ is shown.

Let us consider the screening in a more detail. A plasmon is a propagating wave of charge density with which a high frequency electric field, oscillating in space and time, is related. The electric field is generated by fluctuations of the electron density and it is spread in all directions. In the absence of plasmon, an overall electrical neutrality of the system assures vanishing of the electric field. If one assumes that the lines in a grid are much narrower than the grid period (or, when $\alpha \to 0$), one can neglect the presence of the grid at all and find the plasmon electric field spreading through the top surface of sample to the air (Fig. 2a). In this case, we say that the plasmon field is not screened. However, when $\alpha$ is not negligible, the spread of the electric field is partially blocked. Let us assume that $\alpha \to 1$, or that the structure is totally covered by a metallic plate which means that the lines of the plasmon electric field end at the plate and do not reach the air. In this case, one says that the plasmon field is totally screened (Fig. 2b).

To understand, why is it so, let us recall that screening of the electric field means a decrease of its strength. So, why the presence of the gate decreases the plasmon field? A typical problem of screening is encountered when one considers a positive charge surrounded by an electron plasma. In such a case, an attractive force increases the concentration of free electrons in the vicinity of the positive charge which results in a faster than $r^{-2}$ decrease of the electric field with the distance $r$. Another example is screening in a dielectric medium: an electric field of a charge causes polarization of the surrounding medium which is described by a dielectric constant $\epsilon > 1$ and leads to a weaker field than in the vacuum. Generally, screening takes place each time a system can arrange positions of charges according to the force originating from the electric field. In the case of a grid, electrons in the metallic stripes move in such a way that the electric field is decreased. Of course, screening by the dielectric polarization of the heterostructure is always present.

There are other consequences resulting from the presence of the metallic gate or a grid. First, the value of restoring electric force which is responsible for plasma oscillations is weaker in the gated case, since the field itself is smaller. It means that the frequency of oscillations of gated plasmons is smaller than that of ungated ones (one can easily find a mechanical analogy showing that the frequency of oscillations decreases with the decrease of restoring forces). Second, when the gate or grid is present, oscillations of the charge density in the 2DEP cause oscillations in the near-by metal which is a source of damping.

The essence of usage of a grid is to generate appropriate wave vectors carrying large enough $k$-components in the near field. This, however, can be achieved without a periodic structure. In high-quality heterostructures, one can observe excitation of plasmons even on samples without any special treatment of the surface. In such a case, the plasmons are launched due to diffraction of light just on mesa borders and $k$-vectors of plasmons are determined by the width of mesa, $W$, or its another geometrical detail. The wave vectors of plasmons in a given sample are equal to $k_j = 2\pi j/L$, where $j$ is the number of mode and $L$ is the length responsible for generation of plasmons (typically, but not always, it is equal to $\Lambda$ or $W$). One can ask, what $L$ should be taken if the sample is a mesa of a complicated shape covered by a grid. The answer depends on the frequency of excitation and other experimental details, and we will come back to this problem in Section 5.

## 3. Dispersion relations

A dispersion relation, i.e., a relation between the frequency $\omega$ and the wave vector $k$ is one of the basic characteristics of an excitation. In fact, it fully determines its propagation, giving the phase velocity $v_\varphi = \omega/k$ and group velocity $v_g = \partial\omega/\partial k$. Derivation of plasmons' dispersion relations has been described in many papers and books which show different aspects of propagation of plasmons in the presence (or not) of the magnetic field, in bulk (or lower-dimension) materials, in isotropic (or anisotropic) structures, etc.

It is not our aim to give here even a short review on this research. Let us, however, point at some useful references. Ando, Fowler and Stern[21] cite the main theoretical and experimental work carried out before 80-ties while Kushwaha[18] includes also a detailed description of research done during subsequent two decades. For a deeper study oriented towards semiconductor plasma physics we recommend books by Dressel and Grüner[22] and Wolf and Platzman.[23] Original papers which are essential to the following discussion are Refs. 24–28,30,31.

A general formula which allows to determine the dispersion relation of 2D plasmons in a magnetic field $B$ is given by Eq. (1):

$$\frac{\omega^2 - \omega_{LO}^2}{\omega^2 - \omega_{TO}^2} - \frac{\omega_{p,j}^2}{X_j^2} \sum_{l=1}^{\infty} \frac{4l^2 J_l^2(X_j)}{\omega^2 - (l\omega_c)^2} = 0, \tag{1}$$

where $\omega$ is the frequency of the incident wave and $j$ numbers plasmon modes. A scheme of derivation of this equation is given in Ref. 32 in a concise form and in Ref. 33 in a very detailed way. A wave vector $k_j$, which does not enter explicitly Eq. (1), is hidden in an $\omega_{p,j}(k_j)$ dependence and the parameter $X_j$. $\omega_c$ is the cyclotron frequency, $J_l$ are Bessel functions of the first kind and order $l$, $\omega_{LO}$ and $\omega_{TO}$ are frequencies of the longitudinal and transverse optical phonons, respectively, and $X_j = 2\pi r_c/\lambda_j = r_c k_j$ is so-called non-local parameter determined by the ratio of the cyclotron radius $r_c$ to the plasmon wavelength $\lambda_j$.

Let us discuss the basic aspects of Eq. (1). General considerations[29] show that $\omega_{p,j}(k_j)$ dependence is:

$$\omega_{p,j}^2 = \frac{n_s e^2 k_j}{2\varepsilon_0 \varepsilon_{eff}(k_j d) m^*}, \tag{2}$$

where $n_s$ is the 2DEP concentration, $e$ and $m^*$ are the electron charge and effective mass, respectively, and $d$ is the thickness of the barrier separating the 2DEP from the air (e.g., an AlGaAs layer in the case of GaAs/AlGaAs structure). A distinction between gated and ungated plasma is hidden in the effective dielectric constant $\varepsilon_{eff}$ which is given by

$$\varepsilon_{eff,g}(k_j) = \frac{1}{2}[\varepsilon_s + \varepsilon_b \coth(k_j d)] \tag{3}$$

in the case of gated plasma[30] or

$$\varepsilon_{eff,ug}(k_j) = \frac{1}{2}\left[\varepsilon_s + \varepsilon_b \frac{1 + \varepsilon_b \tanh(k_j d)}{\varepsilon_b + \tanh(k_j d)}\right] \tag{4}$$

in the case of ungated plasma.[31] In the above expressions, $\varepsilon_s$ and $\varepsilon_b$ are the dielectric constants of the substrate and barrier, respectively. These equations are valid if one neglects the dynamics of the lattice, i.e., one considers plasmon frequencies which are substantially different from the frequency of optical phonons. If the lattice dynamics has to be taken into account, one has to replace dielectric constants given by Eq. (3) and (4) with their high-frequency counterparts:

$$\varepsilon_{s,b} = \varepsilon_{s,b}^{\infty} \frac{\omega^2 - \omega_{LO}^2}{\omega^2 - \omega_{TO}^2}. \tag{5}$$

We assume that the frequency of longitudinal ($\omega_{LO}$) and transverse ($\omega_{TO}$) optical phonon is the same for both the barrier and the substrate. Including this interaction, one gets the first term at the left hand side of Eq. (1). If the lattice dynamics can be neglected, this term is replaced by 1.

Resonant denominators in Eq. (1) appear while solving of a linearized Vlasov equation which allows to determine a perturbation of the equilibrium distribution function.[34] This expression shows that the plasma can support resonances with the frequency equal to multiples of the cyclotron frequency $\omega_c$. The physical reason of appearance of these resonances is the fact that if a plasmon wave is present then the cyclotron orbit of the electron is never closed because of the electron drift in the plasmon electric field. In the geometry considered here, i.e., with the propagation of plasmons strictly perpendicular to the static magnetic field, nonlocal corrections lead to appearance of so called Bernstein modes. Their frequency is almost equal to the harmonics of the cyclotron resonance (for details, see Ref. 23). An interesting feature of Bernstein modes is their avoided crossings with magnetoplasmon modes.[28] Observation of such avoided crossings is in fact practically the only way to indicate the presence of Bernstein modes in a 2DEP.

When both the nonlocal corrections and the lattice dynamics are not important, the dispersion relation of magnetoplasmons reduces to the most often considered form:

$$\omega^2 = \omega_{p,j}^2 + \omega_c^2. \tag{6}$$

Discussing the experimental results obtained on heterostructures based on GaAs, CdTe and GaN we will show that the form of the dispersion relation which is appropriate in each case does depend on the material considered.

Having chosen an appropriate equation, Eq. (1) or Eq. (6), we are next looking for the values of $k_j$ which satisfy this equation at given value of $\omega$ and at given magnetic field $B$. In the case of Eq. (6), there is at most one real solution for $k_j$. In the case of Eq. (1), the number of $k_j$'s depends on the number of terms taken into consideration in the infinite series, i.e., on the range of a resulting algebraic equation. Next, using Eq. (2) we can determine the frequency of $j$'th plasmon mode.

## 4. Experimental aspects of magnetoplasmon spectroscopy

We are interested in two types of measurements. One is done with a Fourier spectrometer. In this case, the sample is irradiated with broadband incoherent light generated by a thermal source (typically - a mercury lamp). The sample is placed in a constant magnetic field and a registered spectrum gives the signal as a function of the photon frequency. The signal can be either due to transmission or reflection (in this case one uses an additional detector) or due to photocurrent or photovoltage (in this case, the sample itself generates the measured signal). A corresponding dynamic method of measurements is a THz time-domain spectroscopy (as an example of such an experiment, see Ref. 35) in which one uses pulses of broadband THz radiation.

The other type of experiment is carried out with monochromatic THz excitation and with a sweep of the magnetic field. The registered spectrum shows the measured signal as a function of the magnetic field. In the following, we will show results of both types of measurements.

Although these two methods seem to be equivalent, they are not. First, if we are looking for an excitation which energy does not depend on the magnetic field, no signature of such a resonance will be present in spectra obtained with monochromatic illumination but should be clearly manifested in Fourier spectra. On the other hand, illumination with spectrally broad radiation in a Fourier spectrometer excites simultaneously many transitions which can lead to smearing of spectroscopic details. An example of such a situation was encountered in measurements on a GaAs/AlGaAs sample: while under monochromatic excitation we observed several plasmon modes, only one was observed on the same sample in a Fourier spectroscopy experiment.[36]

Next experimental problem is related to the magnetoresistance and even to the electron localization. If one measures the photocurrent as a function of magnetic field on a detector or on the sample, possible influence of the magnetoresistance on the shape of the spectrum must be taken into account. In the case of low-electron-mobility samples, a strong decrease of the conductivity with increasing magnetic field can drastically modify the shape of the spectrum.[38] Such a decrease is often caused by a magnetic-field-induced localization of conducting electrons in fluctuations of the electrostatic potential.[39] On the other hand, in high-electron-mobility samples (typically, the case of a degenerate 2DEP), a strong geometrical magnetoresistance appears if one studies a sample which is much shorter than wide, e.g., a field-effect transistor which source-drain separation is much smaller than its width.[40,41] Generally, the measured signal must be corrected with taking into account possible influence of magnetoresistance.

We studied GaAs/AlGaAs samples presented in Fig. 3 (for more details, see Refs. 36,37). The thickness of Au metalization in the case of the uniform and meander gates was 15 nm, and 25 nm in the case of the comb gate, and $\alpha = 0.5$. In the case of CdTe/CdMgTe quantum wells, we studied two types of samples. One of them was a square mesa of 1.6 mm x 1.6 mm on which an Al grid was lithographically defined. The period of the grid was either 2 $\mu$m or 3 $\mu$m, $\alpha = 0.5$, and the metalization thickness was 110 nm. The other was a quantum point contact (QPC) in which the quantum constriction was 460 nm-wide and separated two regions of the width of 2.4 $\mu$m (see Fig. 4). Similar grids were deposited on GaN/AlGaN heterostructure (see Ref. 33,42). In this case, however, $\alpha$ was varied between 0.72 and 0.9.

A typical experimental arrangement is shown in Fig. 5 and presents the case of a monochromatic excitation. A molecular laser (optically pumped with a $CO_2$ laser) can be replaced by other monochromatic sources (we used also electronic sources covering the frequency range 0.1 - 0.66 THz) or by a Fourier spectrometer. THz radiation is waveguided through a stainless steel tube from the source to the sample. A cold black polyethylene filter is placed near the sample to block both the visible radiation and thermal radiation of warm parts of the cryostat. The intensity of radiation exciting the sample is measured on the beam reflected from a chopper wheel. The chopper is a source of a reference frequency signal synchronizing the lock-in amplifiers. The sample is placed in the center of a superconducting coil. A bolometer placed under the sample allows to measure a transmitted signal. The sample is biased and variation of the current as a function of the magnetic field is determined from a voltage drop registered on a load resistor. The whole system is controlled by a computer.

## 5. Data analysis

As one can conclude from the above description, similar grid-gated samples were fabricated on each type of heterostructure and it is interesting to

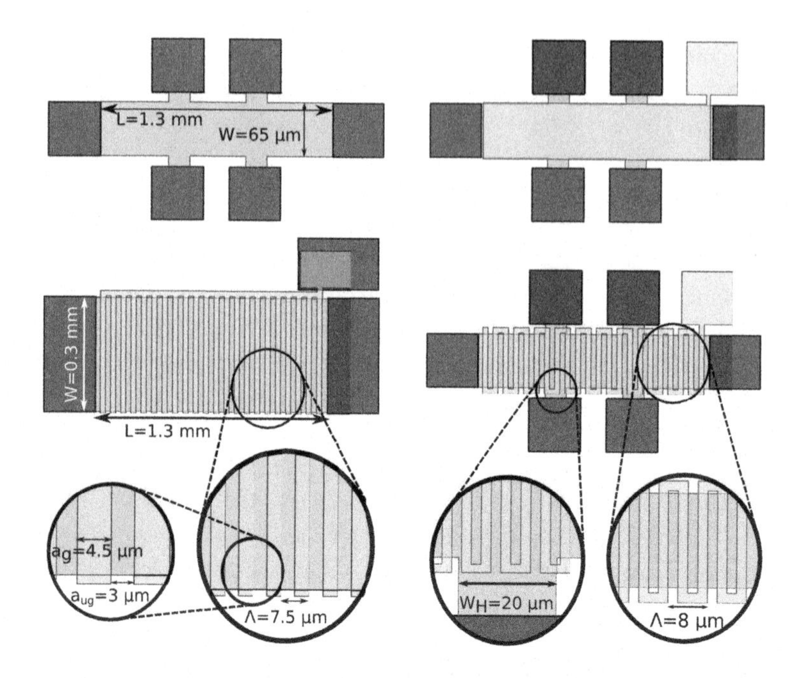


Fig. 3. The scheme of samples processed on GaAs/AlGaAs heterostructure. Dimensions of mesa and gates are given in the figure.

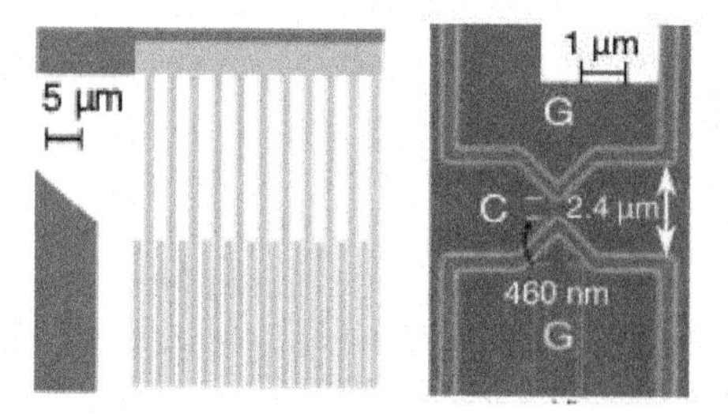


Fig. 4. Left: geometry of a grid deposited on a CdTe/CdMgTe quantum well. Grey areas on right and top are contact metalization. The geometrical aspect ratio of the grid was equal to 0.5. Right: geometry of a quantum point contact on a CdTe/CdMgTe quantum well. The current can flow from left to right through the contact (C). Top and bottom are are lateral gates (G) which polarization controls the quantum constriction. White lines are borders of trenches defining the mesa. After Ref. 32.

compare obtained results. Generally, observed plasmonic excitations follow the theoretical models presented in Section 3, however each material shows its peculiar features.

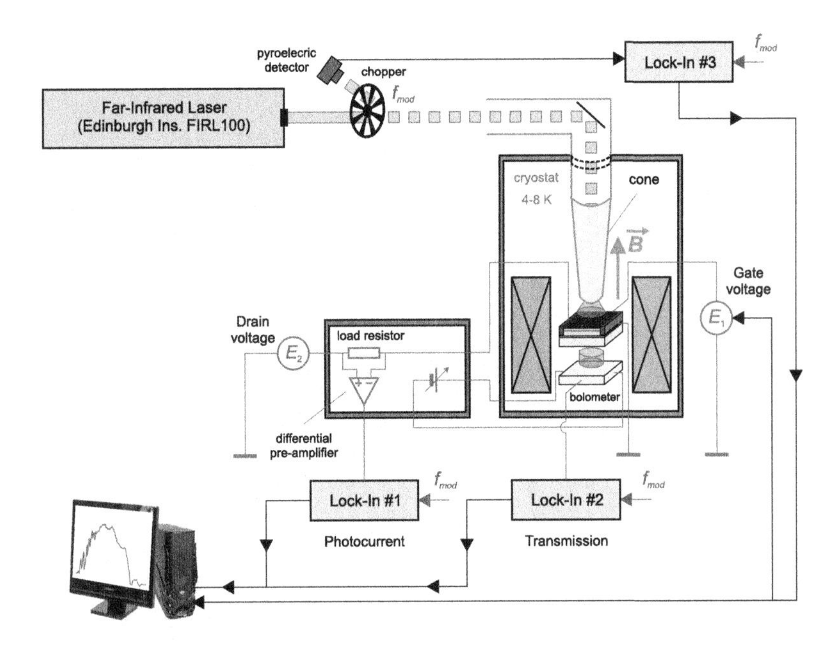


Fig. 5. The scheme of the experimental set up for the case of a simultaneous measurements of the photocurrent and transmission. $f_{mod}$ is the frequency of on/off modulation of the beam with a mechanical chopper and is used to synchronize lock-ins.

Magnetospectroscopy data presented in Fig. 6 were obtained with monochromatic excitation of different wavelengths on the GaAs/AlGaAs heterostructure. Based on this result, the goal is to obtain a corresponding dispersion relation of plasmons, i.e., the $\omega_{p,j}(k_j)$ dependence. The experiment gives, however, only the values of magnetic field of subsequent plasmon modes (marked by vertical bars in Fig. 6). We proceed in the following way. The first approximation to the dispersion relation is given by Eq. (6). Using this equation one can determine the frequency of plasmon modes at zero magnetic field as a function of their number. The result is shown in Fig. 7 if one uses the upper horizontal axis.

Now, the question is, how to determine plasmons' wave vectors. At the end of Section 3 we said that one should solve an appropriate equation. It is true, but in the case of multiple resonances observed, one has to find a common rule defining all $k$-vectors of subsequent modes. So, obtained $k_j$ values must obey the relation $k_j = 2\pi j/L$, where $j$ indices forms a regular series, and $L$ is a dimension related to the geometry of the sample.

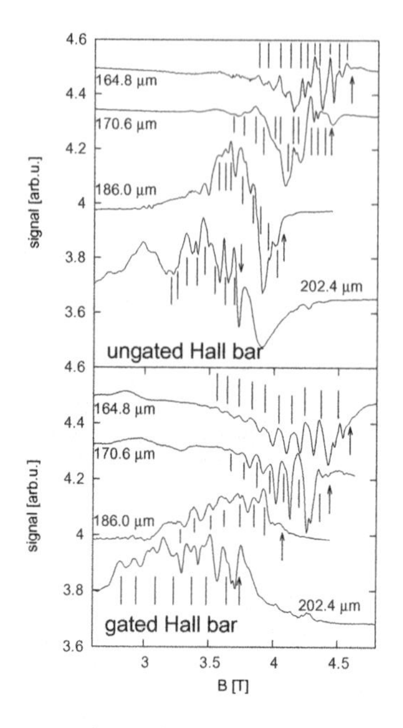


Fig. 6. Photovoltage spectra obtained on the meander-gated and ungated Hall bars shown in Fig. 3. The wavelength of the molecular laser radiation is given next to each spectrum. Spectra are normalized and vertically shifted for clarity. Magnetoplasmon resonances are marked with vertical bars and the cyclotron resonance magnetic field with arrows. After Ref. 36.

We take into account that: a) the wave vector must be quantized by one of dimensions defining the mesa or the periodicity of the grid; b) we do not know if we observe all modes ($j$ = 1, 2, 3, ...), only even ($j$ = 2, 4, 6, ...), odd ($j$ = 1, 3, 5, ...) or yet another sequence; c) we have to decide, which effective dielectric constant should be substituted into Eq. (2): that for gated (Eq. (3)) or ungated (Eq. (4)) plasma or an interpolated value.

In all our considerations we applied a phenomenological formula

$$\varepsilon_{eff} = \alpha\varepsilon_{eff,g} + (1-\alpha)\varepsilon_{eff,ug}, \tag{7}$$

which worked very well in that sense that the values of $\alpha$ determined by fitting were very close to the geometrical aspect ratio of the grid considered. An evolution of the dispersion relation with $\alpha$ is shown in the upper panel in Fig. 7.

Additional parameters needed are the plasma concentration $n_s$ which we obtain from magnetotransport measurements and the electron effective

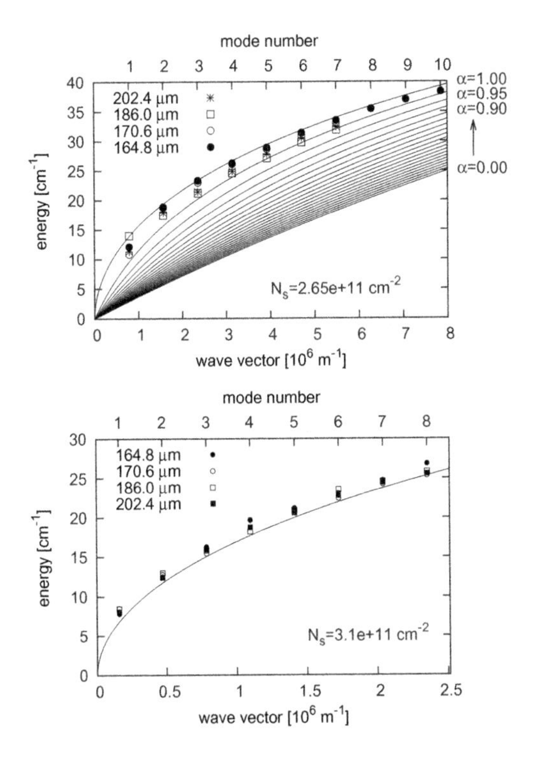


Fig. 7. Points: frequency of plasmon modes at zero magnetic field recalculated from spectra shown in Fig. 6 with Eq. (6) as a function of the number of mode (upper horizontal axis) and the wave vector (lower horizontal axis) at different indicated excitation laser photon wavelengths. Solid lines: theoretical calculations according to Eq. 6. Left panel: meander-gated sample, solid lines for $0 < \alpha < 1$, bottom to top; right panel: ungated sample, solid line: $\alpha = 1$. After Ref. 36.

mass which we obtain from the position of the cyclotron resonance (the best way) or take from the literature data, if there is not a better solution. Taking the value of the effective mass as good as possible is quite important in this analysis since in a 2DEP it depends on the plasma concentration (the effect of the nonparabolicity of the conduction band) and the width of the quantum well. Taking into account a high spectral resolution of presented measurements, one should take great care about the values of parameters used in calculations. Later on we shall come back to this problem in the case of the polaron effect in CdTe/CdMgTe quantum wells.

The only way to solve this multi-parameter task is to fit the experimental results with different possible dispersion relations resulting from a particular choice of the dimension quantizing the $k$-vectors, a sequence of $j$ indices and $\alpha$. There are, of course, strong hints from the experiment. For instance, a square-root like form of $\omega_{p,j}(k_j)$ shown in Fig. 7 suggests an ungated dispersion; the dimensions of mesa are known, so there are only a few possible values of the wave vector of the fundamental mode (with $j$=1). Anyway, several possibilities are to be verified and the final result chosen should agree well with the experimental data.

This is also the time to decide whether one can work with the simplest form of the magnetoplasmon dispersion, given by Eq. (6), neglecting nonlocal corrections and the plasmon - phonon interaction or a more general approach, Eq. (1), should be applied. If the frequency of excitation $\omega$ is far from optical phonon frequencies, one can safely neglect the plasmon - phonon interaction. In our experiments, this is the case of GaAs/AlGaAs and GaN/AlGaN heterostructures, but not CdTe/CdMgTe quantum wells. The necessity to include nonlocal corrections depends on the value of the parameter $X_j$ which can be determined only when $k_j$ values are known.

Having the final result of the fitting, i.e., a series of $k_j$ vectors and the $\omega_{p,j}(k_j)$ dependence, one has to verify that the obtained results are physically justified. This is quite important since plasmons in real samples sometimes manifest in an unexpected way. The results presented above serve as a good example. First, it appeared that in ungated sample, plasmon wave vectors were quantized by a 20 $\mu$m wide "arm" which connected the main channel to a lateral voltage contact, and not by the mesa width of 65 $\mu$m. In this case, odd $j$ = 1, 3, 5, ... were chosen. This is a reasonable choice since one should expect a uniform THz field distribution over a 20 $\mu$m distance which imposes identical boundary conditions on both edges of the "arm" and promotes excitation of odd plasmon harmonics. Second, the dispersion of the meander-gated sample showed to be practically that of ungated plasma (with $\alpha \approx 1$) in spite of the geometrical aspect ratio of the grid equal to 0.5. The reason is a small thickness of the gate metalization of only 15 nm which is too thin to effectively block spreading of the plasmon electric field. Nevertheless, in this sample, the wave vectors were quantized by the grid period equal to 8 $\mu$m and all odd and even modes, $j$ = 1, 2, 3..., were observed. Dispersion relations obtained for other GaAs/AlGaAs samples shown in Fig. 3 and at lower, sub-THz, frequency of excitations were determined in Ref. 37.

The CdTe/CdMgTe case is interesting because of a strong electron - phonon interaction which makes it necessary to take the first term of the l.h.s. of Eq. (1) without any approximations: the frequency of excitation $\omega$

was in our measurements close to $\omega_{LO}$ and $\omega_{TO}$. In fact, Fourier transmission measurements showed a clear increase of the electron effective mass with the magnetic field, as is presented in Fig. 8. In our experiments on

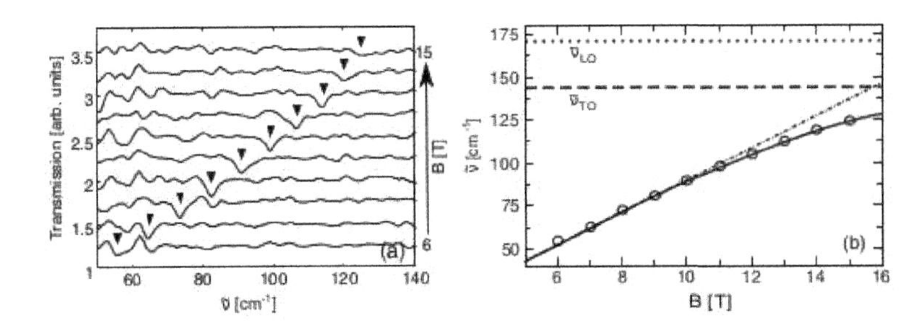


Fig. 8. Left: transmission spectra at magnetic fields form 6 T to 15 T for an unprocessed wafer of CdTe/CdMgTe quantum well at every 1 T (bottom to top). Solid triangles mark the position of the cyclotron resonance. Right: dependence of the cyclotron resonance energy (in $cm^{-1}$) on the magnetic field. Horizontal lines indicate the energies of the LO and TO phonons. After Ref. 43.

CdTe/CdMgTe described in Ref. 32,43, we used the excitation radiation of 2.25 THz and 3.11 THz which corresponded to the cyclotron magnetic field equal to about 9 T and 11.5 T, respectively (see Fig. 9). In such a case, it was necessary to introduce into the dispersion relation, Eq. (2), a dependence of the effective mass on the plasmon mode number, $j$. Although it seems to be small, it changes the calculated plasmon energy by about 5%. Obviously, working with even higher energy of excitation would make this correction more and more important.

A really big correction in the case of CdTe/CdMgTe results from taking into account the plasmon - phonon interaction, as it is shown in Fig. 9 with solid and dotted lines.

Nonlocal corrections seem to be important in the case of CdTe/CdMgTe samples since the values of $X_j$ are equal to about 0.3 for the highest $j$. However, the corrections are essential in this case at magnetic fields between about 1 T and 3 T as it is shown in Fig. 10 while the position of magnetoplasmon resonances were found above 8 T in the case of grid-gated sample and above 7 T in the case of the QPC (for a detailed description of excitation of magnetoplasmons in the QPC, see Ref. 32).

Thus, to observe avoided crossings of the harmonics of the cyclotron resonance with magnetoplasmon modes, one has to choose an appropriate

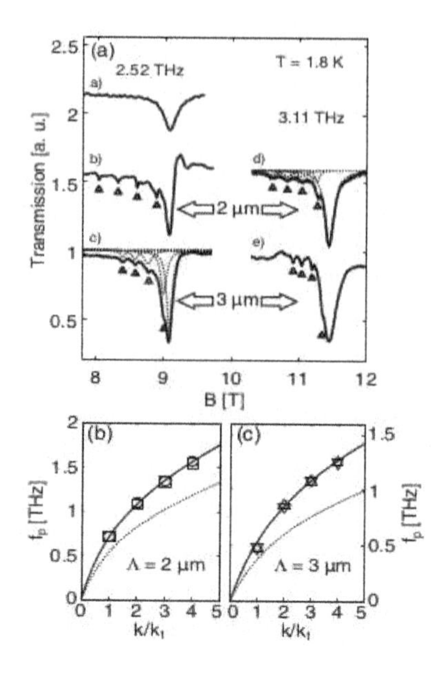


Fig. 9. a) Transmission spectra as a function of the magnetic field on grid-gated CdTe/CdMgTe samples with the grid of the period $\Lambda = 2$ $\mu$m (b and c) or 3 $\mu$m (d and e) at excitation frequency of 2.52 THz and 3.11 THz. The curve a) showing only the cyclotron resonance transition was obtained on a sample without the grid. Solid triangles mark the position of magnetoplasmon modes. b) and c): Plasmon frequency dependence on the wave vector normalized to $k_1 = 2\pi/\Lambda$. Points: experimental data; solid and dotted line: theoretical dispersion relations with and without the plasmon - LO phonon interaction, respectively. After Ref. 32.

value of the frequency of excitation. This suggests that it should be easier to look for such an interaction with Fourier spectroscopy than with laser spectroscopy, just because experimentally it is much easier to fine-tune the magnetic field than to find a THz source with precise frequency tuning, especially above 1 THz.

As an example, Fig. 11 shows normalized negative transmission spectra registered at magnetic fields up to 15 T on a grid-gated GaN/AlGaN sample with $\Lambda = 2.5$ $\mu$m and $\alpha = 0.95$. The positions of maxima, indicated with solid triangles, correspond to the frequency of magnetoplasmon excitation. These positions are next replotted in a frequency - magnetic field coordinates in the right panel of Fig. 12. One can clearly see that there are avoided crossings between second and third harmonic of the cyclotron resonance and the second and third magnetoplasmon mode.

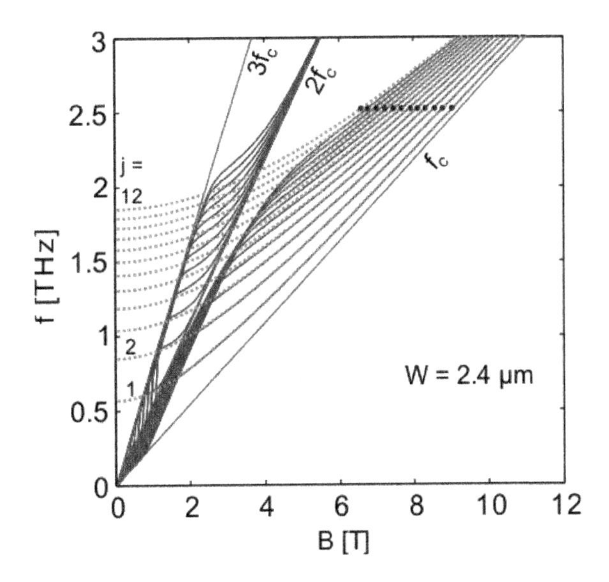


Fig. 10. Solid (dotted) lines: calculated dispersion relation for a CdTe/CdMgTe quantum well with plasmon-phonon interaction and taking into account (neglecting) nonlocal corrections, according to Eq. 1. Points: experimentally determined position of plasmon resonances of the QPC with excitation of 2.52 THz. $W$ is the QPS mesa's width. Straight lines: harmonics of the cyclotron resonance.

The modes which are observed in the vicinity of the avoided crossings are called Bernstein modes. To model these details and explain positions of multiple plasmon resonances presented in Fig. 12 one has to use Eq. (1) to calculate dispersion relations of several magnetoplasmon modes. In calculations, the infinite series in Eq. (1) was truncated to the first nine terms. Final results are shown with solid lines.

The Fourier spectroscopy gave the positions of magnetoplasmon maxima at zero magnetic field which are indicated on the vertical axis of Fig. 12. As one can notice, these positions correspond to a high accuracy to the calculated dispersion relations extrapolated to the zero magnetic field. One can easily plot the frequency of subsequent modes as a function of the wave vector which is in this case equal to $2\pi j/\Lambda$, $j = 1, 2, 3, ...$ (the left panel of Fig. 12). Next, a solid line corresponding to the dispersion relation given by Eq. 2 was plotted with $\alpha = 0.92$, to a high accuracy equal to the geometrical aspect ratio of the grid.

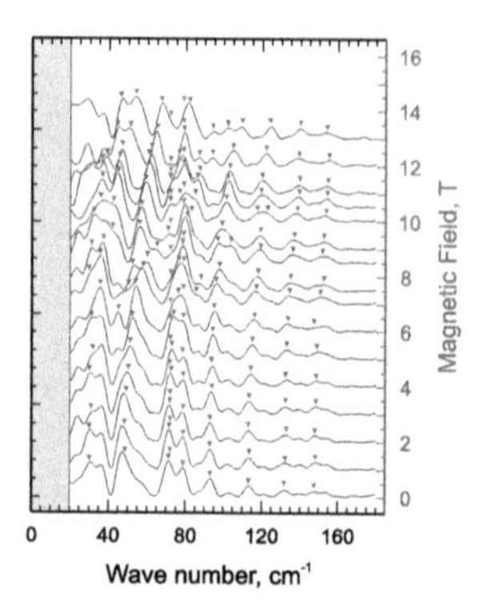


Fig. 11. Negative transmission spectra of a GaN/AlGaN heterostructure with $\Lambda$ = 2.5 $\mu$m and $\alpha = 0.92$ at magnetic fields form 0 to 15 T, every Tesla, bottom to top. Spectra are normalized and vertically shifted for clarity. Solid triangles show positions of maxima corresponding to plasmon resonances. A gray bar at the left shows the experimentally inaccessible frequency band. After Ref. 33.

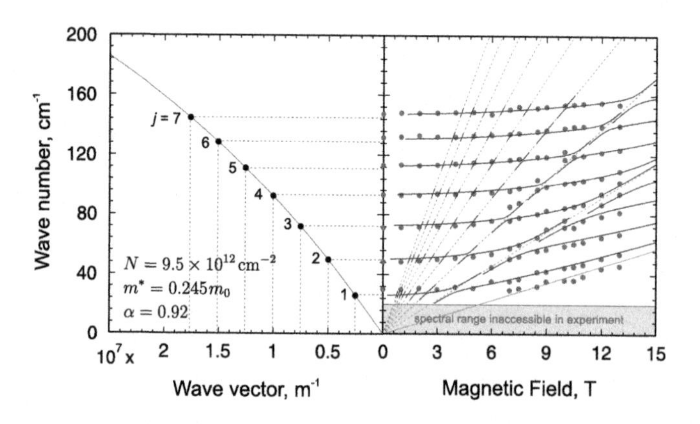


Fig. 12. Right: positions of magnetoplasmon resonances indicated in Fig. 11. Straight dotted lines: harmonics of the cyclotron resonance. Solid lines: magnetoplasmon dispersions for the lowest 7 modes calculated with Eq. (1). A grey bar at the bottom shows the experimentally inaccessible frequency band. After Ref. 33.

## 6. Closing remarks

It is our hope that the present paper will be useful to scientists who would like to begin the research on plasmon excitations in two-dimensional structures. We have tried to explain the basic notions which are essential for understanding the language used to describe plasma excitations. We have concentrated on the problem of a proper choice of the dispersion relation and we have described in detail the procedure leading from the experimentally measured spectra to the $\omega_p(k_j)$ dependence.

The research on a two-dimensional electron plasma at low temperatures and magnetic fields is still a vivid branch of semiconductor physics. New achievements and ideas open yet unexplored directions of research. Among them, one could point at a study on the ratchet effect in heterostructures with non-centrosymmetric grids,[44] an ultra-strong light-matter coupling at THz frequencies[35,45,46] or relativistic plasma excitations.[47] For sure, there is still a lot to do in the field of a two-dimensional plasma.

## Acknowledgement

A financial support from a Polish National Science Centre UMO-2015/17/B/ST7/03630 grant is acknowledged.

## References

1. S. J. Allen, D. C. Tsui, and R. A. Logan, "Observation of the Two-Dimensional Plasmon in Silicon Inversion Layers", *Phys. Rev. Lett.* **38**, 980 (1977).
2. M. A. Zudov, R. R. Du, J. A. Simmons, J. L. Reno, "Shubnikov - de Haas-like oscillations in millimeter wave photoconductivity in a high - mobility two-dimensional electron gas", *Phys. Rev. B* **64**, 201311 (2001).
3. R. G. Mani, J. H. Smet, K. von Klitzing, V. Narayanamurti, W. B. Johnson, and V. Umansky, "Zero-resistance states induced by electromagnetic-wave excitation in GaAs/AlGaAs heterostructures", *Nature* **420**, 646 (2002).
4. I. A. Dmitriev, A. D. Mirlin, D. G. Polyakov, and M. A. Zudov, "Nonequilibrium phenomena in high Landau levels", *Rev. Mod. Phys.* **84**, 1709 (2012).
5. M. A. Zudov, Comment on "Theory of microwave-induced zero-resistance states in two-dimensional electron systems" and on "Microwave-induced zero-resistance states and second-harmonic generation in an ultraclean two-dimensional electron gas", *Phys. Rev. B* **92**, 047301 (2015).
6. S. A. Mikhailov, Reply to "Comment on "Theory of microwave-induced zero-resistance states in two-dimensional electron systems" and on "Microwave-induced zero-resistance states and

second-harmonic generation in an ultraclean two-dimensional electron gas", *Phys. Rev. B* **92**, 047302 (2015).

7. Y. M. Beltukov, and M. I. Dyakonov, "Microwave-Induced Resistance Oscillations as a Classical Memory Effect", *Phys. Rev. Lett.* **116**, 176801 (2016).
8. I. V. Kukushkin, J. H. Smet, V. W. Scarola, V. Umansky, and K. von Klitzing, "Dispersion of the Excitations of Fractional Quantum Hall States", *Science* **324**, 1044 (2009).
9. B. A. Piot, J. Kunc, M. Potemski, D. K. Maude, C. Betthausen, A. Vogl, D. Weiss, G. Karczewski, and T. Wojtowicz, "Fractional quantum Hall effect in CdTe", *Phys. Rev. B* **82**, 081307 (2010).
10. C. Betthausen, P. Giudici, A. Iankilevitch, C. Preis, V. Kolkovsky. M. Wiater, G. Karczewski, B. A. Piot, J. Kune, M. Potemski, T. Wojtowicz, and D. Weiss, "Fractional quantum Hall effect in a dilute magnetic semiconductor", *Phys. Rev. B* **90**, 115302 (2014).
11. M. Czapkiewicz, V. Kolkovsky, P. Nowicki, M. Wiater, T. Wojciechowski, T. Wojtowicz, and J. Wróbel, "Evidence for charging effects in CdTe/CdMgTe quantum point contacts", *Phys. Rev. B* **86**, 165415 (2012).
12. S. D. Ganichev, S. A. Tarasenko, V. V. Bel'kov, P. Olbrich, W. Eder, D. R. Yakovlev, V. Kolkovsky, W. Zaleszczyk, G. Karczewski, T. Wojtowicz, and D. Weiss, "Spin currents in diluted magnetic semiconductors", *Phys. Rev. Lett.* **102**, 156602 (2009).
13. P. Olbrich, C. Zoth, P. Lutz, C. Drexler,V. V. Bel'kov, Ya. V. Terent'ev, S. A. Tarasenko, A. N. Semenov, S. V. Ivanov, D. R. Yakovlev, T. Wojtowicz, U. Wurstbauer, D. Schuh, and S. D. Ganichev, "Spin-polarized electric currents in dilutedmagnetic semiconductor heterostructures induced by terahertz and microwave radiation", *Phys. Rev. B* **86**, 085310 (2012).
14. R. Rungsawang, F. Perez, D. Oustinov, J. Gómez, V. Kolkovsky, G. Karczewski, T. Wojtowicz, J. Madéo, N. Jukam, S. Dhillon, and J. Tignon, "Terahertz Radiation from Magnetic Excitations in Diluted Magnetic Semiconductors", *Phys. Rev. Lett.* **110**, 177203 (2013).
15. M. J. Manfra, K. W. Baldwin, A. M. Sergent, K. W. West, R. J. Molnar, and J. Caissie, "Electron mobility exceeding 160000 $cm^2/Vs$ in AlGaN/GaN heterostructures grown by molecular-beam epitaxy", *Appl. Phys. Lett.* **85**, 5394(2004).
16. C. Skierbiszewski, K. Dybko, W. Knap, M. Siekacz, W. Krupczyński, G. Nowak, M. Boćkowski, J. Łusakowski, Z. R. Wasilewski, D. Maude,

T. Suski, and S. Porowski, "High mobility two-dimensional electron gas in AlGaN/GaN heterostructures grown on bulk GaN by plasma assisted molecular beam epitaxy", *Appl. Phys. Lett.* **86**, 102106 (2005).
17. W. Knap, V. I. Fal'ko, E. Frayssinet, P. Lorenzini, N. Grandjean, D. Maude, G. Karczewski, B. L. Brandt, J. Łusakowski, I. Grzegory, M. Leszczyński, P. Prystawko, C. Skierbiszewski, S. Porowski, X. Hu, G. Simin, M. Asif Khan, and M. S. Shur, "Spin and interaction effects in Shubnikov–de Haas oscillations and the quantum Hall effectin GaN/AlGaN heterostructures", *Journal of Physics: Condensed Matter* bf 16, 3421, (2004).
18. M. S. Kushwaha, "Plasmons and magnetoplasmons in semiconductor heterostructures", *Surface Science Reports* **41**, 1 (2001).
19. R. Krahne, M. Hochgräfe, Ch. Heyn, D. Heitmann, M. Hauser, and K. Eberl, "Excitation of two-dimensional plasmons with cross-grating couplers", *Phys. Rev. B* **62**, 15345 (2000).
20. U. Mackens, D. Heitmann, L. Prager, J. P. Kotthaus, and W. Beinvogl, "Minigaps in the Plasmon Dispersion of a Two-Dimensional Electron Gas with Spatially Modulated Charge Density", *Phys. Rev. Lett.* **53**, 1485 (1984).
21. T. Ando, A. B. Fowler, and F. Stern, "Electronic properties of two-dimensional systems", *Rev. Mod. Phys.* **54**, 437 (1982).
22. M. Dressel and G. Grüner, "Electrodynamics of Solids" (Cambridge University Press, 2002).
23. P. M. Platzman and P. A. Wolf, "Waves and interaction in solid state plasmas", *Solid State Physics: Supplement 13* (Academic Press, 1973).
24. K. W. Chiu and J. J. Quinn, *Phys. Rev. B* **9**, 4724 (1974).
25. M. H. Cohen, M. J. Harrison, and W. A. Harrison, "Magnetic-field dependence of the ultrasonic attenuation in metals", *Phys. Rev.* **117**, 937 (1960).
26. A. V. Chaplik, and D. Heitmann, "Geometric resonances in two-dimensional magnetoplasmons", *J. Phys. C:Solid State Phys.* **18**, 3357 (1985).
27. A. V. Chaplik, "Possible crystallization of charge carriers in low-density inversion layers", *Soviet Physics JETP* **35**, 395 (1972).
28. I. B. Bernstein, "Waves in a plasma in a magnetic field", *Phys. Rev.* **109**, 10 (1958).
29. F. Stern, "Polarizability of a Two-Dimensional Electron Gas", *Phys. Rev. Lett.* **18**, 546 (1967).

30. A. Eguiluz, T. K. Lee, J. J. Quinn, and K. W. Chiu, "Interface excitations in metal-insulator-semiconductor structures", *Phys. Rev. B* **11**, 4989 (1975).
31. V. V. Popov, O. V. Polischuk, and M. S. Shur, "Resonant excitation of plasma oscillations in a partially gated two-dimensional electron layer", *J. Appl. Phys.* **98**, 033510 (2005).
32. I. Grigelionis, K. Nogajewski, G. Karczewski, T. Wojtowicz, M. Czapkiewicz, J. Wróbel, H. Boukari, H. Mariette, and J. Łusakowski, "Magnetoplasmons in high electron mobility CdTe/CdMgTe quantum wells", Phys. Rev. B **91**, 075424 (2015).
33. K. Nogajewski, Ph.D. thesis, University of Warsaw, https://depotuw.ceon.pl/handle/item/310 (2013).
34. A. Hirose, "Lecture Notes on Plasma Waves", http://physics.usask.ca/ hirose/P862/notes.htm
35. G. Scalari, C. Maissen, D. Turčinková, G. Hagenmüller, S, De Liberato, C. Ciuti, C. Reicht, D. Schuh, W. Wegscheider, M. Beck, and J. Feist, "Ultrastrong Coupling of the Cyclotron Transistion of a 2D Electron Gas to a THz Metamaterial", *Science* **335**, 1323 (2012).
36. M. Białek, A. M. Witowski, M. Orlita, M. Potemski, M. Czapkiewicz, J. Wróbel, V. Umansky, M. Grynberg, and J. Łusakowski, *J. Appl. Phys.* **115**, 214503 (2014).
37. M. Białek, M. Czapkiewicz, J. Wróbel, V. Umansky, and J. Łusakowski, "Plasmon dispersions in high electron mobility terahertz detectors", *Appl. Phys. Lett.* **104**, 263514 (2014).
38. K. Karpierz, J. Łusakowski, J. Kossut, and M. Grynberg, "Potential fluctuations in CdTe epitaxial layer studied by shallow donor spectroscopy in the far infrared", *J. Phys.: Cond. Matter* **20**, 195217 (2008).
39. J. Łusakowski, and A. Łusakowski, "Magnetoconductivity and potential fluctuations in semi-insulating GaAs", *J. Phys.: Condens. Matter* **16**, 2661 (2004).
40. M. Sakowicz, R. Tauk, J. Łusakowski, A. Tiberj, W. Knap, Z. Bougrioua, M. Azize, P. Lorenzini, K. Karpierz, and M. Grynberg, "Low temperature electron mobility and concentration under the gate of AlGaN/GaN field effect transistors", *J. Appl. Phys.* **100**, 113726 (2006).
41. J. Łusakowski, W. Knap, Y. Meziani, J.- P. Cesso, A. el Fatimy, R. Tauk, N. Dyakonova, G. Ghibaudo, F. Boeuf and T. Skotnicki, "Electron mobility in quasi ballistic Si MOSFETs", *Solid State Electronics* **50**, 632-636 (2006).

42. K. Nogajewski, J. Łusakowski, W. Knap, V. V. Popov, F. Teppe, S. L. Rumyantsev, and M. S. Shur, "Localized and collective magnetoplasmon excitations in AlGaN/GaN-based grating-gate terahertz modulators", *Appl. Phys. Lett.* **99**, 213501 (2011).
43. I. Grigelionis, M. Białek. M. Grynberg, M. Czapkiewicz, V. Kolkovskiy, M. Wiater, T. Wojciechowski, J. Wróbel, T. Wojtowicz, N. Diakonova, W. Knap, and J. Łusakowski, "Terahertz magnetospectroscopy of a point contact based on CdTe/CdMgTe quantum well", *J. Nanophotonics* **9**, 93082 (2015).
44. P. Olbrich, E. L. Ivchenko, R. Ravash, T. Feil, S. D. Danilov, J. Allerdings, D. Weiss, D. Schuh, W. Wegscheider, and S. D. Ganichev, "Ratchet Effects Induced by Terahertz Radiation in Heterostructures with a Lateral Periodic Potential", *Phys. Rev. Lett.* **103**, 090603 (2009).
45. Y. Todorov, A. M. Andrews, I. Sagnes, R. Colombelli, P. Klang, G. Strasser, and C. Sirtori, "Strong Light-Matter Coupling in Subwavelength Metal-Dielectric Microcavities at Terahertz Frequencies", *Phys. Rev. Lett.* **102**, 186402 (2009).
46. V. M. Muravev, P. A. Gusikhin, I. V. Andreev, and I. V. Kukushkin, "Ultrastrong coupling of high-frequency two-dimensional cyclotron plasma mode with a cavity photon", *Phys. Rev. B* **87**, 045307 (2013).
47. V. M. Muravev, Gusikhin P. A., Andreev I. V., Kukushkin I. V., "Novel relativistic plasma excitations in a gated two-dimensional electron system", *Phys. Rev. Lett.* **114**, 106805 (2015).

# Spin-charge Coupling Effects in a Two-dimensional Electron Gas

Roberto Raimondi

*Dipartimento di Matematica e Fisica, Università Roma Tre,*
*Via della Vasca Navale 84, Rome, 00146, Italy*
*E-mail: roberto.raimondi@uniroma3.it*
*www.uniroma3.it*

Cosimo Gorini

*Institut für Theoretische Physik, Universität Regensburg*
*93040 Regensburg, Germany*
*E-mail: cosimo.gorini@physik.uni-regensburg.de*

Sebastian Tölle

*Institut für Physik, Universität Augsburg*
*Universitätsstr. 1, 86135 Augsburg, Germany*
*E-mail: sebastian.toelle@physik.uni-augsburg.de*

In these lecture notes we study the disordered two-dimensional electron gas in the presence of Rashba spin-orbit coupling, by using the Keldysh non-equilibrium Green function technique. We describe the effects of the spin-orbit coupling in terms of a SU(2) gauge field and derive a generalized Boltzmann equation for the charge and spin distribution functions. We then apply the formalism to discuss the spin Hall and the inverse spin galvanic (Edelstein) effects. Successively we show how to include, within the generalized Boltzmann equation, the side jump, the skew scattering and the spin current swapping processes originating from the extrinsic spin-orbit coupling due to impurity scattering.

*Keywords*: Spin-orbit coupling, electronic transport, many-body Green function, disordered systems.

## 1. Introduction

These lecture notes are based mainly on the work by Gorini et al. of Ref. 1, where by means of a gradient expansion a generalized Boltzmann equation with SU(2) gauge fields was obtained for the disordered Rashba model. The inclusion of the extrinsic spin-orbit coupling (SOC) from impurities in the SU(2) formalism was later considered in the work by Raimondi et al. in Ref. 2. Hence, the aim of these lecture notes is to provide a self-contained and pedagogical introduction to the disordered two-dimensional electron gas (2DEG) with both intrinsic (Rashba) and extrinsic SOC within the SU(2) gauge-field approach.

The layout of these lecture notes is the following. In Section 2 we write down the quantum kinetic equation for the fermion Green function in the presence of U(1), associated with the electromagnetic field, and SU(2) gauge fields. The standard model of disorder is introduced in Section 3. Whereas in Section 2 we derive the hydrodynamic SU(2) derivative of the Boltzmann equation, in Section 3 we obtain an expression for the collision integral describing the scattering from impurities. In Section 4 we apply the formalism to the disordered Rashba model and derive the Bloch equation for the spin density, describe the Dyakonov-Perel spin relaxation, and discuss a thermally induced spin polarization. In Section 5 we introduce SOC from impurity scattering and analyze the so-called side jump mechanism, which manifests as a correction to the velocity operator. In Section 6 we discuss the skew scattering mechanism. Both intrinsic and extrinsic SOCs contribute to the spin relaxation. Extrinsic SOC gives rise to Elliott-Yafet spin relaxation, which is covered in Section 7, together with the complete form of the Boltzmann equation. Section 8 states our conclusions. Throughout we use units such that $\hbar = c = 1$.

## 2. The kinetic equation and the SU(2) covariant Green function

We begin by defining the Keldysh Green function (for an introduction see, e.g., the book by Rammer[3])

$$\check{G} = \begin{pmatrix} G^R & G^K \\ 0 & G^A \end{pmatrix}, \tag{1}$$

where the retarded $G^R$, Keldysh $G^K$ and advanced $G^A$ components are given by

$$\begin{aligned}
G^R(\mathbf{r}_1, t_1; \mathbf{r}_2, t_2) &= -i\Theta(t_1 - t_2)\langle \psi(\mathbf{r}_1, t_1)\psi^\dagger(\mathbf{r}_2, t_2) + \psi^\dagger(\mathbf{r}_2, t_2)\psi(\mathbf{r}_1, t_1)\rangle \\
G^K(\mathbf{r}_1, t_1; \mathbf{r}_2, t_2) &= -i\langle \psi(\mathbf{r}_1, t_1)\psi^\dagger(\mathbf{r}_2, t_2) - \psi^\dagger(\mathbf{r}_2, t_2)\psi(\mathbf{r}_1, t_1)\rangle \\
G^A(\mathbf{r}_1, t_1; \mathbf{r}_2, t_2) &= i\Theta(t_2 - t_1)\langle \psi(\mathbf{r}_1, t_1)\psi^\dagger(\mathbf{r}_2, t_2) + \psi^\dagger(\mathbf{r}_2, t_2)\psi(\mathbf{r}_1, t_1)\rangle.
\end{aligned}$$

In the above definitions $\psi(\mathbf{r}_1, t_1)$ and $\psi^\dagger(\mathbf{r}_2, t_2)$ are Heisenberg field operators for fermions and $\Theta(t)$ the Heaviside step function. In the following we will be concerned with spin one-half fermions. As a consequence, all entries of $\check{G}$ become two by two matrices.

To derive a kinetic equation, it is useful to introduce Wigner mixed coordinates. To this end we perform a Fourier transform with respect to both space $(\mathbf{r}_1 - \mathbf{r}_2)$ and time $(t_1 - t_2)$ relative coordinates

$$\check{G}(\mathbf{p}, \epsilon; \mathbf{r}, t) = \int \mathrm{d}(t_1 - t_2) \int \mathrm{d}(\mathbf{r}_1 - \mathbf{r}_2) \check{G}(\mathbf{r}_1, t_1; \mathbf{r}_2, t_2) e^{i[\epsilon(t_1 - t_2) - \mathbf{p}\cdot(\mathbf{r}_1 - \mathbf{r}_2)]}. \tag{2}$$

The first step in the standard derivation of the kinetic equation is the left-right subtracted Dyson equation

$$\left[\check{G}_0^{-1}(x_1, x_3) \overset{\otimes}{,} \check{G}(x_3, x_2)\right] = 0, \tag{3}$$

where we have used space-time coordinates $x_1 \equiv (\mathbf{r}_1, t_1)$ *etc.* In Eq. (3), the symbol $\otimes$ implies integration over $x_3$ and matrix multiplication both in Keldysh and spin (if any) spaces. Furthermore

$$\check{G}_0^{-1}(x_1, x_3) = (i\partial_{t_1} - H)\,\delta(x_1 - x_3), \tag{4}$$

where $H$ is the Hamiltonian operator. In these lecture notes we do not consider electron-electron interaction. Quite generally the Hamiltonian operator takes the form

$$H = \frac{(-i\nabla_{\mathbf{r}} + e\mathbf{A}(\mathbf{r}, t))^2}{2m} - e\Phi(\mathbf{r}, t) + V(\mathbf{r}). \tag{5}$$

Here $e = |e|$ and we have assumed negatively charged particles. In Eq. (5) the scalar and vector potential have a two by two matrix structure, which can be shown by expanding them in the basis of the Pauli matrices

$$\Phi = \Phi^0\sigma^0 + \Phi^a\frac{\sigma^a}{2}, \quad \mathbf{A} = \mathbf{A}^0\sigma^0 + \mathbf{A}^a\frac{\sigma^a}{2}, \quad a = x, y, z, \tag{6}$$

and summation over the repeated indices is understood. The $\sigma^0$-components are the electromagnetic scalar and vector potentials associated with the U(1) gauge invariance. $V(\mathbf{r})$ describes the disorder potential due to impurities and defects. In this section we set $V(\mathbf{r}) = 0$ and postpone its discussion to the following section. The $\sigma^a$-components are an SU(2) gauge field, whose scalar and vector components can be used to respectively describe a Zeeman/exchange term and SOC – in our case Rashba SOC, as will be shown in Sec. 4. For the time being, we do not consider a specific form of the SU(2) gauge field $(\Phi^a, \mathbf{A}^a)$.

The goal of a kinetic equation is to describe non-equilibrium phenomena. In general this is a formidable task. However, for close-to-equilibrium phenomena or for non-equilibrium ones occurring on scales large compared to microscopic ones, it is possible to derive an effective kinetic equation by means of the so-called gradient expansion. The idea is based on the observation that under specific circumstances the Green function varies fast with respect to the relative coordinate $x_1 - x_2$ and much more slowly with respect to the center-of-mass one $(x_1 + x_2)/2$. In equilibrium, for a translationally invariant system, the Green function does not depend on the center-of-mass coordinate at all.

To understand how the gradient expansion works, consider the convolution of two quantities

$$(A \otimes B)(x_1, x_2) = \int \mathrm{d}x_3 A(x_1, x_3) B(x_3, x_2),$$

which can be equivalently expressed as a function of center-of-mass and relative coordinates

$$\int \mathrm{d}x_3 A\left(\frac{x_1 + x_3}{2}, x_1 - x_3\right) B\left(\frac{x_3 + x_2}{2}, x_3 - x_2\right).$$

Next, replace $x_1 + x_3 = x_1 + x_2 + x_3 - x_2$ and $x_3 + x_2 = x_1 + x_2 - (x_1 - x_3)$ in the first argument of $A$ and $B$, respectively. By Taylor expanding $A$ with respect to $x_3 - x_2$ in its first argument and $B$ with respect to $x_1 - x_3$ in its first argument, after Fourier transforming according to Eq. (2), one gets

$$A(x,p)B(x,p) + \frac{i}{2}(\partial_\mu A(x,p))(\partial_p^\mu B(x,p)) - \frac{i}{2}(\partial_p^\mu A(x,p))(\partial_\mu B(x,p)), \quad (7)$$

where we have introduced the compact (relativistic) space-time notations

$$x^\mu = (t, \mathbf{r}),\ x_\mu = (-t, \mathbf{r}),\ p^\mu = (\epsilon, \mathbf{p}),\ p_\mu = (-\epsilon, \mathbf{p}) \quad (8)$$

and

$$\partial^\mu \equiv \frac{\partial}{\partial x_\mu},\ \partial_\mu \equiv \frac{\partial}{\partial x^\mu},\ \partial_p^\mu \equiv \frac{\partial}{\partial p_\mu},\ \partial_{p,\mu} \equiv \frac{\partial}{\partial p^\mu} \quad (9)$$

in a such a way that the product $p^\mu x_\mu = -\epsilon t + \mathbf{p} \cdot \mathbf{r}$ has the correct Lorentz metrics. Equation (3) acquires then the form

$$-i\left[\check{G}_0^{-1}, \check{G}\right] + \frac{1}{2}\left\{(\partial^\mu \check{G}_0^{-1}), (\partial_{p,\mu}\check{G})\right\} - \frac{1}{2}\left\{(\partial_p^\mu \check{G}_0^{-1}), (\partial_\mu \check{G})\right\} = 0. \quad (10)$$

The Hamiltonian (5) is invariant under a gauge transformation $O(x)$, which locally rotates the spinor field

$$\psi'(x) = O(x)\psi(x),\ \psi'^\dagger(x) = \psi^\dagger(x)O^\dagger(x),\ O(x)O^\dagger(x) = 1. \quad (11)$$

The Green function, however, is not locally covariant, i.e. its transformation depends on two distinct space-time points

$$\check{G}(x_1, x_2) \to O(x_1)\check{G}(x_1, x_2)O^\dagger(x_2). \quad (12)$$

Physical observables, which are locally covariant, are obtained by considering the Green function in the limit of coinciding space-time points. It is then useful to introduce a *locally* covariant Green function

$$\check{\check{G}}(x_1, x_2) = U_\Gamma(x, x_1)\check{G}(x_1, x_2)U_\Gamma(x_2, x) \tag{13}$$

where

$$U_\Gamma(x, x_1) = \mathcal{P}\exp\left(-i\int_{x_1}^{x} eA^\mu(y)\mathrm{d}y_\mu\right). \tag{14}$$

The line integral of the gauge field is referred to as the Wilson line. In Eq. (14) $\mathcal{P}$ is a path-ordering operator and $A^\mu = (\Phi, \mathbf{A})$, $A_\mu = (-\Phi, \mathbf{A})$. Since the Wilson line transforms covariantly

$$U_\Gamma(x, x_1) \to O(x)U_\Gamma(x, x_1)O^\dagger(x_1), \tag{15}$$

one easily sees that the covariant Green function $\check{\check{G}}$ transforms in a locally covariant way

$$\check{\check{G}}(x_1, x_2) \to O(x)\check{\check{G}}(x_1, x_2)O^\dagger(x). \tag{16}$$

When $x_1 = x_2 = x$, the locally covariant Green function coincides with the original Green function. By inverting Eq. (13), one can, via Eq. (10), obtain an equation for the locally covariant Green function. Due to the non-Abelian character of the gauge field, Eq. (13) is not easy to handle. In the spirit of the gradient approximation, we assume that $\partial^\mu\partial_{\mu,p} \ll 1$. In addition we also assume that $eA^\mu\partial_{\mu,p} \ll 1$. This assumption can be justified on physical grounds once an explicit form is assigned to $A$. Under these assumptions, Eq. (13) becomes

$$\check{\check{G}} = \check{G} - \frac{1}{2}\{eA^\mu\partial_{p,\mu}, \check{G}\} \tag{17}$$

and its inverse

$$\check{G} = \check{\check{G}} + \frac{1}{2}\{eA^\mu\partial_{p,\mu}, \check{\check{G}}\}. \tag{18}$$

By using the decomposition

$$\delta(x_1 - x_2) = \int \frac{\mathrm{d}^{d+1}p}{(2\pi)^{d+1}} e^{ip^\mu(x_{1,\mu} - x_{2,\mu})}, \tag{19}$$

one obtains

$$\check{G}_0^{-1}(x, p) = \epsilon - \frac{(\mathbf{p} + e\mathbf{A}(x))^2}{2m} + e\Phi(x) = \epsilon - \frac{p^2}{2m} - V^\mu eA_\mu - \frac{e^2\mathbf{A}^2}{2m}, \tag{20}$$

from which

$$\partial_p^\mu \check{G}_0^{-1}(x,p) = -V^\mu - \left(\partial_p^\mu V^\nu\right) eA_\nu,$$
$$\partial^\mu \check{G}_0^{-1}(x,p) = -V^\nu \partial^\mu eA_\nu. \tag{21}$$

In the above $V^\mu = (1, \mathbf{p}/m)$ is the $d$-current operator, $d$ being the space dimensionality. In the second equation of (21) we neglected the term $\partial^\mu e^2 \mathbf{A}^2/2m = e\mathbf{A}\cdot\partial^\mu e\mathbf{A}$ because it gives a small correction to $\mathbf{p}\cdot\partial^\mu e\mathbf{A}$ when $\mathbf{p}\sim\mathbf{p}_F$. Performing the *shift* transformation of Eq. (17) in Eq. (20) gives

$$\check{\tilde{G}}_0^{-1} = \epsilon - \frac{p^2}{2m}. \tag{22}$$

We have now all the necessary ingredients to obtain the equation for $\check{\tilde{G}}$. We begin by considering the first term of Eq. (10). By applying the *shift* transformation of Eq. (17) and expressing $\check{G}$ in terms of $\check{\tilde{G}}$ via Eq. (18), we obtain

$$-i\left(\left[\check{G}_0^{-1}, \check{\tilde{G}}\right] - \frac{1}{2}\left\{eA_\mu, \partial_p^\mu\left[\check{G}_0^{-1}, \check{\tilde{G}}\right]\right\} + \frac{1}{2}\left[\check{G}_0^{-1}, \left\{eA_\mu, \partial_p^\mu \check{\tilde{G}}\right\}\right]\right). \tag{23}$$

By means of the identity $\{A,[B,C]\} - [B,\{A,C\}] = \{[A,B],C\}$ we get

$$ieV^\mu\left(\left[A_\mu, \check{\tilde{G}}\right] + \frac{1}{2}\left\{e\left[A_\mu, A_\nu\right], \partial_p^\nu \check{\tilde{G}}\right\}\right). \tag{24}$$

As for the second term of Eq. (10)

$$\frac{1}{2}\left\{(\partial^\mu \check{G}_0^{-1}), (\partial_{p,\mu}\check{G})\right\} \to -\frac{e}{2}V^\mu\left\{\left(\partial^\nu A_\mu\right), \partial_{p,\nu}\check{\tilde{G}}\right\}, \tag{25}$$

where the last step follows by considering the first order of the gradient expansion. Finally, for the last term of Eq. (10)

$$-\frac{1}{2}\left\{(\partial_p^\mu \check{G}_0^{-1}), (\partial_\mu \check{G})\right\} \to V^\mu\left[\partial_\mu \check{\tilde{G}} + \frac{1}{2}\left\{\left(e\partial_\mu A^\nu\right), \partial_{p,\nu}\check{\tilde{G}}\right\}\right], \tag{26}$$

where we omitted terms $\sim \partial_\mu \partial_p^\nu \check{\tilde{G}}$ within the first order accuracy of the gradient expansion. By collecting the results of Eqs.(24-26), the equation for $\check{\tilde{G}}$ reads

$$V^\mu\left[\tilde{\partial}_\mu \check{\tilde{G}} + \frac{1}{2}\left\{eF_{\mu\nu}, \partial_p^\nu \check{\tilde{G}}\right\}\right] = 0, \tag{27}$$

where we have introduced the covariant derivative

$$\tilde{\partial}_\mu \check{\tilde{G}} = \partial_\mu \check{\tilde{G}} + i\left[eA_\mu, \check{\tilde{G}}\right] \tag{28}$$

and the field strength

$$F_{\mu\nu} = \partial_\mu A_\nu - \partial_\nu A_\mu + ie\left[A_\mu, A_\nu\right]. \tag{29}$$

It is useful at this stage to separate the space and time parts and rewrite Eq. (27) as

$$\left(\tilde{\partial}_t + \frac{\mathbf{p}}{m}\cdot\tilde{\nabla}_{\mathbf{r}}\right)\check{\tilde{G}} - \frac{e}{2}\left\{\frac{\mathbf{p}}{m}\cdot\mathbf{E}, \partial_\epsilon\check{\tilde{G}}\right\} + \frac{1}{2}\left\{\mathbf{F}, \nabla_{\mathbf{p}}\check{\tilde{G}}\right\} = 0, \tag{30}$$

where the generalized Lorentz force reads

$$\mathbf{F} = -e\left(\mathbf{E} + \frac{\mathbf{p}}{m}\times\mathbf{B}\right) \tag{31}$$

with the U(1)×SU(2) fields given by

$$\begin{aligned} \mathbf{E} &= -\partial_t\mathbf{A} - \nabla_{\mathbf{r}}\Phi + ie\left[\Phi, \mathbf{A}\right], \\ B_i &= \frac{1}{2}\varepsilon_{ijk}F^{jk}. \end{aligned} \tag{32}$$

Equation (30) is the quantum kinetic equation. One can integrate over the energy $\epsilon$, corresponding to the equal-time limit, in order to obtain a semiclassical kinetic equation. We define the distribution function as

$$f(\mathbf{p},\mathbf{r},t) \equiv \frac{1}{2}\left[1 + \int\frac{d\epsilon}{2\pi i}\tilde{G}^K(\mathbf{p},\epsilon;\mathbf{r},t)\right], \tag{33}$$

which is a matrix in spin space, $f = f^0\sigma^0 + f^a\sigma^a$, $a = x, y, z$. By taking the Keldysh component of Eq. (30) we get

$$\left(\tilde{\partial}_t + \frac{\mathbf{p}}{m}\cdot\tilde{\nabla}_{\mathbf{r}}\right) f(\mathbf{p},\mathbf{r},t) + \frac{1}{2}\left\{\mathbf{F}\cdot\nabla_{\mathbf{p}}, f(\mathbf{p},\mathbf{r},t)\right\} = 0. \tag{34}$$

We have then obtained a generalization of the Boltzmann equation, where space and time derivatives are replaced by the covariant ones and the generalized Lorentz force appears. We may then introduce the density and current by integrating over the momentum

$$\rho(\mathbf{r},t) = \sum_{\mathbf{p}} f(\mathbf{p},\mathbf{r},t),\ \mathbf{J}(\mathbf{r},t) = \sum_{\mathbf{p}}\frac{\mathbf{p}}{m}f(\mathbf{p},\mathbf{r},t). \tag{35}$$

Hence the integration over the momentum of Eq. (34) leads to a continuity-like equation

$$\tilde{\partial}_t\rho(\mathbf{r},t) + \tilde{\nabla}_{\mathbf{r}}\cdot\mathbf{J}(\mathbf{r},t) = 0. \tag{36}$$

We will use the above equation in Section 4, when discussing the spin Hall and inverse spin galvanic/Edelstein effects in the Rashba model.

## 3. The standard model of disorder and the diffusive approximation

In this section we consider the effect of disorder due to impurity scattering. According to the standard model of disorder[4] the potential $V(\mathbf{r})$ is assumed to be a random variable with distribution

$$\langle V(\mathbf{r})\rangle = 0,\ \langle V(\mathbf{r})V(\mathbf{r}')\rangle = n_i v_0^2 \delta(\mathbf{r}-\mathbf{r}'). \tag{37}$$

In the above $n_i$ is the impurity density and $v_0$ is the scattering amplitude. Higher momenta can be present, and indeed they will be needed when considering skew-scattering processes, but it is not necessary to specify them for the time being. Disorder effects can be taken into account in perturbation theory via the inclusion of a self-energy. Equation (3) becomes

$$\left[\check{G}_0^{-1}(x_1,x_3)\overset{\otimes}{;}\check{G}(x_3,x_2)\right] = \left[\check{\Sigma}(x_1,x_3)\overset{\otimes}{;}\check{G}(x_3,x_2)\right]. \tag{38}$$

The lowest order self-energy due to disorder is given in Fig. 1 and its expression reads

$$\check{\Sigma}_0(p,x) = n_i v_0^2 \sum_{\mathbf{p}'} \check{G}(p,x). \tag{39}$$

Notice that the integration is only on the space component of the $d$-momentum $p^\mu = (\epsilon, \mathbf{p}')$. This is a result of the fact that the scattering is elastic. In order to use the above self-energy, we must transform it to the locally covariant formalism according to the transformation of Eq. (17) and express $\check{G}$ in terms of $\check{\tilde{G}}$ via Eq. (18). This procedure is the same we have followed in the previous section to transform the kinetic equation from the form of Eq. (10) to the form Eq. (27). Since the procedure will also be used later on, let us show it in detail in this simple case. First we notice that

$$U_\Gamma(x,x_1)\left[\check{\Sigma}(x_1,x_3)\overset{\otimes}{;}\check{G}(x_3,x_2)\right]U_\Gamma(x_2,x) = \left[\check{\tilde{\Sigma}}(x_1,x_3)\overset{\otimes}{;}\check{\tilde{G}}(x_3,x_2)\right] \tag{40}$$

after using the unitarity of the Wilson line by inserting

$$U_\Gamma(x_3,x)U_\Gamma(x,x_3) = 1$$

between the self-energy and the Green function. The locally covariant self-energy reads

$$\check{\tilde{\Sigma}}_0 = n_i v_0^2 \sum_{\mathbf{p}'} \left(\check{\tilde{G}}_{\mathbf{p}'} + \frac{1}{2}\left\{A^\mu(\partial_{p',\mu} - \partial_{p,\mu}), \check{\tilde{G}}_{\mathbf{p}'}\right\}\right) = n_i v_0^2 \sum_{\mathbf{p}'} \check{\tilde{G}}_{\mathbf{p}'}. \tag{41}$$

In the above the derivative with respect to $\epsilon$ cancels in the two terms. The derivative with respect to $\mathbf{p}$ vanishes because there is no dependence on

$\mathbf{p}$. Finally, the derivative with respect to $\mathbf{p}'$ can be integrated giving at most a constant, which can be discarded. As a result, the locally covariant self-energy has the same functional form of the original self-energy. The

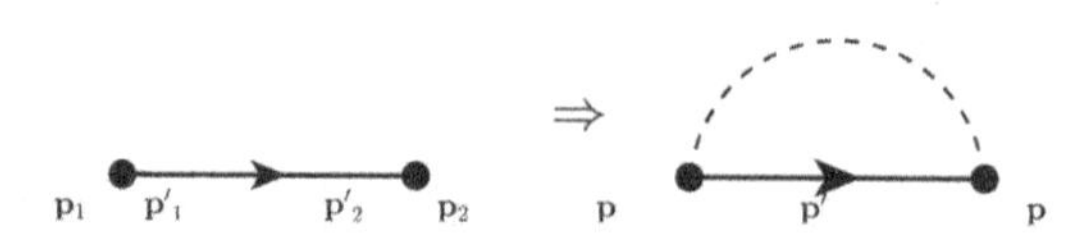


Fig. 1. Self-energy diagram to second order in the impurity potential (black dot vertex). The diagram on the left is before the impurity average, which is carried in the diagram on the right as a dashed line connecting the two impurity insertions. Notice that the impurity average in momentum space yields $\langle V(\mathbf{p}_1 - \mathbf{p}'_1)V(\mathbf{p}'_2 - \mathbf{p}_2)\rangle = n_i v_0^2 \delta(\mathbf{p}_1 - \mathbf{p}'_1 + \mathbf{p}'_2 - \mathbf{p}_2)$.

Keldysh component of the collision integral then reads

$$\tilde{I} = -i\left[\check{\tilde{\Sigma}}, \check{\tilde{G}}_{\mathbf{p}}\right]^K = -in_i v_0^2 \sum_{\mathbf{p}'} \left((\tilde{G}^R_{\mathbf{p}'} - \tilde{G}^A_{\mathbf{p}'})\tilde{G}^K_{\mathbf{p}} - (\tilde{G}^R_{\mathbf{p}} - \tilde{G}^A_{\mathbf{p}})\tilde{G}^K_{\mathbf{p}'}\right). \quad (42)$$

Note that the SU(2) shifted retarded and advanced Green functions have no spin structure and therefore commute with the Keldysh Green function. For weak scattering, one can ignore the broadening of the energy levels in the retarded and advanced Green function and use

$$\tilde{G}^R_{\mathbf{p}} - \tilde{G}^A_{\mathbf{p}} = -2\pi i\delta(\epsilon - \epsilon_{\mathbf{p}}),\ \tilde{G}^K_{\mathbf{p}} = -2\pi i\delta(\epsilon - \epsilon_{\mathbf{p}})(1 - 2f(\mathbf{p}, \mathbf{r}, t)). \quad (43)$$

As a result, Eq. (34) is no longer collisionless and becomes

$$\left(\tilde{\partial}_t + \frac{\mathbf{p}}{m}\cdot\tilde{\nabla}_{\mathbf{r}}\right) f(\mathbf{p}, \mathbf{r}, t) - \frac{e}{2}\left\{\left(\mathbf{E} + \frac{\mathbf{p}}{m}\times\mathbf{B}\right)\cdot\nabla_{\mathbf{p}}, f(\mathbf{p}, \mathbf{r}, t)\right\} = I[f], \quad (44)$$

with the collision integral being

$$I[f] = -2\pi n_i v_0^2 \sum_{\mathbf{p}'} \delta(\epsilon_{\mathbf{p}} - \epsilon_{\mathbf{p}'})(f(\mathbf{p}, \mathbf{r}, t) - f(\mathbf{p}', \mathbf{r}, t)). \quad (45)$$

It is appropriate in the final part of this section to obtain the solution of the Boltzmann equation Eq. (44) in the diffusive approximation. First we notice that, by integration over the momentum $\mathbf{p}$, the collision integral $I$ vanishes reproducing the continuity equation (36) with density and current defined in Eq. (35). In the diffusive approximation we expand the distribution function in spherical harmonics

$$f(\mathbf{p}, \mathbf{r}, t) = \langle f\rangle + 2\hat{\mathbf{p}}\cdot\mathbf{f} + \ldots \quad (46)$$

and keep terms up to the p-wave symmetry. In the above $\langle \dots \rangle$ indicates the integration over the directions of the momentum. We perform the evaluation in two space dimensions having in mind the application of the theory to the 2DEG. By defining the momentum relaxation time[a]

$$\frac{1}{\tau} = 2\pi N_0 n_i v_0^2 \tag{47}$$

with the density of states $N_0 = m/(2\pi)$ the collision integral becomes

$$I[f] = -\frac{1}{\tau} 2\hat{\mathbf{p}} \cdot \mathbf{f}. \tag{48}$$

In the diffusive approximation we consider $\omega\tau \ll 1$ and $v_F q \tau \ll 1$, where $\omega$ and $q$ are typical energy and momentum scales. For instance, $\omega$ can be the magnitude of an externally applied magnetic field. We multiply Eq. (44) by $\hat{\mathbf{p}}$ and integrate over the angle $\phi$ with $\hat{\mathbf{p}} = (\cos\phi, \sin\phi)$. We get

$$-\frac{1}{\tau}\mathbf{f} = \frac{p}{2m}\tilde{\nabla}_{\mathbf{r}}\langle f \rangle - \frac{e}{2}\langle\left\{\hat{\mathbf{p}}\mathbf{E}\cdot\nabla_{\mathbf{p}}, \langle f \rangle\right\}\rangle - \frac{e}{2m}\langle\left\{\hat{\mathbf{p}}(\mathbf{p}\times\mathbf{B}\cdot\nabla_{\mathbf{p}}), 2\hat{\mathbf{p}}\cdot\mathbf{f}\right\}\rangle. \tag{49}$$

The first term, keeping in mind Eq. (35) for the current, represents the *diffusive* contribution including the additional part due to the SU(2) gauge field. Due to the covariant nature of the derivative, such a term differs from zero even in uniform circumstances. The second term yields the usual *drift* contribution, whereas the third one gives rise to a Hall contribution. The gradient with respect to the momentum can be split as $\nabla_{\mathbf{p}} = \hat{\mathbf{p}}\partial_p - \hat{\boldsymbol{\phi}}\partial_\phi/p$ where $\hat{\boldsymbol{\phi}} = (-\sin\phi, \cos\phi)$. Then we get

$$\mathbf{f} = -\frac{\tau p}{2m}\tilde{\nabla}_{\mathbf{r}}\langle f \rangle + \frac{e\tau}{4}\left\{\mathbf{E}, \partial_p\langle f \rangle\right\} + \frac{e\tau}{2m}\left\{\mathbf{B}\times, \mathbf{f}\right\}. \tag{50}$$

By using the definitions of density and current in Eq. (36), we may write the expression for the number and spin components as

$$n = \mathrm{Tr}\left[\rho\right],\ \mathbf{J}^0 = \mathrm{Tr}\left[\mathbf{J}\right],\ s^a = \frac{1}{2}\mathrm{Tr}\left[\sigma^a\rho\right],\ \mathbf{J}^a = \frac{1}{2}\mathrm{Tr}\left[\sigma^a\mathbf{J}\right]. \tag{51}$$

To begin with, let us consider the drift term

$$\mathbf{J}_{dr} = \sum_{\mathbf{p}} \frac{p}{m}\frac{e\tau}{4}\left\{\mathbf{E}, \partial_p\langle f \rangle\right\} = eN_0\int d\epsilon_{\mathbf{p}}\ D(\epsilon_{\mathbf{p}})\frac{1}{2}\left\{\partial_{\epsilon_{\mathbf{p}}}\langle f \rangle, \mathbf{E}\right\} = -\frac{e}{2}\left\{\sigma(\mu), \mathbf{E}\right\} \tag{52}$$

where $\epsilon_{\mathbf{p}} = p^2/2m$, $D(\epsilon_{\mathbf{p}}) = \tau\epsilon_{\mathbf{p}}/m$, $\mu = \rho/N_0$ is a spin-dependent chemical potential, and $\sigma(\mu) = N_0 D(\mu)$. In equilibrium, $\rho_{eq} = N_0\epsilon_F + N_0\Phi$ and its

[a] Its expression can also be obtained, for instance, by the Fermi golden rule.

eigenvalues determine the chemical potentials for *up* and *down* electrons. Then

$$\mathbf{J}^0_{dr} = -eN_0 D^0 \mathbf{E}^0 - \frac{e}{2} N_0 D^a \mathbf{E}^a, \ \mathbf{J}^a_{dr} = -\frac{e}{4} N_0 D^0 \mathbf{E}^a - \frac{e}{2} N_0 D^a \mathbf{E}^0, \quad (53)$$

with $D^0$ and $D^a$ defined by $D(\mu) = D^0 + D^a \sigma^a$. By expanding around $\epsilon_F$, one has $D^0 \approx D(\epsilon_F)$ and $D^a \approx \tau s^a/(N_0 m)$, and therefore

$$\sigma(\mu) = N_0 D(\epsilon_F)\sigma^0 + \frac{\tau}{m} s^a \sigma^a. \quad (54)$$

The diffusion term is obtained by integrating over the momentum the first term of Eq. (50)

$$\mathbf{J}_{dif} = -N_0 \int d\epsilon_{\mathbf{p}} \ D(\epsilon_{\mathbf{p}}) \tilde{\nabla}_{\mathbf{r}} \langle f \rangle = -\frac{1}{2} \left\{ D(\mu), \tilde{\nabla}_{\mathbf{r}} \rho \right\}. \quad (55)$$

The above form of the diffusion term is determined by requiring that in equilibrium it must cancel the drift term according to the Einstein argument. Then

$$\mathbf{J}^0_{dif} = -D^0 \nabla_{\mathbf{r}} n - 2D^a \left[ \tilde{\nabla}_{\mathbf{r}} s \right]^a, \ \mathbf{J}^a_{dif} = -\frac{1}{2} D^a \nabla_{\mathbf{r}} n - D^0 \left[ \tilde{\nabla}_{\mathbf{r}} s \right]^a. \quad (56)$$

Finally the Hall term yields

$$\mathbf{J}^0_{Hall} = \frac{e\tau}{m} \mathbf{B}^0 \times \mathbf{J}^0 + \frac{e\tau}{m} \mathbf{B}^a \times \mathbf{J}^a, \ \mathbf{J}^a_{Hall} = \frac{e\tau}{m} \mathbf{B}^0 \times \mathbf{J}^a + \frac{e\tau}{4m} \mathbf{B}^a \times \mathbf{J}^0. \ (57)$$

To summarize, we may write the particle and spin currents as

$$\begin{aligned} \mathbf{J}^0 &= -eN_0 D^0 \mathbf{E}^0 - \frac{e}{2} N_0 D^a \mathbf{E}^a - D^0 \nabla_{\mathbf{r}} n - 2D^a \left[ \tilde{\nabla}_{\mathbf{r}} s \right]^a \\ &+ \frac{e\tau}{m} \mathbf{B}^0 \times \mathbf{J}^0 + \frac{e\tau}{m} \mathbf{B}^a \times \mathbf{J}^a \end{aligned} \quad (58)$$

and

$$\begin{aligned} \mathbf{J}^a &= -\frac{e}{4} N_0 D^0 \mathbf{E}^a - \frac{e}{2} N_0 D^a \mathbf{E}^0 - \frac{1}{2} D^a \nabla_{\mathbf{r}} n - D^0 \left[ \tilde{\nabla}_{\mathbf{r}} s \right]^a \\ &+ \frac{e\tau}{m} \mathbf{B}^0 \times \mathbf{J}^a + \frac{e\tau}{4m} \mathbf{B}^a \times \mathbf{J}^0. \end{aligned} \quad (59)$$

The above two equations, together with the continuity-like Eq. (36), will be used in the next section to analyze the spin Hall and Edelstein effect in the disordered Rashba model.

## 4. The disordered Rashba model

The Rashba Hamiltonian reads[5]

$$H = \frac{p^2}{2m} + \alpha p_y \sigma^x - \alpha p_x \sigma^y. \tag{60}$$

The only non zero components of the SU(2) gauge field are

$$eA_x^y = -2m\alpha, \; eA_y^x = 2m\alpha. \tag{61}$$

As shown in the previous sections (cf. Eq. (36)), the spin density obeys a continuity-like equation

$$\tilde{\partial}_t s^a + \tilde{\nabla}_{\mathbf{r}} \cdot \mathbf{J}^a = 0, \tag{62}$$

which is deceptively simple. The notable fact in the present theory is that the covariant derivatives defined in Eq. (28) appear also at the level of the effective phenomenological equations, providing an elegant and compact way to derive the equation of motion for the spin density. The explicit expressions of the space and time covariant derivatives of a generic observable $\mathcal{O}^a$ read[b]

$$\tilde{\partial}_t \mathcal{O}^a = \partial_t \mathcal{O}^a + \epsilon_{abc} \, e\Phi^b \mathcal{O}^c \tag{63}$$

$$\tilde{\nabla}_i \mathcal{O}^a = \nabla_i \mathcal{O}^a - \epsilon_{abc} \, eA_i^b \mathcal{O}^c. \tag{64}$$

Equation (62) becomes

$$\partial_t s^a + \epsilon_{abc} \, e\Phi^b s^c + \nabla_i J_i^a - \epsilon_{abc} \, eA_i^b J_i^c = 0, \tag{65}$$

showing that the equation for the spin is not a simple continuity equation, as expected from the non conservation of spin. The second term in Eq. (65) is the standard *precession* term. The last term of (65) can be made explicit by providing the expression for the spin current $J_i^a$, where the lower (upper) index indicates the space (spin) component. The expression of $J_i^a$ was derived via a microscopic theory in the diffusive regime in Eq. (59). The explicit expression reads

$$J_i^a = -\frac{e\tau}{m} s^a E_i + D\epsilon_{abc} eA_i^b s^c - \frac{e\tau}{4m} \epsilon_{ijk} J_j \, B_k^a - \frac{eDN_0}{2} E_i^a, \tag{66}$$

where $D = D(\epsilon_F)$ is the diffusion constant. Let us apply Eqs. (65-66) to the Rashba model defined by Eq. (61) in the presence of an applied electric field along the x direction $E_x$. To linear order in the electric field, the first term of Eq. (66) does not contribute in a paramagnetic system. Also by

[b] $\epsilon_{abc}$ is the fully antisymmetric Ricci tensor.

using Eq. (61) in the expressions of the fields of Eq. (32) we get that the SU(2) electric field vanishes $E_i^a = 0$ and that the only non zero component of the SU(2) magnetic field reads

$$eB_z^z = -(2m\alpha)^2. \tag{67}$$

Because the electric field is uniform, we may ignore the space derivative and obtain the explicit form of Eq. (65)

$$\partial_t s^x = -2m\alpha J_x^z \tag{68}$$
$$\partial_t s^y = -2m\alpha J_y^z \tag{69}$$
$$\partial_t s^z = +2m\alpha J_y^y + 2m\alpha J_x^x \tag{70}$$

with the associated expressions for the spin currents

$$J_x^z = 2m\alpha D s^x \tag{71}$$
$$J_y^z = 2m\alpha D s^y + \theta_{SH}^{int} J_x^0 \tag{72}$$
$$J_x^x = J_y^y = -2m\alpha D s^z, \tag{73}$$

where $J_x^0 = -e\sigma(\epsilon_F)E_x$ is the charge current and $\theta_{SH}^{int}$ is the spin Hall angle for intrinsic SOC

$$\theta_{SH}^{int} = -m\tau\alpha^2. \tag{74}$$

Insertion of Eqs. (71-73) into (68-70) gives the Bloch equations

$$\partial_t s^x = -\frac{1}{\tau_{DP}} s^x \tag{75}$$
$$\partial_t s^y = -\frac{1}{\tau_{DP}} (s^y - s_0) \tag{76}$$
$$\partial_t s^z = -\frac{2}{\tau_{DP}} s^x \tag{77}$$

where the Dyakonov-Perel relaxation time is given by

$$\tau_{DP}^{-1} = (2m\alpha)^2 D \tag{78}$$

and the current-induced spin polarization is given by

$$s_0 = -eN_0\alpha\tau E_x. \tag{79}$$

In the static limit the solution of the Bloch equations yields an in-plane spin polarization perpendicular to the electric field $s^y = s^0$, with $s^x = s^z = 0$. This is known as the Edelstein[6,7] or inverse spin-galvanic effect[8,9]. The vanishing of the time derivative implies, via Eq. (69), the vanishing of the spin current $J_y^z$ associated to the spin Hall effect. This vanishing occurs thanks to the exact compensation of the two contributions appearing in Eq. (72).[c]

The above analysis can be extended in the presence of a thermal gradient. More precisely, we derive $s^y$ in terms of a stationary thermal gradient along the $x$-direction, $\nabla_x T$, in the absence of any additional external fields.[11]

However, in an experiment one would still measure an electric field $E_x$ due to a gradient in the chemical potential, $\nabla_x \mu$, resulting from an imbalance of the charge carriers due to the thermal gradient. For this, we shall first consider the trace of Eq. (50):

$$\mathbf{f}^0 = -\frac{\tau p}{2m} \nabla \langle f^0 \rangle, \tag{80}$$

where we can approximate $\nabla \langle f^0 \rangle$ as the Fermi function $f^{eq}$, giving us

$$f_x^0 = -\frac{\tau p}{2m} \left( \frac{\epsilon - \mu}{T} \nabla_x T + \nabla_x \mu \right) \left( -\frac{\partial f^{eq}}{\partial \epsilon} \right) \tag{81}$$

for the $x$-component of Eq. (80). With use of the Sommerfeld expansion

$$\int d\epsilon g(\epsilon) \left( -\frac{\partial f^{eq}}{\partial \epsilon} \right) = g(\mu) + \frac{\pi^2}{6} (k_B T)^2 \left. \frac{\partial^2 g}{\partial \epsilon^2} \right|_{\epsilon = \mu}, \tag{82}$$

where $g(\epsilon)$ is an arbitrary energy dependent function, we end up with the particle current in the $x$-direction as follows:

$$J_x^0 = -\frac{2\tau N_0}{m} \left[ \frac{\pi^2}{3} \frac{(k_B T)^2}{T} \nabla_x T + \mu \nabla_x \mu \right]. \tag{83}$$

We shall consider an open circuit along $x$-direction, i.e., a vanishing particle current $J_x^0 = 0$. Then, we can express the electric field one would measure in an experiment as

$$E_x = \frac{1}{e} \nabla_x \mu = S \nabla_x T, \tag{84}$$

where $S = -(\pi k_B)^2 T / (3e\mu)$ is the Seebeck coefficient. After having analyzed the charge sector, we consider next the spin sector in order to get an expression for $s^y$. We start by multiplying Eq. (50) with $\sigma^z$ and perform the trace. The $y$-component reads

$$f_y^z = 2p\tau\alpha \langle f^y \rangle + \frac{e\tau}{m} B_z^z f_x^0 . \tag{85}$$

---

[c]To make contact with the diagrammatic Kubo formula approach, we notice that the second term of Eq. (72) corresponds to a bubble-like diagram, whereas the first term describes the so-called vertex corrections.[10]

Note that the form of Eq. (69) doesn't change and since we assume a stationary case we have

$$J_y^z = 0 \Leftrightarrow f_y^z = 0\,. \tag{86}$$

This implies that there is no spin Nernst effect as no spin Hall effect in the disordered Rashba model. From Eq. (85), together with Eq. (81) we therefore find

$$\langle f^y \rangle = \frac{2\alpha m}{p} f_x^0 = -\alpha\tau \left( \frac{\epsilon - \mu}{T} \nabla_x T + \nabla_x \mu \right) \left( -\frac{\partial f^{eq}}{\partial \epsilon} \right) . \tag{87}$$

From the form of the latter equation and with use of the Sommerfeld expansion, Eq. (82), it is clear that we can express the $y$ spin polarization as

$$s^y = P_{sT} \nabla_x T + P_{sE} E_x\,, \tag{88}$$

where $P_{sT}$ can be written in a Mott-like form:

$$P_{sT} = -S\mu \frac{\partial P_{sE}}{\partial \mu}. \tag{89}$$

Here, we find $P_{sE} = -\alpha\tau e N_0$, consistent with Eq. (76). This results in a vanishing $P_{sT}$ since $P_{sE}$ is independent of $\mu$. We express $E_x$ in terms of $\nabla_x T$ with use of Eq. (84) and end up with

$$s^y = -\alpha\tau e N_0 S \nabla_x T, \tag{90}$$

describing the thermal Edelstein effect in the disordered Rashba model.

## 5. The impurity-induced spin-orbit coupling: Swapping and side jump mechanisms

In this and following sections we consider the *extrinsic* SOC due to impurity scattering described by the Hamiltonian

$$H_{ext,so} = -\frac{\lambda_0^2}{4} \boldsymbol{\sigma} \times \nabla V(\mathbf{r}) \cdot \mathbf{p}, \tag{91}$$

where $\lambda_0$ is the effective Compton wave length[12,13]. In developing the perturbation theory in the impurity potential we must now use the lowest order scattering amplitude

$$S_{\mathbf{p}',\mathbf{p}''} = V_{\mathbf{p}'-\mathbf{p}''} \left[ 1 - \frac{i\lambda_0^2}{4} \mathbf{p}' \times \mathbf{p}'' \cdot \boldsymbol{\sigma} \right] \tag{92}$$

with

$$\langle V_{\mathbf{q}_1} V_{\mathbf{q}_2} \rangle = n_i v_0^2 \delta(\mathbf{q}_1 + \mathbf{q}_2). \tag{93}$$

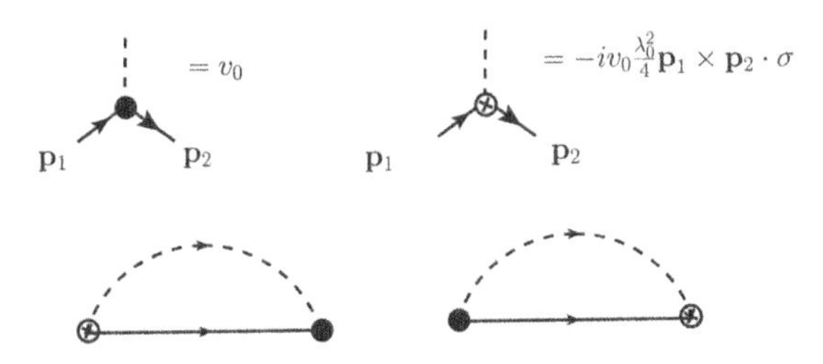


Fig. 2. In the top line the impurity insertion without (black dot vertex) and with (crossed dot vertex) spin-orbit coupling. In the bottom line the two diagrams to first order in the spin-orbit coupling $\lambda_0^2$.

To zeroth order in $\lambda_0^2$, we have the diagram of Fig. 1, which has been studied in the previous section. To first order in $\lambda_0^2$, we must consider the two diagrams of Fig. 2. Here an empty crossed dot stands for the part of the scattering amplitude with the spin-orbit coupling. Let us evaluate these diagrams step by step. Before the impurity average (indicated as $\langle \dots \rangle$), the expression for the two diagrams reads

$$\begin{aligned}\check{\Sigma}_{1,\mathbf{p}',\mathbf{p}''} = &-i\frac{\lambda_0^2}{4}\sum_{\mathbf{p}_1,\mathbf{p}_2}\langle V_{\mathbf{p}'-\mathbf{p}_1}\left(\mathbf{p}' \times \mathbf{p}_1 \cdot \boldsymbol{\sigma}\check{G}_{\mathbf{p}_1,\mathbf{p}_2}\right.\\ &\left.+ \check{G}_{\mathbf{p}_1,\mathbf{p}_2}\mathbf{p}_2 \times \mathbf{p}'' \cdot \boldsymbol{\sigma}\right) V_{\mathbf{p}_2-\mathbf{p}''}\rangle. \end{aligned} \tag{94}$$

In the above, $\check{G}_{\mathbf{p}_1,\mathbf{p}_2}$ is the Fourier transform with respect to the space arguments $\mathbf{r}_1$ and $\mathbf{r}_2$ of $\check{G}(\mathbf{r}_1,\mathbf{r}_2)$. We do not mention explicitly here the time arguments for the sake of simplicity. Performing the impurity average one obtains $\mathbf{p}' - \mathbf{p}_1 = \mathbf{p}'' - \mathbf{p}_2$. It is convenient then to define momenta as

$$\mathbf{p} = \frac{\mathbf{p}' + \mathbf{p}''}{2}, \ \mathbf{q} = \mathbf{p}' - \mathbf{p}'' = \mathbf{p}_1 - \mathbf{p}_2, \ \tilde{\mathbf{p}} = \frac{\mathbf{p}_1 + \mathbf{p}_2}{2} \tag{95}$$

in such a way that $\mathbf{p}$ and $\tilde{\mathbf{p}}$ correspond to the momentum of the mixed Wigner representation introduced previously. The momentum $\mathbf{q}$ instead is the variable conjugated to the center-of-mass coordinate $\mathbf{r}$ by Fourier transform. We then get the impurity-averaged expression of the two first-order diagrams

$$\begin{aligned}\check{\Sigma}_1(\mathbf{p},\mathbf{q}) = &-i\frac{\lambda_0^2}{4}n_i v_0^2\sum_{\tilde{\mathbf{p}}}\left[\left(\mathbf{p}+\frac{\mathbf{q}}{2}\right)\times\left(\tilde{\mathbf{p}}+\frac{\mathbf{q}}{2}\right)\cdot\boldsymbol{\sigma}\check{G}(\tilde{\mathbf{p}},\mathbf{q})\right.\\ &\left.+ \check{G}(\tilde{\mathbf{p}},\mathbf{q})\left(\tilde{\mathbf{p}}-\frac{\mathbf{q}}{2}\right)\times\left(\mathbf{p}-\frac{\mathbf{q}}{2}\right)\cdot\boldsymbol{\sigma}\right]. \end{aligned} \tag{96}$$

The above expression can be divided into three terms

$$\check{\Sigma}_{1,a}(\mathbf{p},\mathbf{q})) = -i\frac{\lambda_0^2}{4}n_i v_0^2 \sum_{\tilde{\mathbf{p}}} \left[\mathbf{p}\times\tilde{\mathbf{p}}\cdot\boldsymbol{\sigma}, \check{G}(\tilde{\mathbf{p}},\mathbf{q})\right] \tag{97}$$

$$\check{\Sigma}_{1,b}(\mathbf{p},\mathbf{q})) = -i\frac{\lambda_0^2}{8}n_i v_0^2 \sum_{\tilde{\mathbf{p}}} \left\{\mathbf{p}\times\mathbf{q}\cdot\boldsymbol{\sigma}, \check{G}(\tilde{\mathbf{p}},\mathbf{q})\right\} \tag{98}$$

$$\check{\Sigma}_{1,c}(\mathbf{p},\mathbf{q})) = i\frac{\lambda_0^2}{8}n_i v_0^2 \sum_{\tilde{\mathbf{p}}} \left\{\tilde{\mathbf{p}}\times\mathbf{q}\cdot\boldsymbol{\sigma}, \check{G}(\tilde{\mathbf{p}},\mathbf{q})\right\}. \tag{99}$$

One can Fourier transform back to the center-of-mass coordinate $\mathbf{r}$ and re-label $\tilde{\mathbf{p}} \to \mathbf{p}'$

$$\check{\Sigma}_{1,a}(\mathbf{p},\mathbf{r})) = -i\frac{\lambda_0^2}{4}n_i v_0^2 \sum_{\mathbf{p}'} \left[\mathbf{p}\times\mathbf{p}'\cdot\boldsymbol{\sigma}, \check{G}(\mathbf{p}',\mathbf{r})\right] \tag{100}$$

$$\check{\Sigma}_{1,b}(\mathbf{p},\mathbf{r})) = -\nabla_{\mathbf{r}}\cdot\frac{\lambda_0^2}{8}n_i v_0^2 \sum_{\mathbf{p}'} \left\{\boldsymbol{\sigma}\times\mathbf{p}, \check{G}(\mathbf{p}',\mathbf{r})\right\} \tag{101}$$

$$\check{\Sigma}_{1,c}(\mathbf{p},\mathbf{r})) = \nabla_{\mathbf{r}}\cdot\frac{\lambda_0^2}{8}n_i v_0^2 \sum_{\mathbf{p}'} \left\{\boldsymbol{\sigma}\times\mathbf{p}', \check{G}(\mathbf{p}',\mathbf{r})\right\}. \tag{102}$$

Equations (100-102) are the final expression for the diagrams of the second line of Fig. 2. The first term $\check{\Sigma}_{1,a}$, as we will see, describes the swapping of spin currents under scattering.[14] The other two terms, $\check{\Sigma}_{1,b}$ and $\check{\Sigma}_{1,c}$, are written as a divergence. As a consequence, when they are inserted in the collision integral of the kinetic equation, they lead to a correction of the velocity operator and hence describe the so-called *side-jump* mechanism.[15]

Until now we have not yet considered the effect of the gauge fields on the extrinsic SOC. To do it, we must transform the above derived self-energy to the locally covariant form. Let us first examine the U(1) gauge field corresponding to a static electric field $\mathbf{E}^0 = -\nabla_{\mathbf{r}}\Phi^0(\mathbf{r})$. The shift of the gradient of the Green function yields

$$\begin{aligned}
\widetilde{\nabla_{\mathbf{r}}\check{G}(\mathbf{p},\mathbf{r})} &= \nabla_{\mathbf{r}}\check{G}(\mathbf{p},\mathbf{r}) - \frac{1}{2}\left\{e\Phi^0\partial_\epsilon, \nabla_{\mathbf{r}}\check{G}(\mathbf{p},\mathbf{r})\right\} \\
&= \nabla_{\mathbf{r}}\check{\tilde{G}}(\mathbf{p},\mathbf{r}) + \frac{1}{2}\nabla_{\mathbf{r}}\left\{e\Phi^0\partial_\epsilon, \check{\tilde{G}}(\mathbf{p},\mathbf{r})\right\} - \frac{1}{2}\left\{e\Phi^0\partial_\epsilon, \nabla_{\mathbf{r}}\check{\tilde{G}}(\mathbf{p},\mathbf{r})\right\} \\
&= \nabla_{\mathbf{r}}\check{\tilde{G}}(\mathbf{p},\mathbf{r}) - e\mathbf{E}^0\partial_\epsilon\check{\tilde{G}}(\mathbf{p},\mathbf{r}).
\end{aligned} \tag{103}$$

As a result we get from Eqs.(101-102) two more terms

$$\delta\check{\Sigma}_{1,b}(\mathbf{p},\mathbf{r})) = e\mathbf{E}^0 \cdot \frac{\lambda_0^2}{8} n_i v_0^2 \sum_{\mathbf{p}'} \left\{ \boldsymbol{\sigma} \times \mathbf{p}, \partial_\epsilon \check{\tilde{G}}(\mathbf{p}',\mathbf{r}) \right\} \tag{104}$$

$$\delta\check{\Sigma}_{1,c}(\mathbf{p},\mathbf{r})) = -e\mathbf{E}^0 \cdot \frac{\lambda_0^2}{8} n_i v_0^2 \sum_{\mathbf{p}'} \left\{ \boldsymbol{\sigma} \times \mathbf{p}', \partial_\epsilon \check{\tilde{G}}(\mathbf{p}',\mathbf{r}) \right\}. \tag{105}$$

Let us transform Eq. (100) to the locally covariant form

$$\begin{aligned}
\check{\tilde{\Sigma}}_{1,a}(\mathbf{p},\mathbf{r}) = & -i\frac{\lambda_0^2}{4} n_i v_0^2 \sum_{\mathbf{p}'} \left[ \mathbf{p} \times \mathbf{p}' \cdot \boldsymbol{\sigma}, \check{\tilde{G}}(\mathbf{p}',\mathbf{r}) \right] \\
& + i\frac{\lambda_0^2}{4} n_i v_0^2 \sum_{\mathbf{p}'} \frac{1}{2} \left[ \mathbf{p} \times \mathbf{p}' \cdot \boldsymbol{\sigma}, \left\{ e\mathbf{A} \cdot \nabla_{\mathbf{p}'}, \check{\tilde{G}}(\mathbf{p}',\mathbf{r}) \right\} \right] \\
& - i\frac{\lambda_0^2}{4} n_i v_0^2 \sum_{\mathbf{p}'} \frac{1}{2} \left\{ e\mathbf{A} \cdot \nabla_{\mathbf{p}}, \left[ \mathbf{p} \times \mathbf{p}' \cdot \boldsymbol{\sigma}, \check{\tilde{G}}(\mathbf{p}',\mathbf{r}) \right] \right\},
\end{aligned}$$

which can be rewritten as

$$\begin{aligned}
\check{\tilde{\Sigma}}_{1,a}(\mathbf{p},\mathbf{r}) = & -i\frac{\lambda_0^2}{4} n_i v_0^2 \sum_{\mathbf{p}'} \left[ \mathbf{p} \times \mathbf{p}' \cdot \boldsymbol{\sigma}, \check{\tilde{G}}(\mathbf{p}',\mathbf{r}) \right] \\
& + i\frac{\lambda_0^2}{4} n_i v_0^2 \sum_{\mathbf{p}'} \frac{1}{2} \left[ e\mathbf{A}; \left\{ \boldsymbol{\sigma} \times \mathbf{p}, \check{\tilde{G}}(\mathbf{p}',\mathbf{r}) \right\} \right] \\
& - i\frac{\lambda_0^2}{4} n_i v_0^2 \sum_{\mathbf{p}'} \frac{1}{2} \left\{ \boldsymbol{\sigma} \times \mathbf{p}'; \left[ e\mathbf{A}, \check{\tilde{G}}(\mathbf{p}',\mathbf{r}) \right] \right\}.
\end{aligned} \tag{106}$$

As a result, the contribution of the diagrams of Fig. 2 can be written as

$$\check{\tilde{\Sigma}}_{1,a}^{SCS} = -i\frac{\lambda_0^2}{4} n_i v_0^2 \sum_{\mathbf{p}'} \left[ \mathbf{p} \times \mathbf{p}' \cdot \boldsymbol{\sigma}, \check{\tilde{G}}(\mathbf{p}',\mathbf{r}) \right] \tag{107}$$

$$\check{\tilde{\Sigma}}_{1,b}^{SJ} = -\frac{\lambda_0^2}{8} n_i v_0^2 \sum_{\mathbf{p}'} \tilde{\nabla}_{\mathbf{r}} \left\{ \boldsymbol{\sigma} \times \mathbf{p}, \check{\tilde{G}}(\mathbf{p}',\mathbf{r}) \right\} \tag{108}$$

$$\check{\tilde{\Sigma}}_{1,c}^{SJ} = \frac{\lambda_0^2}{8} n_i v_0^2 \sum_{\mathbf{p}'} \left\{ \boldsymbol{\sigma} \times \mathbf{p}', \tilde{\nabla}_{\mathbf{r}} \check{\tilde{G}}(\mathbf{p}',\mathbf{r}) \right\}, \tag{109}$$

where $\tilde{\nabla}_{\mathbf{r}} = \nabla_{\mathbf{r}} - e\mathbf{E}^0 \partial_\epsilon + i\,[e\mathbf{A},\dots]$. A few comments are needed at this point. As already mentioned, the term $\check{\tilde{\Sigma}}_{1,a}^{SCS}$ describes the spin current swapping (SCS) defined by

$$J_i^a = \kappa \left[ J_a^i - \delta_{ia} \sum_l J_l^l \right]. \tag{110}$$

This is evident in the fact that this term contains the vector product $\mathbf{p}\times\mathbf{p}'$ of the momenta before and after the scattering from the impurity. The presence of both momenta yields the coupling of the currents of incoming and outgoing particles. The other two terms $\check{\tilde{\Sigma}}^{SJ}_{1,b}$ and $\check{\tilde{\Sigma}}^{SJ}_{1,c}$ describe the so-called side jump (SJ) effect. This is evident in both terms, which show the operator $\boldsymbol{\sigma}\times(\mathbf{p}-\mathbf{p}')$. By a semiclassical analysis one can show that $\Delta\mathbf{r}\equiv-(\lambda_0^2/4)(\mathbf{p}'-\mathbf{p})\times\boldsymbol{\sigma}$ is the side jump shift caused by the SOC to the scattering trajectory of a wave packet. The SJ terms are composed of three parts. The first part is the one under the space derivative sign and can be written as $-\nabla_{\mathbf{r}}\cdot\delta\mathbf{J}^{SJ}$, i.e. it describes a modification of the current operator. Eventually this term yields the first *one-half* of the side jump. The second part, proportional to the electric field, describes how the energy of a scattering particle is affected by the effective dipole energy $\sim e\mathbf{E}^0\cdot\Delta\mathbf{r}$ due to the side jump shift of the trajectory. This is the other *one-half* contribution to the side jump. The third part reconstructs the full covariant derivative in the presence of a SU(2) gauge field $\mathbf{A}=\mathbf{A}^a\sigma^a/2$. To first order accuracy in the gradient expansion, we have replaced the $\breve{G}$ with $\check{\tilde{G}}$ in the terms where the gradient or the gauge field appear. To make the above comments explicit, we start by noticing that only the Keldysh component appears in Eq. (107) because both $\tilde{G}^R$ and $\tilde{G}^A$ are proportional to $\sigma^0$ and commute with all the Pauli matrices. By considering that the Keldysh component of the collision integral requires $\tilde{\Sigma}^R\tilde{G}^K-\tilde{G}^K\tilde{\Sigma}^A-(\tilde{G}^R-\tilde{G}^A)\tilde{\Sigma}^K$, one obtains[16] from Eq. (107)

$$I^{SCS}[f]=-i\frac{\lambda_0^2}{4}n_iv_0^2\sum_{\mathbf{p}'}\delta(\epsilon_{\mathbf{p}}-\epsilon_{\mathbf{p}'})\left[\boldsymbol{\sigma}\cdot\mathbf{p}\times\mathbf{p}',f_{\mathbf{p}'}\right]. \tag{111}$$

Similarly, for the side jump we define

$$I^{SJ}=-i\int\frac{d\epsilon}{4\pi i}\left(-(\tilde{G}^R-\tilde{G}^A)\tilde{\Sigma}^K+\tilde{\Sigma}^R\tilde{G}^K-\tilde{G}^K\tilde{\Sigma}^A\right)\equiv I^{(a)}+I^{(b)}. \tag{112}$$

Because of the integration over the angle, the retarded and advanced components of Eq. (109) vanish. The retarded component of Eq. (108) reads

$$\tilde{\Sigma}^{SJ,R}_{1,b}=-i2\pi\frac{\lambda_0^2}{8}n_iv_0^2\sum_{\mathbf{p}'}\delta(\epsilon_{\mathbf{p}}-\epsilon_{\mathbf{p}'})\tilde{\nabla}_{\mathbf{r}}\boldsymbol{\sigma}\times\mathbf{p} \tag{113}$$

and $\tilde{\Sigma}^{SJ,A}_{1,b}=-\tilde{\Sigma}^{SJ,R}_{1,b}$. As a result, with $h_{\mathbf{p}}\equiv 1-2f_{\mathbf{p}}$ for brevity,

$$I^{(b)}=-\frac{\lambda_0^2}{16}2\pi n_iv_0^2\sum_{\mathbf{p}'}\delta(\epsilon_{\mathbf{p}}-\epsilon_{\mathbf{p}'})\Big\{\tilde{\nabla}_{\mathbf{r}}(\boldsymbol{\sigma}\times\mathbf{p}),h_{\mathbf{p}}\Big\}. \tag{114}$$

By using again the identity $\{A,[B,C]\}-[B,\{A,C\}]=\{[A,B],C\}$, the Keldysh component of Eq. (109) reads

$$\tilde{\Sigma}^{SJ,K}_{1,c}=\frac{\lambda_0^2}{8}n_i v_0^2\sum_{\mathbf{p}'}\left(\tilde{\nabla}_{\mathbf{r}}\left\{\boldsymbol{\sigma}\times\mathbf{p}',\tilde{G}^K_{\mathbf{p}'}\right\}-\left\{\tilde{\nabla}_{\mathbf{r}}(\boldsymbol{\sigma}\times\mathbf{p}'),\tilde{G}^K_{\mathbf{p}'}\right\}\right). \tag{115}$$

By combining the last result with Eq. (108) and Eq. (114), one obtains finally

$$I^{SJ}[f]=-\tilde{\nabla}_{\mathbf{r}}\cdot\frac{\lambda_0^2}{8}n_i v_0^2\sum_{\mathbf{p}'}\delta(\epsilon_{\mathbf{p}}-\epsilon_{\mathbf{p}'})\left\{\boldsymbol{\sigma}\times(\mathbf{p}'-\mathbf{p}),f_{\mathbf{p}'}\right\} \tag{116}$$
$$+\frac{\lambda_0^2}{8}n_i v_0^2\sum_{\mathbf{p}'}\delta(\epsilon_{\mathbf{p}}-\epsilon_{\mathbf{p}'})\left(\left\{\tilde{\nabla}_{\mathbf{r}}(\boldsymbol{\sigma}\times\mathbf{p}'),f_{\mathbf{p}'}\right\}-\left\{\tilde{\nabla}_{\mathbf{r}}(\boldsymbol{\sigma}\times\mathbf{p}),f_{\mathbf{p}}\right\}\right).$$

The first term on the right hand side, under the covariant space derivative, defines the modification of the current operator due to the SOC. We emphasize that such anomalous part of the current is subject to the full covariant derivative. Hence the last term in Eq. (65) remains the same. Notice also that the second term in $I^{SJ}[f]$, although it does not contribute to the continuity equation, is necessary to make sure that the equilibrium distribution function solves the kinetic equation.

## 6. The impurity-induced spin-orbit coupling: Skew scattering

In this section we discuss skew scattering by considering the diagrams of Fig. 3. To understand the meaning of these diagrams, recall that in general, in the presence of SOC, the scattering amplitude reads

$$S=A+\hat{\mathbf{p}}\times\hat{\mathbf{p}}'\cdot\boldsymbol{\sigma}B, \tag{117}$$

where $\hat{\mathbf{p}}$ and $\hat{\mathbf{p}}'$ are unit vectors in the direction of the momentum before and after the scattering event. To lowest order in perturbation theory or Born approximation one has $A=v_0$ and $B=-\mathrm{i}(\lambda_0^2p_F^2/4)v_0$ and one recovers Eq. (92). By considering the scattering probability proportional to $|S|^2$, one obtains three contributions given by $|A|^2$, $|B|^2$ and $2\,\mathcal{R}e\,(AB^*)\hat{\mathbf{p}}\times\hat{\mathbf{p}}'\cdot\boldsymbol{\sigma}$. Whereas the first two contributions are spin independent and give the total scattering time, the third one represents the so-called skew scattering term according to which electrons with opposite spin are scattered in opposite directions. Clearly, since $A$ and $B$ are out of phase, there is no skew scattering effect to the order of the Born approximation. For it to appear to first order in the spin-orbit coupling constant $\lambda_0^2$, $A$ has to be evaluated

Fig. 3. Third order in $v_0$ and first order in $\lambda_0^2$ diagrams. The skew scattering contribution arises from the first and last diagram.

beyond the Born approximation. The scattering problem can be cast in terms of the Lippman-Schwinger equation

$$\psi(\mathbf{x}) = e^{\mathrm{i}\mathbf{k}\cdot\mathbf{x}} + \int \mathrm{d}\mathbf{x}'\; G(\mathbf{x}-\mathbf{x}')V(\mathbf{x}')\psi(\mathbf{x}'), \tag{118}$$

where $G(\mathbf{x})$ is the retarded Green function at fixed energy. From (118) we get

$$\psi^{(1)} = v_0 G(\mathbf{x}),\ A^{(1)} = v_0;\ \ \psi^{(2)} = v_0^2 G(\mathbf{0})G(\mathbf{x}),\ A^{(2)} = v_0^2 G(\mathbf{0}). \tag{119}$$

Notice that only the imaginary part of $A^{(2)}$ is needed. By recalling that $\mathcal{I}m\ G(\mathbf{0}) = -\pi N_0$, the spin-orbit independent scattering amplitude $A$ up to second order in $v_0$ reads

$$A = v_0\,(1 - \mathrm{i}\pi N_0 v_0)\,. \tag{120}$$

The skew scattering contribution will then follow by inserting the modified scattering amplitude (120) into the collision integral of the Boltzmann equation. The same result can, of course, be obtained in quantum field theory using the Green function technique. The latter becomes necessary when one wants to consider skew scattering in the presence of Rashba spin-orbit interaction. To this end one has to consider the electron self-energy at least to third order in the scattering potential $v_0$. The diagrams of Fig. 3 yield

$$\check{\Sigma}_a^{SS} = -\frac{i\lambda_0^2}{4}\sum_{\mathbf{p}_1,\mathbf{p}_2,\mathbf{p}_3,\mathbf{p}_4}\langle V_{\mathbf{p}'-\mathbf{p}_1}\check{G}_{\mathbf{p}_1,\mathbf{p}_2}V_{\mathbf{p}_2-\mathbf{p}_3}\check{G}_{\mathbf{p}_3,\mathbf{p}_4}V_{\mathbf{p}_4-\mathbf{p}''}\mathbf{p}_4\times\mathbf{p}''\cdot\boldsymbol{\sigma}\rangle$$

$$\check{\Sigma}_b^{SS} = -\frac{i\lambda_0^2}{4}\sum_{\mathbf{p}_1,\mathbf{p}_2,\mathbf{p}_3,\mathbf{p}_4}\langle V_{\mathbf{p}'-\mathbf{p}_1}\check{G}_{\mathbf{p}_1,\mathbf{p}_2}V_{\mathbf{p}_2-\mathbf{p}_3}\mathbf{p}_2\times\mathbf{p}_3\cdot\boldsymbol{\sigma}\check{G}_{\mathbf{p}_3,\mathbf{p}_4}V_{\mathbf{p}_4-\mathbf{p}''}\rangle$$

$$\check{\Sigma}_c^{SS} = -\frac{i\lambda_0^2}{4}\sum_{\mathbf{p}_1,\mathbf{p}_2,\mathbf{p}_3,\mathbf{p}_4}\langle V_{\mathbf{p}'-\mathbf{p}_1}\mathbf{p}'\times\mathbf{p}_1\cdot\boldsymbol{\sigma}\check{G}_{\mathbf{p}_1,\mathbf{p}_2}V_{\mathbf{p}_2-\mathbf{p}_3}\check{G}_{\mathbf{p}_3,\mathbf{p}_4}V_{\mathbf{p}_4-\mathbf{p}''}\rangle.$$

By requiring the existence of third moments of the random potential $\langle V(\mathbf{q}_1)V(\mathbf{q}_2)V(\mathbf{q}_3)\rangle = n_i v_0^3\delta(\mathbf{q}_1+\mathbf{q}_2+\mathbf{q}_3)$, we perform the impurity average and obtain

$$\check{\Sigma}_a^{SS} = -n_i v_0^3 \frac{i\lambda_0^2}{4} \sum_{\mathbf{p}_1,\mathbf{p}_2,\mathbf{p}_3,\mathbf{p}_4} \delta_{\mathbf{p}'-\mathbf{p}_1+\mathbf{p}_2-\mathbf{p}_3+\mathbf{p}_4-\mathbf{p}''} \check{G}_{\mathbf{p}_1,\mathbf{p}_2} \check{G}_{\mathbf{p}_3,\mathbf{p}_4} \mathbf{p}_4 \times \mathbf{p}'' \cdot \boldsymbol{\sigma}$$

$$\check{\Sigma}_b^{SS} = -n_i v_0^3 \frac{i\lambda_0^2}{4} \sum_{\mathbf{p}_1,\mathbf{p}_2,\mathbf{p}_3,\mathbf{p}_4} \delta_{\mathbf{p}'-\mathbf{p}_1+\mathbf{p}_2-\mathbf{p}_3+\mathbf{p}_4-\mathbf{p}''} \check{G}_{\mathbf{p}_1,\mathbf{p}_2} \mathbf{p}_2 \times \mathbf{p}_3 \cdot \boldsymbol{\sigma} \check{G}_{\mathbf{p}_3,\mathbf{p}_4}$$

$$\check{\Sigma}_c^{SS} = -n_i v_0^3 \frac{i\lambda_0^2}{4} \sum_{\mathbf{p}_1,\mathbf{p}_2,\mathbf{p}_3,\mathbf{p}_4} \delta_{\mathbf{p}'-\mathbf{p}_1+\mathbf{p}_2-\mathbf{p}_3+\mathbf{p}_4-\mathbf{p}''} \mathbf{p}' \times \mathbf{p}_1 \cdot \boldsymbol{\sigma} \check{G}_{\mathbf{p}_1,\mathbf{p}_2} \check{G}_{\mathbf{p}_3,\mathbf{p}_4}.$$

Let us introduce as before, momenta associated to center-of-mass and relative coordinates $\mathbf{p} = (\mathbf{p}' + \mathbf{p}'')/2$, $\mathbf{q} = \mathbf{p}' - \mathbf{p}''$, $\widetilde{\mathbf{p}}_a = (\mathbf{p}_1 + \mathbf{p}_2)/2, \widetilde{\mathbf{q}}_a = \mathbf{p}_1 - \mathbf{p}_2, \widetilde{\mathbf{p}}_b = (\mathbf{p}_3 + \mathbf{p}_4)/2, \widetilde{\mathbf{q}}_b = \mathbf{p}_3 - \mathbf{p}_4$ and we get after integrating over the momentum $\widetilde{\mathbf{q}}_b$

$$\check{\Sigma}_a^{SS} = -n_i v_0^3 \frac{i\lambda_0^2}{4} \sum_{\widetilde{\mathbf{p}}_a,\widetilde{\mathbf{p}}_b,\widetilde{\mathbf{q}}_a} \check{G}_{\widetilde{\mathbf{p}}_a,\widetilde{\mathbf{q}}_a} \check{G}_{\widetilde{\mathbf{p}}_b,\mathbf{q}-\widetilde{\mathbf{q}}_a} \left(\widetilde{\mathbf{p}}_b - \frac{\mathbf{q}}{2} + \frac{\widetilde{\mathbf{q}}_a}{2}\right) \times \left(\mathbf{p} - \frac{\mathbf{q}}{2}\right) \cdot \boldsymbol{\sigma}$$

$$\check{\Sigma}_b^{SS} = -n_i v_0^3 \frac{i\lambda_0^2}{4} \sum_{\widetilde{\mathbf{p}}_a,\widetilde{\mathbf{p}}_b,\widetilde{\mathbf{q}}_a} \check{G}_{\widetilde{\mathbf{p}}_a,\widetilde{\mathbf{q}}_a} \left(\widetilde{\mathbf{p}}_a - \frac{\widetilde{\mathbf{q}}_a}{2}\right) \times \left(\widetilde{\mathbf{p}}_b + \frac{\mathbf{q}}{2} - \frac{\widetilde{\mathbf{q}}_a}{2}\right) \cdot \boldsymbol{\sigma} \check{G}_{\widetilde{\mathbf{p}}_b,\mathbf{q}-\widetilde{\mathbf{q}}_a}$$

$$\check{\Sigma}_c^{SS} = -n_i v_0^3 \frac{i\lambda_0^2}{4} \sum_{\widetilde{\mathbf{p}}_a,\widetilde{\mathbf{p}}_b,\widetilde{\mathbf{q}}_a} \left(\mathbf{p} + \frac{\mathbf{q}}{2}\right) \times \left(\widetilde{\mathbf{p}}_a + \frac{\widetilde{\mathbf{q}}_a}{2}\right) \cdot \boldsymbol{\sigma} \check{G}_{\widetilde{\mathbf{p}}_a,\widetilde{\mathbf{q}}_a} \check{G}_{\widetilde{\mathbf{p}}_b,\mathbf{q}-\widetilde{\mathbf{q}}_a}.$$

We Fourier transform back with respect to the momentum $\mathbf{q}$ and neglect derivatives with respect to $\mathbf{r}$, i.e. we confine to lowest order in the gradient expansion. We then get

$$\check{\Sigma}_a^{SS} = -n_i v_0^3 \frac{i\lambda_0^2}{4} \sum_{\widetilde{\mathbf{p}}_a,\widetilde{\mathbf{p}}_b} \check{G}(\widetilde{\mathbf{p}}_a, \mathbf{r}) \check{G}(\widetilde{p}_b, \mathbf{r}) \widetilde{\mathbf{p}}_b \times p \cdot \boldsymbol{\sigma} \tag{121}$$

$$\check{\Sigma}_b^{SS} = -n_i v_0^3 \frac{i\lambda_0^2}{4} \sum_{\widetilde{p}_a,\widetilde{\mathbf{p}}_b} \check{G}(\widetilde{\mathbf{p}}_a, \mathbf{r}) \widetilde{\mathbf{p}}_a \times \widetilde{\mathbf{p}}_b \cdot \boldsymbol{\sigma} \check{G}(\widetilde{\mathbf{p}}_b, \mathbf{r}) \tag{122}$$

$$\check{\Sigma}_c^{SS} = -n_i v_0^3 \frac{i\lambda_0^2}{4} \sum_{\widetilde{\mathbf{p}}_a,\widetilde{\mathbf{p}}_b} \mathbf{p} \times \widetilde{\mathbf{p}}_a \cdot \boldsymbol{\sigma} \check{G}(\widetilde{\mathbf{p}}_a, \mathbf{r}) \check{G}(\widetilde{\mathbf{p}}_b, \mathbf{r}). \tag{123}$$

When Rashba SOC is present one has to consider the covariant self-energy, as done for the side jump and spin current swapping contribution. To leading order in the gradient expansion, this is done simply by replacing the Green function $\check{G}$ with its covariant expression $\check{\tilde{G}}$. Hence the self-energy responsible for the skew scattering reads

$$\check{\tilde{\Sigma}}_a^{\rm SS} = -in_i v_0^3 \frac{\lambda_0^2}{4} \sum_{\mathbf{p}_a, \mathbf{p}_b} \check{\tilde{G}}(\mathbf{p}_a, \mathbf{r})\, \check{\tilde{G}}(\mathbf{p}_b, \mathbf{r})\, \mathbf{p}_b \times \mathbf{p} \cdot \boldsymbol{\sigma} \tag{124}$$

$$\check{\tilde{\Sigma}}_b^{\rm SS} = -in_i v_0^3 \frac{\lambda_0^2}{4} \sum_{\mathbf{p}_a, \mathbf{p}_b} \check{\tilde{G}}(\mathbf{p}_a, \mathbf{r})\, \mathbf{p}_a \times \mathbf{p}_b \cdot \boldsymbol{\sigma}\, \check{\tilde{G}}(\mathbf{p}_b, \mathbf{r}) \tag{125}$$

$$\check{\tilde{\Sigma}}_c^{\rm SS} = -in_i v_0^3 \frac{\lambda_0^2}{4} \sum_{\mathbf{p}_a, \mathbf{p}_b} \mathbf{p} \times \mathbf{p}_a \cdot \boldsymbol{\sigma}\, \check{\tilde{G}}(\mathbf{p}_a, \mathbf{r})\, \check{\tilde{G}}(\mathbf{p}_b, \mathbf{r}). \tag{126}$$

Since we are considering the effect to first order in $\lambda_0^2$, the covariant Green functions entering Eqs.(124-126) are spin independent and isotropic in momentum space. As a result the retarded and advanced components of the above self-energies vanish, while the Keldysh component survives only for $\check{\tilde{\Sigma}}_a^{\rm SS}$ and $\check{\tilde{\Sigma}}_c^{\rm SS}$. Their joint contribution, after recalling that $\sum_{\mathbf{p}} \tilde{G}^R(\mathbf{p}) = -i\pi N_0$, leads to an extra term on the right hand side of the Boltzmann equation

$$I^{SS}[f] = -2\pi n_i v_0^2 (v_0 \pi N_0) \frac{\lambda_0^2}{4} \sum_{\mathbf{p}'} \delta(\epsilon_{\mathbf{p}} - \epsilon_{\mathbf{p}'}) \left\{\mathbf{p}' \times \mathbf{p} \cdot \boldsymbol{\sigma},\ f_{\mathbf{p}'}\right\}. \tag{127}$$

Finally by collecting the collision integrals Eq. (111) for spin current swapping, Eq. (116) for side jump and Eq. (127) for skew scattering, the Boltzmann equation Eq. (43) reads now

$$\begin{aligned}
&\tilde{\partial}_t f_{\mathbf{p}} + \tilde{\nabla}_{\mathbf{r}} \cdot \left[\frac{\mathbf{p}}{m} f_{\mathbf{p}} + \frac{\lambda_0^2}{8\tau} \langle \{\boldsymbol{\sigma} \times (\mathbf{p}' - \mathbf{p}), f_{\mathbf{p}'}\}\rangle\right] \\
&\quad - \frac{e}{2}\left\{\left(\mathbf{E} + \frac{\mathbf{p}}{m} \times \mathbf{B} \cdot \nabla_{\mathbf{p}}\right), f_{\mathbf{p}}\right\} \\
&\quad = -\frac{1}{\tau}\left(f_{\mathbf{p}} - \langle f_{\mathbf{p}'}\rangle\right) - (\pi v_0 N_0) \frac{\lambda_0^2}{4\tau} \langle \{\mathbf{p}' \times \mathbf{p} \cdot \boldsymbol{\sigma},\ f_{\mathbf{p}'}\}\rangle \\
&\quad - i\frac{\lambda_0^2}{4\tau} \langle [\boldsymbol{\sigma} \cdot \mathbf{p} \times \mathbf{p}', f_{\mathbf{p}'}]\rangle \\
&\quad + \frac{\lambda_0^2}{8\tau} \left\langle \left(\left\{\tilde{\nabla}_{\mathbf{r}}(\boldsymbol{\sigma} \times \mathbf{p}'), f_{\mathbf{p}'}\right\} - \left\{\tilde{\nabla}_{\mathbf{r}}(\boldsymbol{\sigma} \times \mathbf{p}), f_{\mathbf{p}}\right\}\right)\right\rangle
\end{aligned} \tag{128}$$

where, being the scattering elastic, $f_{\mathbf{p}} = f(\epsilon_{\mathbf{p}}, \hat{\mathbf{p}})$ and $f_{\mathbf{p}'} = f(\epsilon_{\mathbf{p}}, \hat{\mathbf{p}}')$ with $\langle \ldots \rangle$ indicating the integration over the directions of $\mathbf{p}'$. Equation (128) is the Boltzmann equation valid to first order in the gradient expansion and up to first order in the extrinsic SOC $\lambda_0$. By setting to zero the Rashba SOC and any exchange field, the covariant derivatives only include the standard U(1) electromagnetic field. One can then derive the standard results for the spin Hall effect and spin current swapping

$$\sigma_{sj}^{sH} = e\frac{\lambda_0^2}{4}n, \; \sigma_{ss}^{sH} = e\frac{\lambda_0^2}{4}np_F^2\tau v_0, \; \kappa = \frac{\lambda_0^2 p_F^2}{4}, \tag{129}$$

where $\sigma_{sj}^{sH}$ and $\sigma_{ss}^{sH}$ are the spin Hall conductivities[2] for the side jump and skew scattering contribution and $\kappa$ is the spin current swapping coefficient.[16] When the Rashba SOC is present, the above equation allows the analysis of the interplay between the intrinsic and extrinsic SOC. However, it turns out that such interplay has some subtle aspects, which have led to the suggestion of an non analytical behavior for vanishing Rashba coupling $\alpha$.[17–19] In the next section we will consider these aspects in detail and show that, indeed, there is no need to invoke a non-analyticity. Rather, one must take into account the fact that the extrinsic SOC introduces a further spin relaxation mechanism.

## 7. The impurity-induced spin-orbit coupling: Elliott-Yafet spin relaxation

The spin-orbit interaction with scattering centers, Eq. (91), gives rise also to spin-flip events leading to the Elliott-Yafet spin relaxation. Such a process is $\mathcal{O}(\lambda_0^4)$ and shown diagrammatically in Fig. 4. Performing the impurity average and defining momenta as in Eq. (95), the Elliott-Yafet self-energy reads

$$\begin{aligned}\check{\Sigma}^{EY}(\mathbf{p},\mathbf{q}) &= -\frac{\lambda_0^4}{16}n_i v_0^2 \sum_{\tilde{\mathbf{p}}} \Big[(\mathbf{p}+\mathbf{q}/2)\times(\tilde{\mathbf{p}}+\mathbf{q}/2)\Big]\cdot\boldsymbol{\sigma}\,\check{G}_{\tilde{\mathbf{p}},\mathbf{q}} \\ &\qquad \boldsymbol{\sigma}\cdot\Big[(\tilde{\mathbf{p}}-\mathbf{q}/2)\times(\mathbf{p}-\mathbf{q}/2)\Big] \\ &\approx \frac{\lambda_0^4}{16}n_i v_0^2 \sum_{\tilde{\mathbf{p}}} (\mathbf{p}\times\tilde{\mathbf{p}})\cdot\boldsymbol{\sigma}\,\check{G}_{\tilde{\mathbf{p}},\mathbf{q}}\,\boldsymbol{\sigma}\cdot(\mathbf{p}\times\tilde{\mathbf{p}}).\end{aligned} \tag{130}$$

In the last line subleading gradient terms $\mathcal{O}(\mathbf{q})$ have been neglected. Considering a 2D system, so that all momenta lie in the $x$-$y$ plane, and transforming back to the center-of-mass coordinate $\mathbf{r}$, one has (renaming $\tilde{\mathbf{p}}\to\mathbf{p}'$)

$$\check{\tilde{\Sigma}}^{EY}(\mathbf{p},\mathbf{r}) = \frac{\lambda_0^4}{16}n_i v_0^2 \sum_{\mathbf{p}'} \sigma^z\,\check{\tilde{G}}(\mathbf{p}',\mathbf{r})\,\sigma^z(\mathbf{p}\times\mathbf{p}')_z^2. \tag{131}$$

Notice that the Elliott-Yafet self-energy is already second order in the (extrinsic) SOC strength, so that its standard and covariant forms coincide up to $\mathcal{O}(\lambda_0^4\,\alpha)$. Such higher-order corrections are not needed here, and therefore covariant quantities were directly introduced in Eq. (131). The

Fig. 4. Self-energy diagram in second order in the spin-orbit impurity potential contributing to the Elliott-Yafet spin relaxation.

expression (131) yields the following extra collision term on the right hand side of the Boltzmann equation Eq. (128):

$$I^{EY}[f] = -\frac{\lambda_0^4}{16} 2\pi n_i v_0^2 \sum_{\mathbf{p}'} \delta(\epsilon_{\mathbf{p}} - \epsilon_{\mathbf{p}'}) \Big[ f_{\mathbf{p}} - \sigma^z f_{\mathbf{p}'} \sigma^z \Big] (\mathbf{p} \times \mathbf{p}')_z^2. \tag{132}$$

The two $\sigma^z$ Pauli matrices flip the in-plane spin components of the distribution function, $\sigma^z \left( f_{\mathbf{p}'}^{x,y} \sigma^{x,y} \right) \sigma^z = -f_{\mathbf{p}'}^{x,y} \sigma^{x,y}$, while leaving the out-of-plane component $f_{\mathbf{p}'}^z \sigma^z$ unchanged. This is a signature that spin-orbit interaction with the random 2D impurity potential conserves the $z$-spin component, while relaxing only the in-plane spins – a result valid strictly in 2D.[d] In order to see this more clearly we follow the procedure of Sec. 3, and compute the modification to the spin equation of motion (62) according to

$$\frac{1}{2} \mathrm{Tr} \Big[ \sigma^a \sum_{\mathbf{p}} \delta \tilde{I}^{EY}[f] \Big] = -\hat{\Gamma}_{EY}^{ab} s^b, \tag{133}$$

where $\hat{\Gamma}_{EY}$ is the Elliott-Yafet spin relaxation matrix. Explicitly

$$\begin{aligned} \frac{1}{2} \mathrm{Tr} \Big[ \sigma^{x,y} \sum_{\mathbf{p}} \delta \tilde{I}^{EY}[f] \Big] &= -\frac{\lambda_0^4}{16} 2\pi n_i v_0^2 \sum_{\mathbf{p},\mathbf{p}'} \delta(\epsilon_{\mathbf{p}} - \epsilon_{\mathbf{p}'}) \Big[ f_{\mathbf{p}}^{x,y} + f_{\mathbf{p}'}^{x,y} \Big] (\mathbf{p} \times \mathbf{p}')_z^2 \\ &\approx -\frac{1}{\tau} \left( \frac{\lambda_0 p_F}{2} \right)^4 s^{x,y} \end{aligned} \tag{134}$$

and

$$\begin{aligned} \frac{1}{2} \mathrm{Tr} \Big[ \sigma^{z} \sum_{\mathbf{p}} \delta \tilde{I}^{EY}[f] \Big] &= -\frac{\lambda_0^4}{16} 2\pi n_i v_0^2 \sum_{\mathbf{p},\mathbf{p}'} \delta(\epsilon_{\mathbf{p}} - \epsilon_{\mathbf{p}'}) \Big[ f_{\mathbf{p}}^{z} - f_{\mathbf{p}'}^{z} \Big] (\mathbf{p} \times \mathbf{p}')_z^2 \\ &= 0, \end{aligned} \tag{135}$$

[d] One may consider in certain circumstances a more general model including also random Rashba SOC.[20]

implying

$$\hat{\Gamma}_{EY} = \frac{1}{\tau_{EY}} \begin{pmatrix} 1 & 0 & 0 \\ 0 & 1 & 0 \\ 0 & 0 & 0 \end{pmatrix}, \quad \frac{1}{\tau_{EY}} = \frac{1}{\tau} \left( \frac{\lambda_0 p_F}{2} \right)^4 . \tag{136}$$

In order to obtain Eq. (134) we employed the diffusive expansion (46) and set

$$\langle f_{\mathbf{p}} \rangle \approx \frac{1}{N_0} \delta(\epsilon_{\mathbf{p}} - \epsilon_F) s^a , \tag{137}$$

which is appropriate for the present paramagnetic case. On the other hand Eq. (135) is obtained at once by noticing that $(\mathbf{p} \times \mathbf{p}')_z^2$ is even under exchange $\mathbf{p} \leftrightarrow \mathbf{p}'$.

Elliott-Yafet relaxation, though typically weak, plays a crucial role in the appropriate description of the spin Hall effect in 2DEGs. Indeed, the first efforts to combine intrinsic and extrinsic contributions to the spin Hall effect lead to a puzzling non-analytical behaviour:[17–19] the spin Hall conductivity for a purely extrinsic sample, with $\alpha = 0$, differed from the value obtained by considering a system with both mechanisms present, where however $\alpha \to 0$. This unphysical behaviour is cured by Elliott-Yafet processes.[21] To be definite, we follow Ref. 2 and consider the coupled dynamics of $J_y^z$ and $s^y$ in the presence of a $x$-pointing electric field (cf. Eq. (69) and Eq. (72)). The spin current is

$$J_y^z = 2m\alpha D s^y + (\sigma_{int}^{sH} + \sigma_{ext}^{sH}) E_x , \tag{138}$$

with $\sigma_{ext}^{sH} = \sigma_{sj}^{sH} + \sigma_{ss}^{sH}$, while from Eqs. (62) and (136) one has for $s^y$

$$\partial_t s^y = -\frac{1}{\tau_{EY}} - 2m\alpha J_y^z . \tag{139}$$

Solving Eqs. (138)-(139) yields[2]

$$J_y^z = \frac{1}{1 + \tau_{EY}/\tau_{DP}} (\sigma_{int}^{sH} + \sigma_{ext}^{sH}) E_x \tag{140}$$

$$s^y = -\frac{2m\alpha\tau_{EY}}{1 + \tau_{EY}/\tau_{DP}} (\sigma_{int}^{sH} + \sigma_{ext}^{sH}) E_x , \tag{141}$$

which are respectively the generalisation of the spin Hall and inverse spin galvanic/Edelstein effects in the presence of intrinsic (Rashba) and extrinsic spin-orbit coupling. These expressions are analytical and reduce to the known results for either $\alpha \to 0$ or $\lambda_0^2 \to 0$. Physically, they show that the behaviour of both effects is determined by the ratio between Dyakonov-Perel and Elliott-Yafet spin relaxation.

## 8. Conclusions

We have employed the Keldysh formalism – in its semiclassical limit – to describe the spin-charge coupled dynamics in a 2DEG with both intrinsic (Rashba) and extrinsic sources of spin-orbit interaction. Such dynamics are rich and typically rather intricated, but we have seen that rewriting spin-orbit coupling in terms of non-Abelian gauge fields leads to a compact and physically transparent set of equations. In particular the latter, obtained from the equations of motion of the locally $SU(2)$-covariant Keldysh Green's function $\check{\tilde{G}}(1,2)$, show that:

- Spin and charge are coupled via the $SU(2)$ field tensor ("spin-electric" and "spin-magnetic" fields);
- The spin obeys an $SU(2)$-covariant continuity equation, appropriately modified when extrinsic spin-orbit is present; this corrects both the definition of the spin current and the collision integral, but preserves their covariance properties;
- The side-jump mechanism is naturally seen as a modification of the velocity operator arising from the $\mathbf{q} \neq 0$ corrections to the Born self-energy;
- Skew scattering and spin current swapping are interference processes proportional to $\mathrm{Re}(AB^*)$ and $\mathrm{Im}(AB^*)$, respectively, where $A$ and $B$ are the scattering amplitudes defined in Eq. (92). Skew scattering arises beyond the Born approximation, which is instead enough to have the spin current swapping. Both processes, in contrast with side jump, are due to $\mathbf{q} = 0$ self-energy terms;
- Elliott-Yafet spin relaxation introduces a typically small though crucial energy scale, which is necessary to cure unphysical non-analytical behaviours of various physical quantities.

The Keldysh non-Abelian approach we have outlined focusing on the paradigmatic Rashba 2DEG can actually be (and indeed has been) employed in a wide range of systems, and offers certain further advantages. Let us briefly mention a few.

- The lack of spin conservation in the presence of spin-orbit coupling may lead to ambiguous definitions of, e.g. spin currents. This was extensively debated from early on[22] and posed problems concerning Onsager reciprocity.[23] Such problems were solved via the non-Abelian formulation,[24,25] by construction devoid of any ambiguity.[26]

- The kinetic equations can include any linear-in-momentum spin-orbit field, e.g. à la Rashba-Dresselhaus. Furthermore, (pseudo) spin-orbit coupling in $N$-band models can be written in terms of $SU(N)$ gauge fields, and formally handled exactly as we did in the single band, $2\times2$ case.
- The formalism can describe the dynamics of exotic systems, such as cold atoms in artificial gauge fields.[27]
- Non-homogeneous and/or dynamical spin-orbit coupling, e.g. gate-controlled[1,28,29] or due to thermal vibrations,[30] has numerous practical and theoretical implications, and is by construction included in the non-Abelian approach. In a similar way, the latter can deal with the spin and charge dynamics induced by time-dependent magnetic textures.[31]

## References


1. C. Gorini, P. Schwab, R. Raimondi and A. L. Shelankov, Non-Abelian gauge fields in the gradient expansion: Generalized Boltzmann and Eilenberger equations, *Phys. Rev. B* **82**, p. 195316 (Nov 2010).
2. R. Raimondi, P. Schwab, C. Gorini and G. Vignale, Spin-orbit interaction in a two-dimensional electron gas: a SU(2) formulation, *Ann. Phys.* **524**, p. 153 (2012).
3. J. Rammer, *Quantum Field Theory of Nonequilibrium States* (Cambridge University Press, Cambridge, 2007).
4. A. A. Abrikosov, L. P. Gorkov and I. E. Dzyaloshinski, *Methods of Quantum Field Theory in Statistical Physics* (Dover Publications, 1975).
5. Y. A. Bychkov and E. I. Rashba, Oscillatory effects and the magnetic susceptibility of carriers in inversion layers, *J. Phys. C* **17**, p. 6039 (1984).
6. V. M. Edelstein, *Solid State Commun.* **73**, p. 233 (1990).
7. A. G. Aronov and Y. B. Lyanda-Geller, *JETP Lett.* **50**, p. 431 (1989).
8. S. D. Ganichev, E. L. Ivchenko, V. V. Belkov, S. A. Tarasenko, M. Sollinger, D. Weiss, W. Wegscheider and W. Prettl, Spin-galvanic effect, *Nature* **417**, p. 153 (2002).
9. S. D. Ganichev, M. Trushin and J. Schliemann, Spin polarisation by current, *ArXiv e-prints* (June 2016).
10. R. Raimondi and P. Schwab, *Phys. Rev. B* **71**, p. 033311 (2005).
11. S. Tölle, C. Gorini and U. Eckern, Room-temperature spin thermoelectrics in metallic films, *Phys. Rev. B* **90**, p. 235117 (Dec 2014).

12. H.-A. Engel, B. I. Halperin and E. Rashba, Theory of Spin Hall Conductivity in n-Doped GaAs, *Phys. Rev. Lett.* **95**, p. 166605 (2005).
13. W.-K. Tse and S. Das Sarma, Spin hall effect in doped semiconductor structures (2006).
14. M. B. Lifshits and M. I. Dyakonov, Swapping spin currents: Interchanging spin and flow directions, *Phys. Rev. Lett.* **103**, p. 186601 (Oct 2009).
15. L. Berger, Side-jump mechanism for the hall effect of ferromagnets, *Phys. Rev. B* **2**, 4559 (Dec 1970).
16. K. Shen, R. Raimondi and G. Vignale, Theory of coupled spin-charge transport due to spin-orbit interaction in inhomogeneous two-dimensional electron liquids, *Phys. Rev. B* **90**, p. 245302 (Dec 2014).
17. W.-K. Tse and S. Das Sarma, *Phys. Rev. B* **74**, p. 245309 (2006).
18. E. M. Hankiewicz and G. Vignale, Phase Diagram of the Spin Hall Effect, *Phys. Rev. Lett.* **100**, p. 026602 (2008).
19. J. L. Cheng and M. W. Wu, Kinetic investigation of the extrinsic spin Hall effect induced by skew scattering, *Journal of Physics: Condensed Matter* **20**, p. 085209 (2008).
20. V. K. Dugaev, E. Y. Sherman, V. I. Ivanov and J. Barnaś, Spin relaxation and combined resonance in two-dimensional electron systems with spin-orbit disorder, *Phys. Rev. B* **80**, p. 081301 (Aug 2009).
21. R. Raimondi and P. Schwab, Tuning the Spin Hall Effect in a Two-Dimensional Electron Gas, *Europhys. Lett.* **87**, p. 37008 (2009).
22. E. I. Rashba, Spin currents in thermodynamic equilibrium: The challenge of discerning transport currents, *Phys. Rev. B* **68**, p. 241315(R) (2003).
23. L. Y. Wang, A. G. Mal'shukov and C. S. Chu, Nonuniversality of the intrinsic inverse spin-Hall effect in diffusive systems, *Phys. Rev. B* **85**, p. 165201 (2012).
24. C. Gorini, R. Raimondi and P. Schwab, Onsager Relations in a Two-Dimensional Electron Gas with Spin-Orbit Coupling, *Phys. Rev. Lett.* **109**, p. 246604 (2012).
25. K. Shen, G. Vignale and R. Raimondi, Microscopic Theory of the Inverse Edelstein Effect, *Phys. Rev. Lett.* **112**, p. 096601 (Mar 2014).
26. I. V. Tokatly, Equilibrium Spin Currents: Non-Abelian Gauge Invariance and Color Diamagnetism in Condensed Matter, *Phys. Rev. Lett.* **101**, p. 106601 (2008).
27. I. V. Tokatly and E. Y. Sherman, Spin evolution of cold atomic gases in $SU(2) \otimes U(1)$ fields, *Phys. Rev. A* **93**, p. 063635 (2016).

28. P. W. Brouwer, J. N. H. J. Cremers and B. I. Halperin, Weak localization and conductance fluctuations of a chaotic quantum dot with tunable spin-orbit coupling, *Phys. Rev. B* **65**, p. 081302(R) (2002).
29. A. G. Mal'shukov, C. S. Tang, C. S. Chu and K. A. Chao, Spin-current generation and detection in the presence of an AC gate, *Phys. Rev. B* **68**, p. 233307 (2003).
30. C. Gorini, U. Eckern and R. Raimondi, Spin Hall Effects Due to Phonon Skew Scattering, *Phys. Rev. Lett.* **115**, p. 076602 (2015).
31. Y. Tserkovnyak and M. Mecklenburg, Electron transport driven by nonequilibrium magnetic textures, *Phys. Rev. B* **77**, p. 134407 (2008).

# Dynamics of Quenched Topological Edge Modes

P. D. Sacramento

*Departamento de Física and CeFEMA*
*Instituto Superior Técnico, Universidade de Lisboa*
*Av. Rovisco Pais, 1049-001 Lisboa, Portugal*
*E-mail: pdss@cefema-gt.tecnico.ulisboa.pt*

A characteristic feature of topological systems is the presence of robust gapless edge states. In this work the effect of time-dependent perturbations on the edge states is considered. Specifically we consider perturbations that can be understood as changes of the parameters of the Hamiltonian. These changes may be sudden or carried out at a fixed rate. In general, the edge modes decay in the thermodynamic limit, but for finite systems a revival time is found that scales with the system size. The dynamics of fermionic edge modes and Majorana modes are compared. The effect of periodic perturbations is also referred allowing the appearance of edge modes out of a topologically trivial phase.

*Keywords*: Topology, time-dependent perturbations.

## 1. Sudden quantum quenches

An example of a time-dependent transformation of the Hamiltonian is a sudden change of its parameters. Let us consider an Hamiltonian defined by an initial set of parameters $\xi_0$ for times $t < t_0$. The single-particle eigenstates of the Hamiltonian are given by

$$H(\xi_0)|\psi_{m_0}(\xi_0)\rangle = E_{m_0}(\xi_0)|\psi_{m_0}(\xi_0)\rangle, \tag{1}$$

where $m_0$ are the quantum numbers. At time $t = t_0$ a sudden transformation of the parameters is performed, $\xi_0 \to \xi_1$. The Hamiltonian eigenstates transform to

$$H(\xi_1)|\psi_{m_1}(\xi_1)\rangle = E_{m_1}(\xi_1)|\psi_{m_1}(\xi_1)\rangle. \tag{2}$$

After this sudden quench the system will evolve in time under the influence of a different Hamiltonian. The time evolution of a single-particle state, with quantum number $m_0$, is given by

$$|\psi^I_{m_0}(t)\rangle = \sum_{m_1} e^{-iE_{m_1}(\xi_1)(t-t_0)} |\psi_{m_1}(\xi_1)\rangle\langle\psi_{m_1}(\xi_1)|\psi_{m_0}(\xi_0)\rangle \tag{3}$$

for times $t \geq t_0$. The survival probability of some initial state $|\psi_{m_0}(\xi_0)\rangle$ is defined by

$$P_{m_0}(t) = |\langle\psi_{m_0}(\xi_0)|\psi^I_{m_0}(t)\rangle|^2. \tag{4}$$

We will be interested in the fate of single particle states after a quantum quench across the phase diagram. We consider a subspace of one excitation such that the total Hamiltonian is given by the ground state energy plus one excited state and assume we remain in the one excitation subspace after the quench. In this work only unitary evolution of single-particle states is considered and effects of dissipation are neglected.

We may as well consider further quenches defined in a sequence of times and sets of parameters as $t_0 < t_1 < t_2 < t_3 < \cdots$ and $\xi_0, \xi_1, \xi_2, \xi_3, \cdots$, respectively. These intervals define regions as $I(t_0 \leq t < t_1), II(t_1 \leq t < t_2), III(t_2 \leq t < t_3), \cdots$. The case of a single quench is clearly obtained taking $t_1 \to \infty$, and so on for further quenches ($t_0 = 0$ is chosen hereafter).

Consider now a case for which we have two quenches in succession. In this case we have that the evolution of the initial state with quantum number $m_0$ is

$$\begin{aligned}
|\psi^{II}_{m_0}(t)\rangle &= e^{-iH(\xi_2)(t-t_1)}|\psi^I_{m_0}(t_1)\rangle \\
&= \sum_{m_2} e^{-iE_{m_2}(\xi_2)(t-t_1)} |\psi_{m_2}(\xi_2)\rangle\langle\psi_{m_2}(\xi_2)|\psi^I_{m_0}(t_1)\rangle \\
&= \sum_{m_2}\sum_{m_1} e^{-iE_{m_2}(\xi_2)(t-t_1)} e^{-iE_{m_1}(\xi_1)t_1} |\psi_{m_2}(\xi_2)\rangle\langle\psi_{m_2}(\xi_2)|\psi_{m_1}(\xi_1)\rangle\langle\psi_{m_1}(\xi_1)|\psi_{m_0}(\xi_0)\rangle
\end{aligned} \tag{5}$$

Choosing $\xi_2 = \xi_0$ we get that for $t_1 \leq t < \infty$ ($t_2 \to \infty$) the overlap with an initial state, $n_0$, is given by

$$\langle\psi_{n_0}(\xi_0)|\psi^{II}_{m_0}(t)\rangle = \sum_{m_1} e^{-iE_{n_0}(\xi_0)(t-t_1)} e^{-iE_{m_1}(\xi_1)t_1} \langle\psi_{n_0}(\xi_0)|\psi_{m_1}(\xi_1)\rangle\langle\psi_{m_1}(\xi_1)|\psi_{m_0}(\xi_0)\rangle \tag{6}$$

Therefore, the probability to find a projection to an initial state, $n_0$, given that the initial state is $m_0$ is given by

$$\begin{aligned} P_{n_0 m_0}(t) &= |\langle \psi_{n_0}(\xi_0)|\psi^{II}_{m_0}(t)\rangle|^2 \\ &= |\sum_{m_1} e^{-iE_{m_1}(\xi_1)t_1} \\ &\quad \langle \psi_{n_0}(\xi_0)|\psi_{m_1}(\xi_1)\rangle\langle \psi_{m_1}(\xi_1)|\psi_{m_0}(\xi_0)\rangle|^2, \end{aligned} \tag{7}$$

which is independent of time.

We may now at some given finite time, $t_2$, change the parameters from $\xi_2 \to \xi_3$. As before we find that for $t_2 \leq t < \infty$ the same probability as in eq. (7) is given by

$$P_{n_0 m_0}(t) = |\langle \psi_{n_0}(\xi_0)|\psi^{III}_{m_0}(t)\rangle|^2 \tag{8}$$

where

$$|\psi^{III}_{m_0}(t)\rangle = e^{-iH(\xi_3)(t-t_2)}|\psi^{II}_{m_0}(t_2)\rangle \tag{9}$$

The probability is now a function of time.

## 2. Models

In this chapter we consider systems that are topologically non-trivial, such as one or two-dimensional topological insulators or topological superconductors. The topological nature of these systems reveals itself both in the topological nature of the groundstate of the infinite system and in the appearance of edge states if the system is finite (bulk-edge correspondence). Different topological invariants may be defined such as winding numbers for the one-dimensional examples considered here and the Chern number for the two-dimensional superconductor considered later. Both the winding numbers and the Chern number may be understood in various ways[1-3] and typically they count the number of edge modes at the interface between the topological system and the vacuum.

Some examples are the models considered in this section which display both trivial and topological phases. The dynamics of the edge modes of the topological phases after a quantum quench is considered in sections 3-5.

### 2.1. *One-band spinless superconductor: The 1D Kitaev model*

The Kitaev one-dimensional superconductor with triplet p-wave pairing is described by the Hamiltonian[4]

$$H = \sum_{j=1}^{\bar{N}} \left[ -\tilde{t} \left( c_j^\dagger c_{j+1} + c_{j+1}^\dagger c_j \right) + \Delta \left( c_j c_{j+1} + c_{j+1}^\dagger c_j^\dagger \right) \right] - \sum_{j=1}^{N} \mu \left( c_j^\dagger c_j - \frac{1}{2} \right) \tag{10}$$

where $\bar{N} = N$ if we use periodic boundary conditions (and $N+1=1$) or $\bar{N} = N-1$ if we use open boundary conditions. Here $N$ is the number of sites. $\tilde{t}$ is the hopping amplitude taken as the unit of energy, $\Delta$ is the pairing amplitude and $\mu$ the chemical potential. The operator $c_j$ destroys a spinless fermion at site $j$.

In momentum space the model is written as

$$\hat{H} = \frac{1}{2} \sum_k \left( c_k^\dagger, c_{-k} \right) H_k \begin{pmatrix} c_k \\ c_{-k}^\dagger \end{pmatrix} \tag{11}$$

where

$$H_k = \begin{pmatrix} \epsilon_k - \mu & i\Delta \sin k \\ -i\Delta \sin k & -\epsilon_k + \mu \end{pmatrix} \tag{12}$$

with $\epsilon_k = -2\tilde{t} \cos k$. Here $c_k$ is the Fourier transform of $c_j$.

In general, a fermion operator may be written in terms of two hermitian operators, $\gamma_1, \gamma_2$, in the following way

$$\begin{aligned} c_{j,\sigma} &= \frac{1}{2} \left( \gamma_{j,\sigma,1} + i\gamma_{j,\sigma,2} \right) \\ c_{j,\sigma}^\dagger &= \frac{1}{2} \left( \gamma_{j,\sigma,1} - i\gamma_{j,\sigma,2} \right) \end{aligned} \tag{13}$$

The index $\sigma$ represents internal degrees of freedom of the fermionic operator, such as spin and/or sublattice index, the $\gamma$ operators are hermitian and satisfy a Clifford algebra

$$\{\gamma_m, \gamma_n\} = 2\delta_{nm}. \tag{14}$$

In the case of the Kitaev model it is enough to consider $c_j = (\gamma_{j,1} + i\gamma_{j,2})/2$, since the fermions are spinless. In terms of these hermitian (Majorana) operators we may write that the Hamiltonian is given by, using open boundary

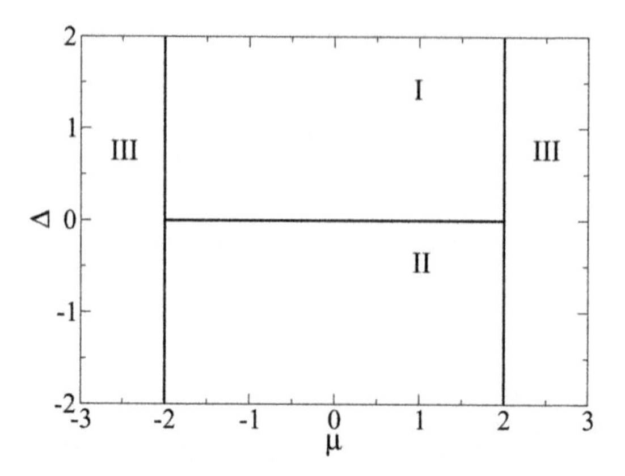


Fig. 1. Phase diagram of $1D$ Kitaev model. In phases I and II there are edge modes.

conditions,

$$H = \frac{i}{2}\sum_{j=1}^{N-1}\left[(-\tilde{t}+\Delta)\gamma_{j,1}\gamma_{j+1,2} + (\tilde{t}+\Delta)\gamma_{j,2}\gamma_{j+1,1}\right] \\ - \frac{i}{2}\sum_{j=1}^{N}\mu\gamma_{j,1}\gamma_{j,2} \tag{15}$$

Taking $\mu = 0$ and selecting the special point $\tilde{t} = \Delta$ the Hamiltonian simplifies considerably to

$$H(\mu = 0, \tilde{t} = \Delta) = i\tilde{t}\sum_{j=1}^{N-1}\gamma_{j,2}\gamma_{j+1,1} = -i\tilde{t}\sum_{j=1}^{N-1}\gamma_{j+1,1}\gamma_{j,2} \tag{16}$$

Note that the operators $\gamma_{1,1}$ and $\gamma_{N,2}$ are missing from the Hamiltonian. Therefore there are two zero energy modes. Defining from these two Majorana fermions a single usual fermion operator (non-hermitian), taking one of the Majorana operators as the real part and the other as the imaginary part, its state may be either occupied or empty with no cost in energy. Defining $d_j = 1/2\left(\gamma_{j,2} + i\gamma_{j+1,1}\right)$ and $d_N = 1/2\left(\gamma_{N,2} + i\gamma_{1,1}\right)$ we can write the Hamiltonian as

$$H = \tilde{t}\sum_{j=1}^{N-1}\left(2d_j^{\dagger}d_j - 1\right) + \epsilon_N\left(2d_N^{\dagger}d_N - 1\right) \tag{17}$$

with $\epsilon_N = 0$. Therefore the fermionic mode $d_N$ does not appear in the Hamiltonian and the state may be occupied or empty ($d_N^{\dagger}d_N = 1, 0$, respectively) with no energy cost. These two states are therefore degenerate in energy and are perfectly localized at the edges of the chain as $\delta$-function peaks (with exponential accuracy as the system size grows).

The phase diagram of the Kitaev model shows three types of phases (see Fig. 1): two topological phases in which there are gapless edge modes, if the system is finite, and two trivial phases with no edge modes. In the various phases the bulk of the system is gapped and at the transition lines the gap closes, allowing the possibility of a change of topology. The transition lines are located at $\Delta = 0, |\mu| \leq 2\tilde{t}$ and at $|\mu| = 2\tilde{t}$ and any $\Delta$.

### 2.2. *Multiband system: 1D Two-band Shockley model*

The Shockley model is a model of a dimerized system of spinless fermions with alternating nearest-neighbor hoppings, given by the Hamiltonian (see for instance Ref. 3)

$$H = \sum_{j=1}^{N} \psi^\dagger(j) \left[ U\psi(j) + V\psi(j-1) + V^\dagger \psi(j+1) \right) \tag{18}$$

where the $2 \times 2$ matrices $U$ and $V$ and the spinor $\psi$ representing two orbitals at site $j$ that are hybridized by the matrices $U$ and $V$ are given by

$$U = \begin{pmatrix} 0 & t_1^* \\ t_1 & 0 \end{pmatrix}; V = \begin{pmatrix} 0 & t_2^* \\ 0 & 0 \end{pmatrix}; \psi(j) = \begin{pmatrix} c_{j,A} \\ c_{j,B} \end{pmatrix}. \tag{19}$$

$t_1$ and $t_2$ are hoppings and $c_{j,A}$ ($c_{j,B}$) destroy spinless fermions at site $j$ belonging to sublattice $A$ ($B$), respectively.

We may as well define Majorana operators as

$$\begin{aligned} c_{j,A} &= \frac{1}{2} \left( \gamma_{j,A,1} + i\gamma_{j,A,2} \right) \\ c_{j,B} &= \frac{1}{2} \left( \gamma_{j,B,1} + i\gamma_{j,B,2} \right) \end{aligned} \tag{20}$$

Here $A$ and $B$ take the role of pseudospins. Taking $t_1$ and $t_2$ real, the

Hamiltonian may be written as

$$
\begin{aligned}
H &= \frac{it_1}{2}\sum_{j=1}^{N}(\gamma_{j,A,1}\gamma_{j,B,2}+\gamma_{j,B,1}\gamma_{j,A,2})\\
&+\frac{t_2}{4}\sum_{j=2}^{N}(\gamma_{j,A,1}\gamma_{j-1,B,1}+\gamma_{j,A,2}\gamma_{j-1,B,2})\\
&+\frac{it_2}{4}\sum_{j=2}^{N}(\gamma_{j,A,1}\gamma_{j-1,B,2}-i\gamma_{j,A,2}\gamma_{j-1,B,1})\\
&+\frac{t_2}{4}\sum_{j=1}^{N-1}(\gamma_{j,B,1}\gamma_{j+1,A,1}+\gamma_{j,B,2}\gamma_{j+1,A,2})\\
&+\frac{it_2}{4}\sum_{j=1}^{N-1}(\gamma_{j,B,1}\gamma_{j+1,A,2}-i\gamma_{j,B,2}\gamma_{j+1,A,1})
\end{aligned}
\tag{21}
$$

Choosing $t_1 = 0$ we find that the Majorana fermions $\gamma_{1,A,1}$, $\gamma_{1,A,2}$, $\gamma_{N,B,1}$ and $\gamma_{N,B,2}$ do not contribute and are zero energy modes. These decoupled zero-energy modes are fermionic in nature, since the decoupled Majoranas are located at the two end sites, $A$ and $B$, respectively. This point is characteristic of the topological phase as long as the bulk gap does not vanish. In the trivial phase there are no decoupled Majorana operators. As discussed for instance in Ref. 3 the two types of phases may also be distinguished by the winding number.

### 2.3. *Multiband system: 1D SSH model with triplet pairing*

This model may be viewed as a dimerized Kitaev superconductor.[5] The dimerization is parametrized by $\eta$ and the superconductivity by $\Delta$.

This model is given by the Hamiltonian

$$
\begin{aligned}
H = -\mu \textstyle\sum_j & \left(c^\dagger_{j,A}c_{j,A}+c^\dagger_{j,B}c_{j,B}\right)\\
-\tilde{t}\textstyle\sum_j & \Big[(1+\eta)c^\dagger_{j,B}c_{j,A}+(1+\eta)c^\dagger_{j,A}c_{j,B}\\
&+\ (1-\eta)c^\dagger_{j+1,A}c_{j,B}+(1-\eta)c^\dagger_{j,B}c_{j+1,A}\Big]\\
+\Delta\textstyle\sum_j & \Big[(1+\eta)c^\dagger_{j,B}c^\dagger_{j,A}+(1+\eta)c_{j,A}c_{j,B}\\
&+\ (1-\eta)c^\dagger_{j+1,A}c^\dagger_{j,B}+(1-\eta)c_{j,B}c_{j+1,A}\Big]
\end{aligned}
\tag{22}
$$

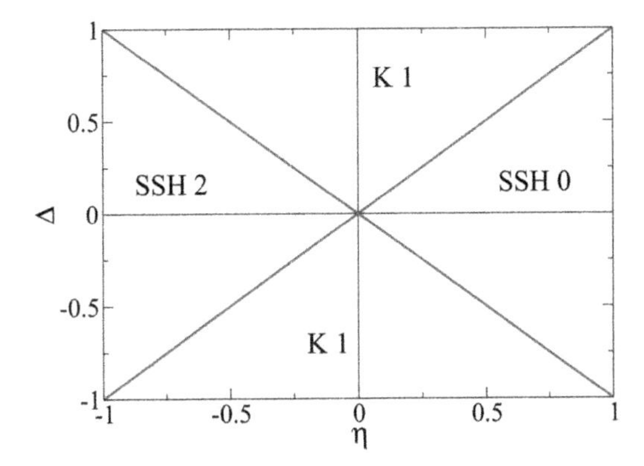


Fig. 2. (Color online) Phase diagram of $1D$ SSH-Kitaev model for $\mu = 0$. The phase SSH0 is trivial and has no edge modes. In the phases K1 there is one edge Majorana mode at each edge and in the phase SSH2 there are fermionic edge modes at each edge.

($\tilde{t}$ is the hopping, $\Delta$ the pairing amplitude and $\mu$ the chemical potential). The model with no superconductivity ($\Delta = 0$) is related to the Shockley model taking $t_1 = \tilde{t}(1+\eta)$ and $t_2 = \tilde{t}(1-\eta)$. The region of $\eta > 0$ corresponds to $t_1 > t_2$ and vice-versa for $\eta < 0$. The Hamiltonian in real space mixes nearest-neighbor sites and also has local terms.

In terms of Majorana operators the Hamiltonian is written as

$$
\begin{aligned}
H = &-\frac{\mu}{2}\sum_{j=1}^{N}(2 + i\gamma_{j,A,1}\gamma_{j,A,2} + i\gamma_{j,B,1}\gamma_{j,B,2}) \\
&- \frac{i\tilde{t}}{2}(1+\eta)\sum_{j=1}^{N}(\gamma_{j,B,1}\gamma_{j,A,2} + \gamma_{j,A,1}\gamma_{j,B,2}) \\
&- \frac{i\tilde{t}}{2}(1-\eta)\sum_{j=1}^{N-1}(\gamma_{j+1,A,1}\gamma_{j,B,2} + \gamma_{j,B,1}\gamma_{j+1,A,2}) \\
&+ \frac{i\Delta}{2}(1+\eta)\sum_{j=1}^{N}(\gamma_{j,A,1}\gamma_{j,B,2} + \gamma_{j,A,2}\gamma_{j,B,1}) \\
&+ \frac{i\Delta}{2}(1-\eta)\sum_{j=1}^{N-1}(\gamma_{j,B,1}\gamma_{j+1,A,2} + \gamma_{j,B,2}\gamma_{j+1,A,1})
\end{aligned} \tag{23}
$$

Consider once again a vanishing chemical potential. Taking $\eta = -1$

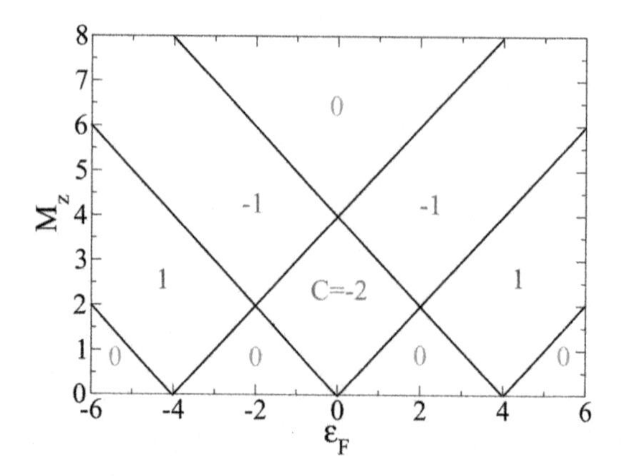


Fig. 3. (Color online) Phase diagram of $2D$ p-wave model as a function of the chemical potential and magnetization. $C$ is the Chern number of each phase associated with the number of protected one-dimensional edge modes at finite magnetization.

and $\Delta = 0$ we have a state similar to the SSH or Shockley models with two fermionic-like zero energy edge states, since the four operators $\gamma_{1,A,1}, \gamma_{1,A,2}; \gamma_{N,B,1}, \gamma_{N,B,2}$ are missing from the Hamiltonian. If we select $\eta = 0$ and $t = \Delta$ is a Kitaev like state since there are two Majorana operators missing from the Hamiltonian, $\gamma_{1,A,1}$ and $\gamma_{N,B,2}$, one from each end. An example of a trivial phase is the point $\eta = 1$ and $\Delta = 0$ in which case there are no zero energy edge states. In Fig. 2 the phase diagram is shown. This model provides a testing ground for the comparison between fermionic and Majorana edge modes. In addition, in some regimes it displays finite energy modes that are localized at the edges of the chain, as obtained before in other multiband models.[6]

### 2.4. *Two-dimensional spinfull triplet superconductor*

Another interesting case is that of a two-dimensional triplet superconductor with $p$-wave symmetry, spin-orbit coupling and a Zeeman term.[7] We write the Hamiltonian for the bulk system in momentum space as

$$\hat{H} = \frac{1}{2}\sum_{\boldsymbol{k}} \left(\psi^\dagger_{\boldsymbol{k}}, \psi_{-\boldsymbol{k}}\right) \begin{pmatrix} \hat{H}_0(\boldsymbol{k}) & \hat{\Delta}(\boldsymbol{k}) \\ \hat{\Delta}^\dagger(\boldsymbol{k}) & -\hat{H}_0^T(-\boldsymbol{k}) \end{pmatrix} \begin{pmatrix} \psi_{\boldsymbol{k}} \\ \psi^\dagger_{-\boldsymbol{k}} \end{pmatrix} \tag{24}$$

where $\left(\psi^\dagger_{\boldsymbol{k}}, \psi_{-\boldsymbol{k}}\right) = \left(\psi^\dagger_{\boldsymbol{k}\uparrow}, \psi^\dagger_{\boldsymbol{k}\downarrow}, \psi_{-\boldsymbol{k}\uparrow}, \psi_{-\boldsymbol{k}\downarrow}\right)$ and

$$\hat{H}_0 = \epsilon_{\boldsymbol{k}}\sigma_0 - M_z\sigma_z + \hat{H}_R\,. \tag{25}$$

Here, $\epsilon_{\boldsymbol{k}} = -2\tilde{t}(\cos k_x + \cos k_y) - \varepsilon_F$ is the kinetic part, $\tilde{t}$ denotes the hopping parameter set in the following as the energy scale ($\tilde{t} = 1$), $\boldsymbol{k}$ is a wave vector in the $xy$ plane, and we have taken the lattice constant to be unity. Furthermore, $M_z$ is the Zeeman splitting term responsible for the magnetization, in $\tilde{t}$ units. The Rashba spin-orbit term is written as

$$\hat{H}_R = \boldsymbol{s} \cdot \boldsymbol{\sigma} = \alpha \left( \sin k_y \sigma_x - \sin k_x \sigma_y \right) , \tag{26}$$

where $\alpha$ is measured in the same units and $\boldsymbol{s} = \alpha(\sin k_y, -\sin k_x, 0)$. The matrices $\sigma_x, \sigma_y, \sigma_z$ are the Pauli matrices acting on the spin sector, and $\sigma_0$ is the $2 \times 2$ identity. The pairing matrix reads

$$\hat{\Delta} = i \left( \boldsymbol{d} \cdot \boldsymbol{\sigma} \right) \sigma_y = \begin{pmatrix} -d_x + id_y & d_z \\ d_z & d_x + id_y \end{pmatrix} . \tag{27}$$

We consider here $d_z = 0$. If the spin-orbit coupling is strong it is energetically favorable that the pairing is of the form $\boldsymbol{d} = d\boldsymbol{s}$.

The energy eigenvalues and eigenfunction may be obtained solving the Bogoliubov-de Gennes equations

$$\begin{pmatrix} \hat{H}_0(\boldsymbol{k}) & \hat{\Delta}(\boldsymbol{k}) \\ \hat{\Delta}^\dagger(\boldsymbol{k}) & -\hat{H}_0^T(-\boldsymbol{k}) \end{pmatrix} \begin{pmatrix} u_n \\ v_n \end{pmatrix} = \epsilon_{\boldsymbol{k},n} \begin{pmatrix} u_n \\ v_n \end{pmatrix} . \tag{28}$$

The 4-component spinor can be written as

$$\begin{pmatrix} u_n \\ v_n \end{pmatrix} = \begin{pmatrix} u_n(\boldsymbol{k}, \uparrow) \\ u_n(\boldsymbol{k}, \downarrow) \\ v_n(-\boldsymbol{k}, \uparrow) \\ v_n(-\boldsymbol{k}, \downarrow) \end{pmatrix} . \tag{29}$$

The superconductor we consider here is time-reversal invariant if the Zeeman term is absent. The system then belongs to the symmetry class DIII where the topological invariant is a $\mathbb{Z}_2$ index.[8] If the Zeeman term is finite, time reversal symmetry (TRS) is broken and the system belongs to the symmetry class D. The topological invariant that characterizes this phase is the first Chern number $C$, and the system is said to be a $\mathbb{Z}$ topological superconductor. The phase diagram is shown in Fig. 3.

Due to the bulk-edge correspondence if the system is placed in a strip geometry and the system is in a topologically non-trivial phase, there are robust edge states, in a number of pairs given by the Chern number, if time reversal symmetry is broken. There are also counterpropagating edge states in the $Z_2$ phases even though the Chern number vanishes, as in the spin Hall effect. In these phases time reversal symmetry is preserved and

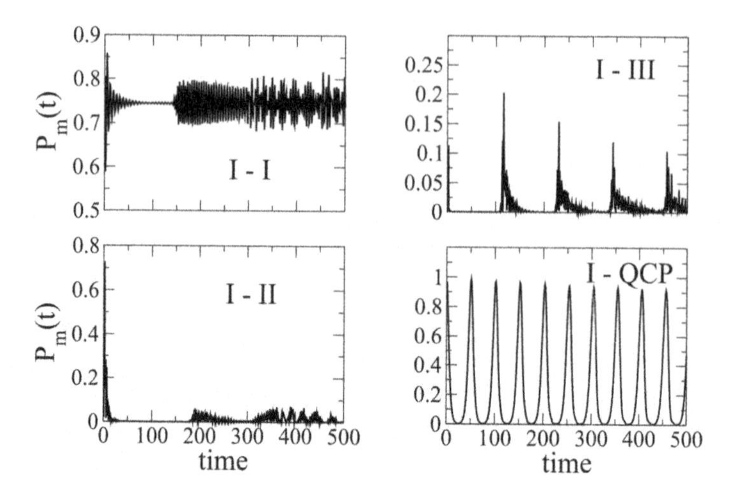


Fig. 4. Survival probability of the Majorana state of the one-dimensional Kitaev model for different transitions across the phase diagram: i) transition within the same topological phase, $I$, $(\mu = 0.5, \Delta = 0.6) \to (\mu = 1.0, \Delta = 0.6)$, ii) transition from the topological phase $I$ to the trivial phase $III$ $(\mu = 0.5, \Delta = 0.6) \to (\mu = 2.2, \Delta = 0.6)$, iii) transition from the topological phase $I$ with positive $\Delta$ to the topological phase $I$ with negative $\Delta$ $(\mu = 0.5, \Delta = 0.6) \to (\mu = 0.5, \Delta = -0.6)$, iv) transition within the same topological phase, $I$, to the quantum critical point $(\mu = 0, \Delta = 0.1) \to (\mu = 0, \Delta = 0)$ where the system is gapless. The system has 100 sites. Reproduced from Ref. 10.

the Kramers pairs of edge states give opposite contributions to the Chern number. Interestingly, turning on the magnetization (Zeeman field) time reversal symmetry is broken and the edge states are no longer topologically protected. However, it was found that, even in regimes where $C = 0$, there are edge states, reminiscent of the edge states of the $Z_2$ phases.

## 3. Dynamics of edge modes of 1$D$ Kitaev model

### 3.1. *Single quench*

The stability of the Majorana fermions in this model has been considered recently.[9] In Fig. 4 we present results for the survival probability of the Majorana mode for different quenches.[10] In the first panel we consider the case of a quench within the same topological phase clearly showing that the survival probability is finite. Since the parameters change, there is a decrease of $P(t)$ as a function of time due to the overlap with *all* the eigenstates of the chain with the new set of parameters, but after some oscillations the survival rate stabilizes at some finite value. As time grows, oscillations appear again centered around some finite value. Therefore the Majorana mode is robust to the quench. In the second panel we consider a quench from the topological phase $I$ to the trivial, non-topological phase $III$. The behavior is quite different. After the quench the survival probability decays fast to nearly zero. After some time it increases sharply and repeats the decay and revival process. Similar results are found for a quench between the two topological phases $I$ and $II$. As discussed in Ref. 9 the revival time scales with the system size. At this instant the wave function is peaked around the center of the system and is the result of a propagating mode across the system with a given velocity and, therefore, scales with the system size. In the infinite system limit the revival time will diverge and the Majorana mode decays and is destroyed. A qualitatively different case is illustrated in the last panel of Fig. 2 where a quench from the topological phase $I$ to the quantum critical point at the origin is considered.

Let us analyse these oscillations in greater detail. Consider first $\mu = 0$ and quenches where one varies $\Delta$, or a fixed $\Delta$ and changing $\mu$. In the case of $\mu = 0$ the critical point is located at $\mu = 0, \Delta = 0$ and in the second case there is a line of critical points at $\mu = 2\tilde{t}$. One finds that there is a point that separates the existence or not of oscillations. If the initial state is close enough to the critical point there are oscillations. Otherwise they are absent. For instance, in the quench from the topological phase $I$ to the critical point at $\mu = 0, \Delta = 0$, the point is located as $N = 100, \Delta = 0.34$, $N = 200, \Delta = 0.18$, $N = 400, 0.05 < \Delta < 0.1$. In the vicinity of the two critical lines of points (around $\mu = 2\tilde{t}, \Delta = 0$), no matter how close the initial point is to the critical line, one does not find oscillations (for further details see Ref. 11).

In Fig. 5 the survival probability, $P(t)$, of a Majorana mode as a function of time, for various critical quenches is presented. In the first panel are shown the oscillations of $P(t)$ as one quenches from a given value of $\Delta$ to the critical point $\mu = 0, \Delta = 0$, maintaining $\mu = 0$. For small deviations of the initial value of $\Delta$ from the critical point, $P(t)$ is close to 1 and as one increases the distance from the critical point the amplitude decreases considerably. The oscillations are quite smooth and clear until the amplitude has decreased enough to reach zero. Beyond this point there is a periodicity but no longer oscillations since there are increasing regions where $P(t)$ basically vanishes. In this case it seems more like the revival times of non-critical quenches, even though the curves are still smooth. Beyond a given value of $\Delta_d$ there is a *period doubling*. Also, after this period

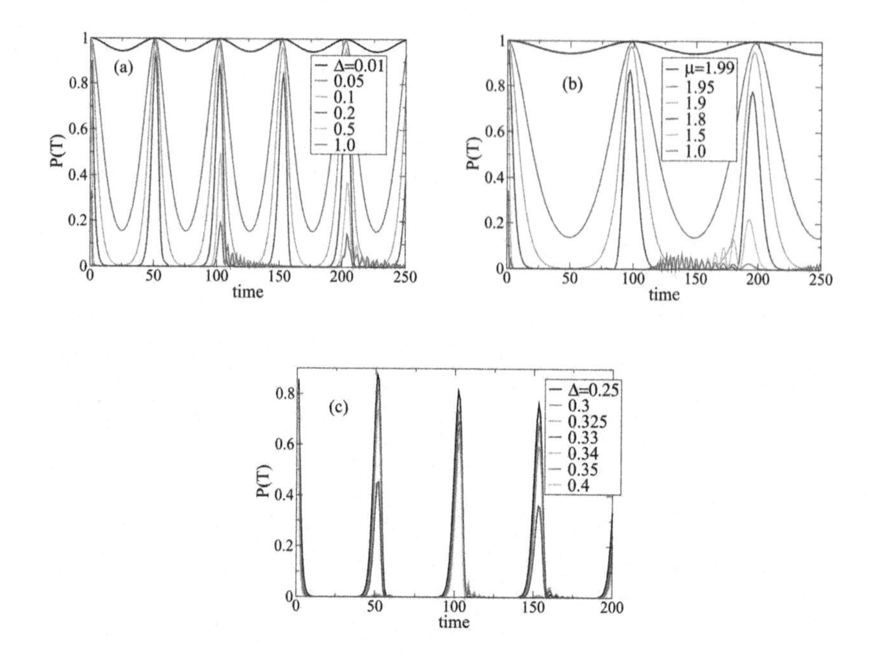


Fig. 5. (Color online) Survival probability, $P(t)$, of a Majorana mode in the $1D$ Kitaev model as one approaches critical points. In panel (c) the crossover to period doubling is shown as one approaches the critical region. The system size is $N = 100$. In (a) $\mu = 0$, in (b) $\Delta = 0.5$, and in (c) $\mu = 0$. Reproduced from Ref. 11.

doubling the survival probability looses its regular periodic behavior and shows more oscillations of smaller periods and amplitude decays that are similar to results previously found in quenches away from critical points.[9,10] In the second panel are shown quenches to the critical line $\mu = 2\tilde{t}$ keeping $\Delta = 0.5$ and decreasing the chemical potential. The behavior is similar to the first panel. In the third panel is shown in greater detail the *crossover* to period doubling for the transition to the critical point. The point of crossover, $\Delta_d$, scales linearly with $1/N$.

The survival probability is determined by the various energies of the final Hamiltonian eigenstates and their overlaps to the initial single-particle state. In Fig. 6 the overlaps between the initial lowest energy state (Majorana mode) and all the final state eigenvectors are shown, as a function of their energies, for $N = 200$. In general, the overlaps are peaked at the lowest energies. There is a clear separation of regimes as one reaches the crossover region where the period doubling occurs. At small values of $\Delta$

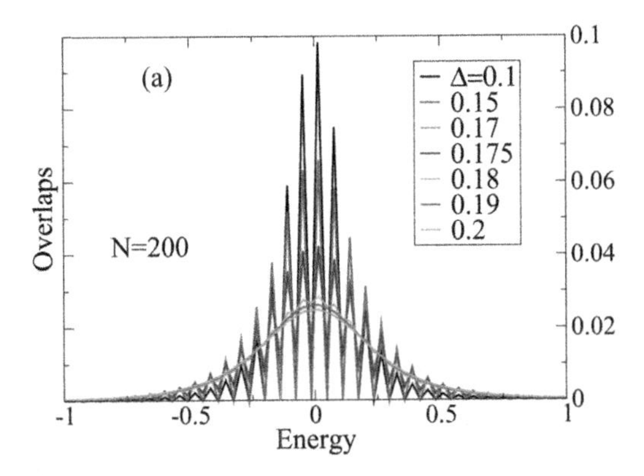


Fig. 6. (Color online) Overlaps for the $1D$ Kitaev model as a function of energy. The critical point is $\mu = 0, \Delta = 0$. Reproduced from Ref. 11.

the overlaps oscillate between finite and zero values. This is a parity effect distinguishing even and odd number of sites. It can be noted that the overlaps are very sharp around the lowest energy states. As the crossover occurs the overlaps are no longer zero at some energy eigenvalues and actually become very smooth. This means that the contributions from the various energy states changes, the time behavior is affected and the clean oscillations are no longer observed. In order to have clean oscillations one needs contributions from few energy levels. A perfect oscillation requires finite overlaps to two states and the frequency of the oscillations is the difference in their energy values. In general, the overlaps have very different magnitudes to the two states and the period of oscillations shown in $P(t)$ depends on their magnitudes. Adding significant contributions from other energy eigenstates leads first to modulated oscillations and then to a complicated time dependence.

The origin of the period doubling is understood in the following way. In Fig. 7 the time evolution of the Majorana state is shown, for a critical quench from a region far from the critical point where the period has doubled, and a quench from the region of oscillations, close to the critical point. In the first case the wave functions at each edge are separated in two energy modes while for the second they are mixed. This is due to the long range correlations close to the critical point that effectively decrease the system size and lead to the coupling of the two edge modes. In the first case the time evolved states from each edge cross each other in a solitonic

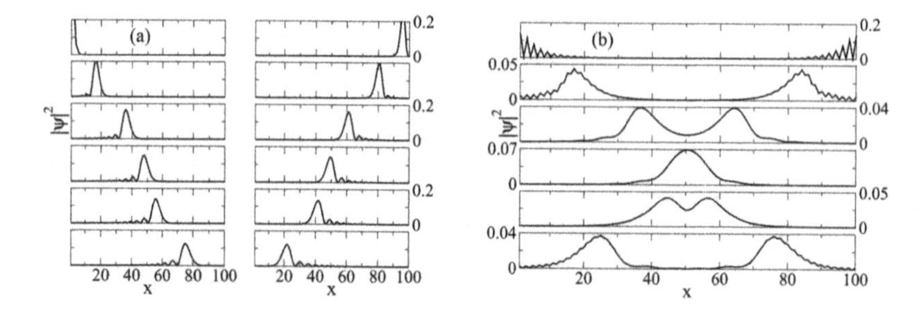


Fig. 7. Solitonic-like vs. constructive interference behavior of the wave functions in the $1D$ Kitaev model. In (a) the initial state is far from the critical point (CP) and in (b) one is close to the CP. In (a) the quench takes place from $\mu = 0, \Delta = 0.5$ to $\mu = 0, \Delta = 0$ and in (b) from $\mu = 0, \Delta = 0.1$ to the same CP. The results are for a system size $N = 100$. In (a) the sequence of times is $t = 2, 10, 20, 26, 30, 40$ and in (b) $t = 2, 10, 20, 25, 30, 40$. Reproduced from Ref. 11.

like behavior while in the second case there is a constructive interference when the peaks of the evolved state meet at the center of the wire. Consistently with the results for the overlaps, in this regime the difference in energy between states with high weight halves, and the period doubles.

Both the revival time and the period of the oscillations are associated with the propagation of the state along the system, with a velocity that, in the case of a free system, travels at a velocity given by the quasiparticle energy slope.[12] In the case of interacting systems, it generalizes to a limiting velocity value, similar to a light-cone propagation.[13–16]

A similar conclusion is obtained performing a Fourier analysis of the time evolution of the survival probability. This is shown in Fig. 8. While for small initial values of $\Delta$ the distribution is quite narrow around low frequencies, it changes significantly as $\Delta$ grows, becoming quite extended. In the Fourier decomposition the amplitudes, $u_n$, are for the frequencies with values $\omega_n = \pi(n-1)/N_t$, where $N_t$ is the number of time points considered.

It is also interesting to study the survival probability of excited states, that in this problem are extended states throughout the chain. Close to the critical point the survival probability of most states is close to 1 except near the low energy modes. Further away from the critical point the deviation of the survival probability from unity extends to higher energy states due to the orthogonality between the eigenstates of the original and final Hamiltonians.[11]

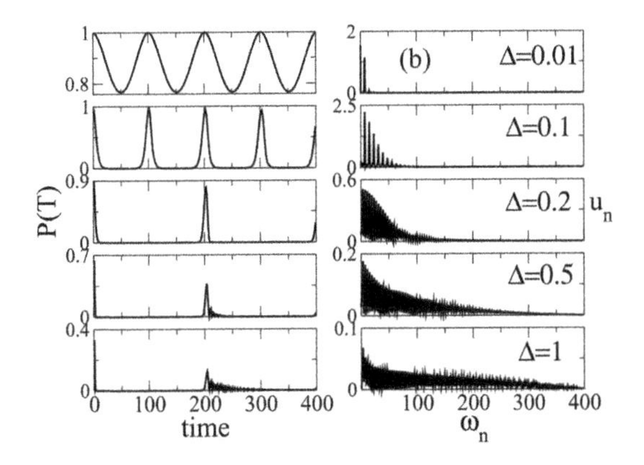


Fig. 8. Survival probability and its Fourier analysis for the $1D$ Kitaev model. Reproduced from Ref. 11.

### 3.2. *Generation of Majorana states*

While quenches in general destabilize the edge states, due to the finiteness of a system, we may generate Majorana states through a sudden quench starting from a trivial phase. Even though in the thermodynamic limit the topological properties can not be changed by a unitary transformation,[10,17] the probability that a given initial state in a trivial phase $III$ may collapse to a Majorana of the final state Hamiltonian in phase $I$ is finite and independent of time. Quenching to a state close to the transition line, the overlaps of several (extended) states are considerable due to the spatial extent of the Majorana states. If the quench is deeper into the topological phase these become more localized and the overlap decreases. Interestingly the larger overlap is found for some higher energy, extended states.

A sequence of quenches allows for the manipulation of the states.[11] A possibility to turn off and on Majoranas can be trivially seen in the following way. Consider starting from a state inside region $I$ of the phase diagram Fig. 1. Perform a critical quench to the line $\Delta = 0$ and then a quench back to the original state. Choosing appropriately $t_1$ we may get a state with no overlap with the initial Majoranas, as illustrated in Fig. 5. So we are back to a topological phase but with no edge states. But Majoranas may be switched back on if at a time $t_2 > t_1$ we perform another quench to a state in region $I$. Due to the quench to $\xi_3$ a finite

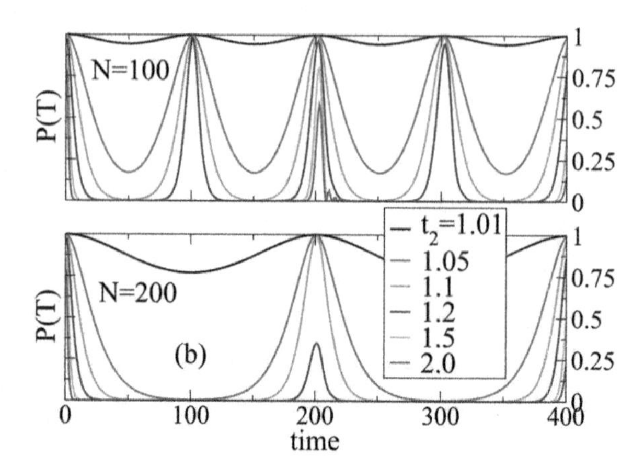


Fig. 9. (Color online) Survival probability of edge modes of the $1D$ Schockley model. Critical quenches are considered from the topological region to the transition point ($t_1 = t_2$). Reproduced from Ref. 11.

probability to find the Majorana state is found[11] even though if no quench from $\xi_2 = \xi_0 \to \xi_3$ was performed, and having chosen appropriately $t_1$, the survival probability of the Majorana states was tuned to vanish. Note that the overlap of Majorana state of $H(\xi_3)$ with a Majorana state of $H(\xi_0)$ is finite, since the states are chosen to be close by.

## 4. Dynamics of $1D$ multiband systems

While in the previous section Majorana edge states of Kitaev's model were considered, edge states in other systems, including topological insulators, have also been considered and show similar properties. In this section we consider two topological systems, the Shockley model[3] which has fermionic edge states and no Majoranas, and the SSH-Kitaev model[5] which displays both types of edge states in different parts of the phase diagram, allowing a comparison of different edge state dynamics.

### 4.1. $1D$ *Shockley model*

In Fig. 9 we show critical quenches to a final state with $t_2 = t_1 = 1$ starting from different initial points in the topological region ($t_2 > t_1$). The cases

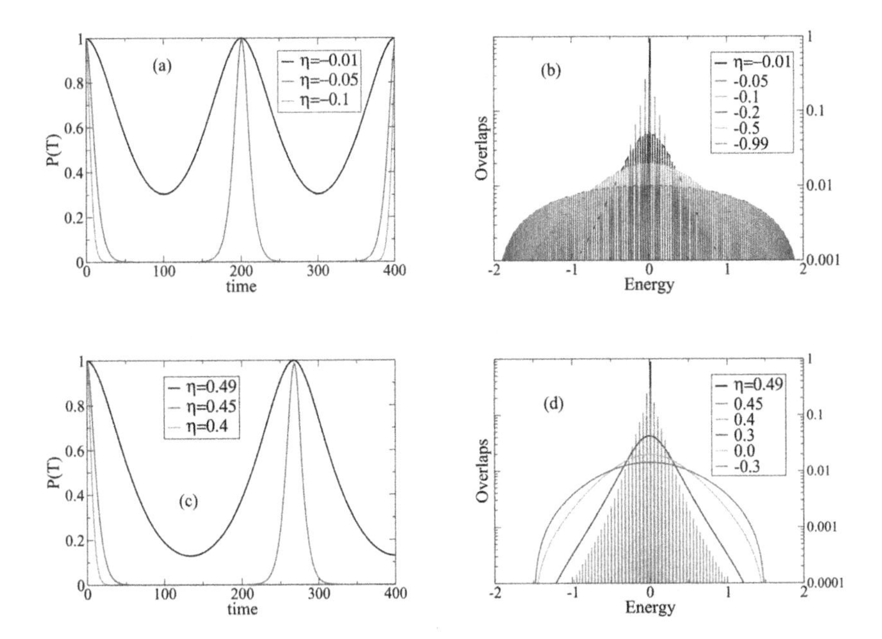


Fig. 10. (Color online) Critical quenches in the $1D$ SSH-Kitaev model: survival probability and overlaps. The CP on the first two panels is $\eta = 0, \Delta = 0$ (SSH2) and on the last two panels the CP is $\eta = 0.5, \Delta = 0.5$ (K1). Reproduced from Ref. 11.

of $N = 100$ and $N = 200$ are shown. As in the Kitaev model the period scales with the system size. The behavior is very similar to the Kitaev model. We see the period doubling for both cases for $t_2 = 1.5$. For $t_2 = 2.0$ the smoothness of the oscillations is replaced by a superposition of many frequencies. From the point of view of edge state dynamics the behavior of Majoranas and fermionic edge states are similar.

Also, moving further away from the critical point a behavior similar to Fig. 4b is seen with a rapid decrease of $P(t)$ and the appearance of revival times.

## 4.2. 1D SSH-Kitaev model

The similarities between Majorana and fermionic edge states are further shown considering the SSH-Kitaev model. In Fig. 2 we showed the phase diagram of the SSH-Kitaev model[5] in the case of $\mu = 0$. In phase K1 we are in the Kitaev regime with one zero energy edge mode at each edge

(Majoranas). In the SSH regimes we are closer to the behavior of the SSH model with fermionic modes. In SSH 0 there are no edge modes. In SSH 2 there are two zero energy fermionic modes.

In Fig. 10 we consider critical quenches to points in the transition between different topological regions. In the top panels we consider $P(t)$ and the overlaps, respectively, of a transition at $\mu = 0$ from the SSH 2 regime to the critical point $\eta = 0, \Delta = 0$ by considering different initial values of $\eta = -0.01, -0.05, -0.1, -0.2, -0.5, -0.99$. In the lower panels we consider critical quenches to the critical point $\eta = 0.5, \Delta = 0.5$ changing the initial value of $\eta$. In both cases note that there is again a change of the distribution of the overlaps from sharp peaks, at small deviations from the critical point, to a broad distribution of the overlaps as one moves sufficiently away from the critical point; again there is a crossover between the two regimes (not shown), as for the Kitaev model. However, the overlaps are not smooth as a function of energy. Note that in the first case $\Delta = 0$, which means that this occurs in the context of the SSH model with no superconductivity. In the second case we have a mixture of SSH and Kitaev model, but the behavior is qualitatively similar in the crossover region. Beyond it we find again the very smooth distributions of the overlaps as in the Kitaev model.

## 5. Dynamics of edge states of 2D triplet superconductor

### 5.1. *Wave-function propagation*

The edge states appear if we consider a strip geometry of finite transversal width, $N_y$, with open boundary conditions (OBC) along $y$ and periodic boundary conditions (PBC) along the longitudinal direction, $x$, of size $N_x$. The diagonalization of this Hamiltonian expressed in real space involves the solution of a $(4N_xN_y) \times (4N_xN_y)$ eigenvalue problem. The energy states include states in the bulk and states along the edges and are written in the form of a 4-component spinor as

$$\begin{pmatrix} u_n \\ v_n \end{pmatrix} = \begin{pmatrix} u_n(j_x, j_y, \uparrow) \\ u_n(j_x, j_y, \downarrow) \\ v_n(j_x, j_y, \uparrow) \\ v_n(j_x, j_y, \downarrow) \end{pmatrix} . \tag{30}$$

Here $j_x, j_y$ are the spatial lattice coordinates along $x$ and $y$, respectively. Focusing our attention on a Majorana mode, we present in Fig. 11 the time evolution of the absolute value of the spinor component $u_n(j_x, j_y, \uparrow)$, as an example, for a time evolution for $(M_z = 2, \mu = -5) \rightarrow (M_z = 0, \mu =$

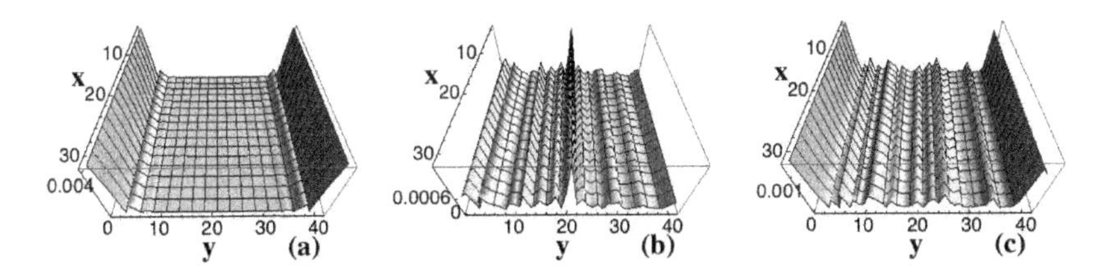


Fig. 11. (Color online) Time evolution of real space $|u_\uparrow|^2$ for $(M_z = 2, \mu = -5) \to (M_z = 0, \mu = -5)$ $C = 1 \to C = 0$ (trivial) for $t = 0, t = 50, t = 62$, shown in (a), (b), (c), respectively. Note that in these transitions there are no edge states in the final states. In the various panels the horizontal axis is the $y$ direction and the vertical direction is the $x$ direction. The system size is $31 \times 41$. Reproduced from Ref. 10.

$-5)$ $C = 1 \to C = 0$ (trivial) for $t = 0, t = 50, t = 62$, shown in (a), (b), (c), respectively. The other spinor components have a qualitatively similar behavior. A set of characteristic time values are selected (time is expressed in units of $1/\tilde{t}$). The initial state shows a mode that is very much peaked at the borders of the system and that decays fast inside the superconductor along the transverse direction. As time evolves the peaks move towards the center until they merge at some later time, dependent of the system transverse size (as for the Kitaev model). After this time the peaks move back from the center, the wave functions become more extended as a mixture to all the eigenstates becomes more noticeable. Eventually at later times the wave function recovers a shape that is close to the initial state and there is a partial revival of the original state. The process then repeats itself but the same degree of coherence is somewhat lost. In Fig. 11 the quenches are carried out between a topological phase and a trivial phase ($C = 1 \to C = 0$) and ($C = -2 \to C = 0$). The behavior is therefore qualitatively the same as for the $1D$ case.

### 5.2. *Evolution of Chern numbers*

The topology of each phase may be characterized by the Chern number, defined over the Brillouin zone of the system.[7] As the system evolves in time, the wave functions change. Solving for the evolution of the wave functions we may calculate the Chern number as a function of time and determine how the topology changes as well. Due to the fluctuating evolution of the overlaps between a given state and all the others in the appropriate subspace, we may expect that wave functions over the Brillouin zone will fluctuate considerably as time goes by.

In Fig. 12 it is shown that the Chern number remains locked to the

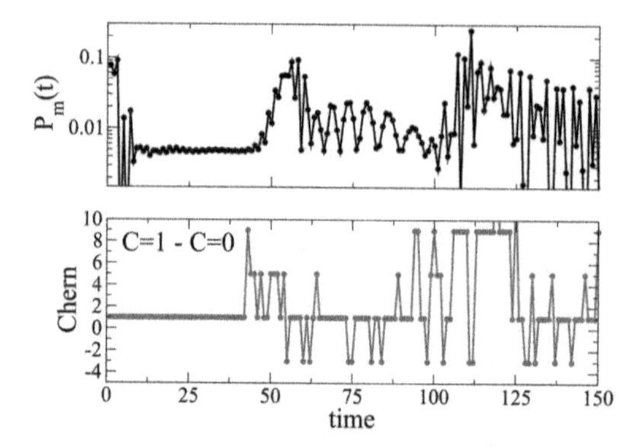


Fig. 12. (Color online) Comparison of time evolution of (a) survival probability and (b) Chern number for the case of strong spin orbit coupling for $(M_z = 2, \mu = -5) \rightarrow (M_z = 0, \mu = -5)$ corresponding to $C = 1 \rightarrow C = 0$. The Chern number remains stable at the initial state value until the Majorana mode reaches the middle point of the system. Beyond this instant the Chern number fluctuates. Reproduced from Ref. 10.

initial state value until the Majorana mode reaches the center point of the system and, therefore, the topology is maintained. Beyond that instant the Chern number starts to fluctuate which indicates that gaps are closing and opening due to the time evolution. In the thermodynamic limit the revival times extend to infinity and the Chern number does not change,[10,17] even though the edge states do decay. However, the Chern number may change due to the finiteness of the system. The values taken by the Chern number at a given time can be quite large. Since the Chern numbers fluctuate considerably it may make sense to look at the time averaged Chern numbers. These average values have a very slow convergence to the value corresponding to the Chern value of the final state and is not conclusive if it fully occurs.

## 6. Periodic driving

A different type of time perturbation that has attracted considerable interest are periodic perturbations. While quenches, either abrupt or slow, in general destabilize the edge states, topological phases can be induced by periodically driving the Hamiltonian of a non-topological system, such as shown before in topological insulators[18–20] and in topological superconductors, with the appearance of Majorana fermions.[21–24] Their appearance in a one-dimensional p-wave superconductor was studied in Ref. 25 and in

Ref. 26 introducing external periodic perturbations; the case of intrinsic periodic modulation was also considered.[27] The periodic driving leads to new topological states,[19] and to a generalization of the bulk-edge correspondence, that reveals a richer structure[28,29] as compared with the equilibrium situation.[8,30] Similarly, in topological superconductors new phases may be induced and manipulated due to the presence of the periodic driving,[25,31,32] such as shining a laser on a topologically trivial system.

## 6.1. *Floquet formalism*

The time evolution of a state under the influence of a time dependent Hamiltonian is given by

$$i\frac{\partial}{\partial t}\psi(k,t) = H(k,t)\psi(k,t) \tag{31}$$

where $k$ is the momentum, $t$ the time and we take $\hbar = 1$. We can decompose the Hamiltonian in two terms: a time independent one, $H(k)$, and an extra term due to the external time-dependent perturbation, that we want to take as periodic with a given frequency, $\omega$,

$$H(k,t) = H(k) + f(\omega t)H_d(k) \tag{32}$$

where $f(\omega t + 2\pi) = f(\omega t)$. Here $H_d(k)$ is of the form of the unperturbed Hamiltonian but with only one non-vanishing term. Looking for a solution of the type

$$\psi(k,t) = e^{-i\epsilon(k)t}\Phi(k,t) \tag{33}$$

and using that $\Phi(k,t) = \Phi(k,t+T)$, where $T$ is the period ($\omega = 2\pi/T$), one gets that

$$\left(H(k,t) - i\frac{\partial}{\partial t}\right)\Phi(k,t) = \epsilon(k)\Phi(k,t) \tag{34}$$

The time-independent quasi-energies $\epsilon(k)$ are the eigenvalues of the operator $H(k,t) - i\frac{\partial}{\partial t}$ and the function $\Phi(k,t)$ the eigenfunction. Note that due to the external time dependent perturbation, energy is not conserved and therefore the original energy bands loose their meaning. Since this function is periodic, we can expand it as

$$\Phi(k,t) = \sum_m \phi_m(k)e^{im\omega t} \tag{35}$$

Inserting this expansion in equation 34 we obtain the time-independent eigensystem

$$\sum_{m'} H_{mm'}(k)\phi_{m'}(k) = \epsilon(k)\phi_m(k) \tag{36}$$

with the quasi-energies the eigenvalues. The Hamiltonian matrix is given by

$$H_{mm'}(k) = \delta_{mm'} m\omega + \frac{1}{T}\int_0^T dt e^{-im\omega t} H(k,t) e^{im'\omega t} \tag{37}$$

Choosing a perturbation of the type $f(\omega t) = \cos(\omega t)$ the second term of the Hamiltonian matrix reduces to $1/2\,(\delta_{m'+1,m} + \delta_{m'-1,m})$.

The time evolution of the state is then obtained solving for the quasi-energies, $\epsilon(k)$, and the functions $\phi_m(k)$ diagonalizing the infinite matrix

$$\begin{pmatrix} \cdots & \cdots & \cdots & \cdots & \cdots \\ \cdots & (m-1)\omega + H(k) & \frac{1}{2}H_d(k) & 0 & \cdots \\ \cdots & \frac{1}{2}H_d(k) & m\omega + H(k) & \frac{1}{2}H_d(k) & \cdots \\ \cdots & 0 & \frac{1}{2}H_d(k) & (m+1)\omega + H(k) & \cdots \\ \cdots & \cdots & \cdots & \cdots & \cdots \end{pmatrix} \tag{38}$$

The matrix can be reduced if the frequency is high enough and then only a few values of $m$ are needed. In the case of a $2D$ triplet superconductor, hopping, chemical potential, spin-orbit coupling or magnetization, will be considered to vary with time. The first three parameters preserve time reversal symmetry while the magnetization naturally breaks time reversal symmetry if the unperturbed Hamiltonian is in a regime with vanishing magnetization. Emphasis will be placed on the effects of varying the chemical potential or the magnetization which are easily tuned externally. In this last case it has been determined before[26] that even though the low energy states have a very low energy, they may not be strictly Majorana fermions since the eigenvalues of the Floquet operator (time evolution operator over one time period) are not strictly $\pm 1$.

Due to the periodicity of the eigenfunctions, $\Phi(k, t+T) = \Phi(k,t)$, the action of the evolution operator, $\mathcal{U}(t)$, on a state over a period, $T$, leads to the same state minus a phase

$$|\psi(T)\rangle = \mathcal{U}(T)|\psi(0)\rangle = e^{-i\epsilon T}|\psi(0)\rangle \tag{39}$$

Therefore, the quasi-energies are defined minus a shift of a multiple of $w = 2\pi/T$, and we can restrict the quasi-energies to the first Floquet zone, defined by the interval $-w/2 \leq \epsilon \leq w/2$. States with quasi-energies $\epsilon =$

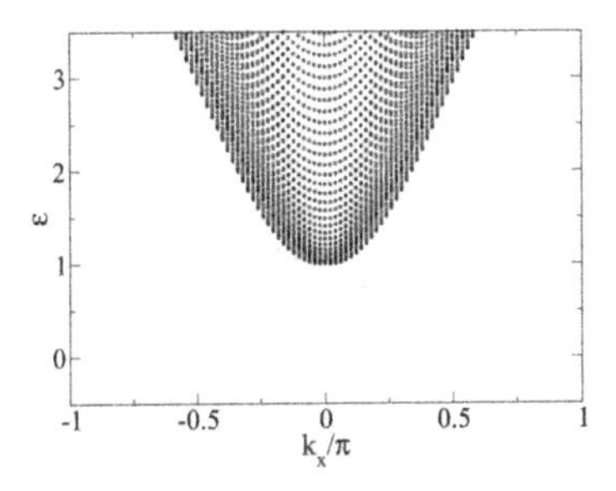


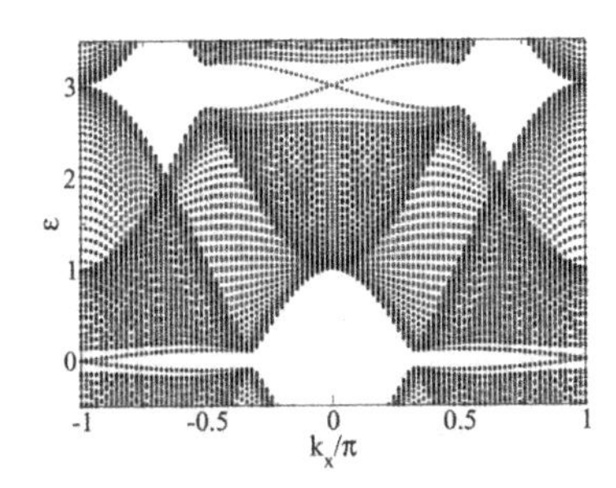


Fig. 13. Energy spectrum of (a) the unperturbed Hamiltonian (or Floquet spectrum for $l = 0$) and (b) Floquet spectrum for $l = 3$ for a $d_x, d_y$ triplet superconductor in a topologically trivial phase $\alpha = 0.1, M_z = 0, \mu = -5, \Delta_s = 0.1, d_z = 0, d = 0.6$ where the magnetization is changed with time with frequency $w = 6$. The periodic driving is $M_{zd} \cos wt$ with $M_{zd} = 4$. Reproduced from Ref. 33.

$w/2$ and $\epsilon = -w/2$ are therefore equivalent and there is a reflection of any bands as one exits the Floquet zone from above (or below) and as one enters from below (or above). Considering the particle-hole symmetry of a superconductor, $\gamma_{-\epsilon} = \gamma_\epsilon^\dagger$ and the equivalence between the energies $\epsilon = -w/2, w/2$ one expects a new type of finite quasi-energy Majorana mode in addition to any zero energy states, the usual Majorana modes.

### 6.2. *Quasi-energy bands of 2D triplet superconductor*

The solutions for the quasi-energies of the perturbed Hamiltonians lead to bands that have a similar structure to the energy bands of the unperturbed Hamiltonian obtained taking a real space description with OBC along $y$ and a momentum description along $x$.

At large frequencies, $w > 4\tilde{t}$, the size of the truncated matrix is relatively small and the quasi-energies and physical properties (calculated over the first Floquet zone) converge fast for small values of $m$. Considering $m = 0$ one reproduces the Hamiltonian of the unperturbed superconductor. The first approximation for the driven system is obtained considering $m = 1, 0, -1$, then $m = 2, 1, 0, -1, -2$ and so on. One may therefore use a short notation for the number of terms considered in the diagonalization of the Hamiltonian matrix by using $l = 0, 1, 2, 3, \ldots$. The unperturbed case is denoted by $l = 0$ and the perturbed cases by $l = 1, 2, \cdots$ (considering that we are using $2l + 1$ states). If the frequency $w$ is small, one needs to consider large values of $l$ and the problem of finding the edge states in a ribbon geometry quickly becomes heavy computationally. Increasing the value of the frequency it is easy to find that it is enough to consider $l = 2$, since taking $l = 3$ leads to very similar results, with a good accuracy.

In Fig. 13 we consider periodic drivings in the magnetization for moderate couplings of $M_{zd} = 4$ and compare to the unperturbed case. We consider frequency $w = 6$. In this case the unperturbed system is in a trivial phase evidenced by the absence of gapless edge states inside the bulk gap. Adding the perturbation edge states appear at low energies and also appear at the border of the Floquet zone around $w/2$ (and $-w/2$). These states are also localized at the edges of the system. In general, edge states appear at the border of the Floquet zone, but as we can see from the figure there is no clearly defined gap throughout the Brillouin zone. Edge states at low energies do not always appear or are mixed with bulk edges. If the driving frequency is smaller or the perturbation has a small amplitude the convergence is slow and in general the quasi-energy spectrum is complex with a strong mixture of the edge and bulk states.[33]

## 6.3. *Currents*

The edge states lead to the appearance of currents. The charge current operator along direction $x$ at a given position $\hat{j}_y$ along $y$ is given by[33]

$$\hat{j}_c(j_y) = \frac{2e}{\hbar} \sum_{k_x} \psi^\dagger_{k_x,j_y} \begin{pmatrix} -\tilde{t}\sin(k_x) & -\frac{i}{2}\alpha\cos(k_x) \\ \frac{i}{2}\alpha\cos(k_x) & -\tilde{t}\sin(k_x) \end{pmatrix} \psi_{k_x,j_y} \tag{40}$$

where $\psi^\dagger_{k_x,j_y} = \left(\psi^\dagger_{k_x,j_y,\uparrow}, \psi^\dagger_{k_x,j_y,\downarrow}\right)$. The current has contributions from the hopping and the spin-orbit terms. The operators are written in real space along $y$ and in momentum space along $x$. One may also define a longitudinal spin current, $\hat{j}_s(j_y)$, taking the difference between the two diagonal components of the charge current. The other terms correspond to spin-flip terms and do not contribute to the $z$ component of the spin current.

The average value of the charge current in the groundstate is given by summing over the single particle occupied states (negative energies) in the

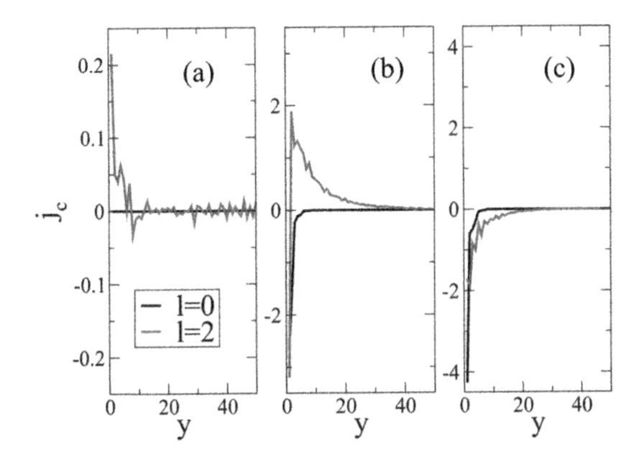


Fig. 14. (Color online) Charge current profiles for the unperturbed ($l = 0$) and perturbed ($l = 2$) $d_x, d_y$ superconductor for (a) $\mu = -5, M_z = 0$ (with $C = 0$), (b) $\mu = -5, M_z = 2$ (with $C = 1$) and (c) $\mu = -1, M_z = 2$ (with $C = -2$) and for $\mu_d = 1$. Only one half of the system is shown since the current profile is anti-symmetric around the middle point. The system size is $N_y = 100$.

usual way

$$
\begin{aligned}
j_c(j_y) = \langle \hat{j}_c(j_y) \rangle = \sum_{k_x, n} \{ & \tilde{t} \sin k_x \left[ \tilde{v}_n(-k_x, j_y, \uparrow) \tilde{v}_n^*(-k_x.j_y, \uparrow) \right. \\
& + \left. \tilde{v}_n(-k_x, j_y, \downarrow) \tilde{v}_n^*(-k_x.j_y, \downarrow) \right] \\
& - \frac{i\alpha}{2} \cos k_x \left[ \tilde{v}_n(-k_x, j_y, \uparrow) \tilde{v}_n^*(-k_x.j_y, \downarrow) \right. \\
& - \left. \tilde{v}_n(-k_x, j_y, \downarrow) \tilde{v}_n^*(-k_x.j_y, \uparrow) \right] \}
\end{aligned} \tag{41}
$$

Here the functions are of the type

$$
\tilde{u}_n(k_x, j_y, \sigma) = \sum_m e^{imwt} u_{n,m}(k_x, j_y, \sigma) \tag{42}
$$

where as usual $\sigma = \uparrow, \downarrow$.

The charge currents at the edges of the unperturbed Hamiltonian are well understood. If there is TRS the currents vanish and if TRS is broken the edge charge currents are finite (finite Chern number). We consider as example a $d_x, d_y$ triplet superconductor. If the magnetization vanishes, the system has TRS and vanishing charge edge currents in the topologically trivial phases. The charge current also vanishes in the $Z_2$ topological phase but the spin edge currents are non-vanishing.

We consider a set of parameters $d = 0.6, \Delta_s = 0.1, d_z = 0, \alpha = 0.6$ and different values for the chemical potential and the magnetization. In Fig. 14 we show results for the profile of the charge current as a function of $y$ for various cases. We compare the unperturbed case with the perturbed one by considering that at $w = 6$ it is enough to truncate the Hamiltonian matrix at $l = 2$. To calculate the currents we sum over the states in the first Floquet zone. Also the results are for time $t = 0$ or any multiple of the time period $T$. As seen in Fig. 14a the periodic driving gives rise to a finite charge current that is absent in the unperturbed trivial phase. In the other two panels the edge states of the unperturbed system carry a finite current which is altered by the periodic driving due to the appearance of extra edge states and also a reshaping of the continuum states; notably there is a change of sign in Fig. 14b.

As shown in Ref. 33 the edge states also generate spin currents. Associated with the spin currents in the topological phases, it has been shown that the eigenstates have non trivial spin polarizations that depend strongly on the momentum.[34] In Ref. 33 the spin polarization of the induced edge states is also considered.

## 7. Conclusions

In this work the robustness of edge modes of topological systems to time-dependent perturbations was considered. The fermionic and Majorana edge mode dynamics of various topological systems were compared, after a global quench of the Hamiltonian parameters takes place. Also, the effect of a periodic perturbation was considered. In general the edge modes are not stable, however in finite systems there is the possibility of revival after a finite time. It was shown that the distinction due to the Majorana nature of the excitations plays a small role in comparison to the details of the energy spectrum and overlaps between states.

Slow transformations were not considered here but allow the occurrence of the Kibble-Zurek mechanism of defect production as one crosses a quantum critical point. In the context of the Creutz ladder, it was shown before that the presence of edge states modifies the process of defect production expected from the Kibble-Zurek mechanism, leading in this problem to a scaling with the change rate with a non-universal critical exponent.[35] A similar result was obtained for the one-dimensional superconducting Kitaev model, where it was shown that, although bulk states follow the Kibble-Zurek scaling, the produced defects for an edge state quench are quite anomalous and independent of the quench rate.[36] Similar results have been found for a $2D$ triplet superconductor.[10] As in the case of sudden quenches, there seems to be no particular signature of the Majorana fermions in comparison to other edge modes. Note that Majoranas are absent in the Creutz ladder.

Fermionic edge states in a topological insulator are now established.[37] Even though Majorana edge states have been extensively studied in the literature their experimental detection has proved challenging. While there is promising evidence of Majorana edge states[38,39] in magnetic chains superimposed on a conventional superconductor, there may be other sources of the edge states observed in the system considered (see for instance Ref. 40 for a discussion and references therein).

One of the methods proposed to detect the presence of Majorana edge states is the measurement of the differential conductance at the interface between a lead and a topological superconductor. If the lead is metallic one expects a zero-bias peak in the differential conductance, if zero-energy modes are present in the superconducting side. In the presence of Majorana modes one expects a vanishing conductance if the number of Majorana modes is even and a quantized value of $2e^2/h$, if the number of modes is odd.[41,42] In the case of the dimerized SSH model here considered, it has been shown[5] that the fermionic edge modes do not contribute to the conductance and, therefore, provides a method to distinguish the various phases and the type of edge modes.

## Acknowledgements

The author acknowledges discussions with Pedro Ribeiro, Antonio Garcia-Garcia, Xiaosen Yang, Maxim Dzero and Henrik Johannesson. Partial support in the form of a BEV by the CNPq at CBPF (Rio de Janeiro) and partial support and hospitality by the Department of Physics of Gothenburg University are acknowledged. Partial support from the Portuguese FCT under Grants No. PEST- OE/FIS/UI0091/2011, No. PTDC/FIS/111348/2009 and UID/CTM/04540/2013 is also acknowledged.

## References

1. M. Z. Hasan and C. L. Kane, *Rev. Mod. Phys.* **82**, 3045 (2010).
2. J. Alicea, *Rep. Prog. Phys.* **75**, 076501 (2012).
3. S. S. Pershoguba and V. M. Yakovenko, *Phys. Rev. B* **86**, 075304 (2012).

4. A. Y. Kitaev, *Phys.-Usp.* **44**, 131 (2001).
5. R. Wakatsuki, M. Ezawa, Y. Tanaka and N. Nagaosa, *Phys. Rev. B* **90**, 014505 (2014).
6. T. O. Puel, P. D. Sacramento and M.A. Continentino, *J. Phys. Cond. Matt.* **27**, 422002 (2015).
7. M. Sato and S. Fujimoto, *Phys. Rev. B* **79**, 094504 (2009).
8. A. P. Schnyder, S. Ryu, A. Furusaki and A. W. W. Ludwig, *Phys. Rev. B* **78**, 195125 (2008; A. P. Schnyder, S. Ryu, A. Furusaki and A. W .W. Ludwig, in *Advances in Theoretical Physics*, edited by Vladimir Lebedev and Mikhail Feigel'man, AIP Conf. Proc. No. 1134 (AIP, Melville, NY, 2009), p. 10; S. Ryu, A. P. Schnyder, A. Furusaki and A. W. W. Ludwig, *New J. Phys.* **12**, 065010 (2010).
9. A. Rajak and A. Dutta, *Phys. Rev. E* **89**, 042125 (2014).
10. P. D. Sacramento, *Phys. Rev. E* **90**, 032138 (2014).
11. P. D. Sacramento, *Phys. Rev. E* **93**, 062117 (2016).
12. J. Happola, G. B. Halasz and A. Hamma, *Phys. Rev. B* **85**, 032114 (2012).
13. E. H. Lieb and D. W. Robinson, *Comm. Math. Phys.* **28**, 251 (1972).
14. S. Bravyi, M. B. Hastings, and F. Verstraete, *Phys. Rev. Lett.* **97**, 050401 (2006).
15. M. Cheneau, P. Barmettler, D. Poletti, M. Endres, P. Schauß, T. Fukuhara, C. Gross, I. Bloch, C. Kollath, and S. Kuhr, *Nature* **481**, 484 (2012).
16. L. Bonnes, F. H. L. Essler, and A. M. Laüchli, *Phys. Rev. Lett.* **113**, 187203 (2014).
17. L. D'Alessio and M. Rigol, *Nat. Comm.* **6**, 8336 (2015).
18. T. Oka and H. Aoki, *Phys. Rev. B* **79**, 081406(R) (2009).
19. N. H. Lindner, G. Refael and V. Galitski, *Nat. Phys.* **7**, 490 (2011).
20. J. I. Inoue and A. Tanaka, *Phys. Rev. Lett.* **105**, 017401 (2010); T. Kitagawa, E. Berg, M. Rudner and E. Demler, *Phys. Rev. B* **82**, 235114 (2010).
21. L. Jiang, T. Kitagawa, J. Alicea, A. R. Akhmerov, D. Pekker, G. Refael, J. I. Cirac, E. Demler, M. D. Lukin and P. Zoller, *Phys. Rev. Lett.* **106**, 220402 (2011).
22. Qing-Jun Tong, Jun-Hong An, Jiangbin Gong, Hong-Gang Luo, and C. H. Oh, *Phys. Rev. B* **87**, 201109(R) (2013).
23. Xiaosen Yang, arXiv:1410.5035.
24. A. Poudel, G. Ortiz, and L. Viola, *Europhys. Lett.* **110**, 17004 (2015).

25. D.E. Liu, A. Levchenko and H.U. Baranger, *Phys. Rev. Lett.* **111**, 047002 (2013).
26. M. Thakurathi, A. A. Patel, D. Sen and A. Dutta, *Phys. Rev. B* **88**, 155133 (2013).
27. M. S. Foster, V. Gurarie, M. Dzero and E. A. Yuzbashyan, *Phys. Rev. Lett.* **113**, 076403 (2014).
28. M. S. Rudner, N. H. Lindner, E. Berg, and M. Levin, *Phys. Rev. X* **3**, 031005 (2013).
29. G. Usaj, P. M. Perez-Piskunow, L. E. F. Foa Torres and C. A. Balseiro, *Phys. Rev. B* **90**, 115423 (2014).
30. J. K. Asbóth, B. Tarasinski and P. Delplace, *Phys. Rev. B* **90**, 125143 (2014).
31. M. Benito, A. Gómez-León, V. M. Bastidas, T. Brandes, and G. Platero, *Phys. Rev. B* **90**, 205127 (2014).
32. Zi-Bo Wang, H. Jiang, H. Liu, and X. C. Xie, *Sol. Stat. Commun.* **215-216**, 18 (2015).
33. P. D. Sacramento, *Phys. Rev. B* **91**, 214518 (2015).
34. A. P. Schnyder, C. Timm and P. M. R. Brydon, *Phys. Rev. Lett.* **111**, 077001 (2013).
35. A. Bermudez, D. Patanè, L. Amico and M. A. Martin-Delgado, *Phys. Rev. Lett.* **102**, 135702 (2009).
36. A. Bermudez, L. Amico and M. A. Martin-Delgado, *New Journ. Phys.* **12**, 055014 (2010).
37. B. Bernevig, T. Hughes and S. Zhang, *Science* **314**, 1757 (2006).
38. V. Mourik, K. Zuo, S. M. Frolov, S. R. Plissard, E. P. A. M. Bakkers, and L. P. Kouwenhoven, *Science* **336**, 1003 (2012).
39. S. Nadj-Perge, I. K. Drozdov, J. Li, H. Chen, S. Jeon, J. Seo, A. H. MacDonald, B. A. Bernevig and A. Yazdani, *Science* **346**, 602 (2014).
40. E. Dumitrescu, B. Roberts, S. Tewari, J. D. Sau, and S. D. Sarma, *Phys. Rev. B* **91**, 094505 (2015).
41. K. T. Law, P. A. Lee and T. K. Ng, *Phys. Rev. Lett.* **103**, 237001 (2009).
42. M. Wimmer, A. R. Akhmerov, J. P. Dahlhaus and C. W. J. Beenakker, *New J. Phys.* **13**, 053016 (2011).

# Effect of Confinement on Melting in Nanopores

M. Sliwinska-Bartkowiak*,‡, M. Jazdzewska*, K. Domin* and K. E. Gubbins†

**Faculty of Physics, Adam Mickiewicz University, ul. Umultowska 85*
*61-614 Poznan, Poland*
*‡E-mail: msb@amu.edu.pl*
*†Department of Chemical and Biomolecular Engineering, North Carolina State University, Raleigh, NC, 27695, USA*
*E-mail:keg@ncsu.edu*

We report experimental measurements of melting behavior of liquids adsorbed in carbon and silica nanopores. Phase transition of $H_2O$, $D_2O$ and $C_6H_5NO_2$ confinement in pores of different inner diameter in the range of 2.2-7.5nm has been characterized by Dielectric Spectroscopy and Differential Scanning Calorimetry methods. The measurements presented in this paper have been made in a wide temperature range from 150K to 300K. We find that the melting point inside pores strongly depend on pores size and fluid-wall interaction. We show that melting temperatures Tmp of confinement system decrease relative to the bulk for weaker fluid-wall than fluid-fluid interaction (silica glasses) and increase in the case of stronger fluid-wall interaction (carbon pores).

*Keywords*: Nanopores, phase transitions, dielectric spectroscopy.

## 1. Introduction

Melting of fluids confined in nano-porous materials has been extensively studied using both experiments and simulations . This effort is relevant to the fundamental understanding of the behavior of nano-dimensional fluids and solids, and the influence of surface forces on such materials. Nanoporous materials, such as activated carbons, carbon nanotubes, silicas, etc., play a prominent role in chemical processing, particularly in separations and as catalysts and catalyst supports. Fluids confined in such porous materials possess many novel properties that can form the basis of future nanotechnologies. Recent studies for pores of simple geometry have shown a rich phase behavior associated with melting in confined systems. The melting temperature may be lowered or raised relative to the bulk melting point, depending on the nature of the adsorbate and the porous material.[1–10] Much of the apparently complex phase behavior is a result of competition between the fluid-wall and fluid-fluid intermolecular interactions. The ratio of the fluid - wall/ fluid-fluid interactions described also as a microscopic wetting parameter $\alpha_w$ of the system[10,11] becomes a criterion of the shift of the melting temperature of the substances placed in pores relatively to the bulk. Early experiments showed a decrease in the melting temperature on confinement in porous solids, and it was widely assumed that such a depression would always occur for confined nano-phases[1,2] More recent experiments and simulations on carbons and mica materials, for which the wetting parameter $\alpha_w$ can be considerably larger than 1, show a significant increase in the melting temperature[3,4] The magnitude of these shifts in the melting temperature depends on the pore width, H, usually becoming larger as H decreases. These results show clearly the important role of both the pore width and wetting parameter. As the pore width H decreases, the shift in the melting temperature relatively to the bulk increases, while the value of the wetting parameter $\alpha_w$ determines the sign of this shift, as well as its magnitude.[12] In this paper we report experimental studies of the melting transition of dipolar liquids: water and nitrobenzene confined in different porous materials such as silica pores (MCM-41, SBA-15, CPG) and carbon nanotubes (CNT) with the cylindrical pores and the pore size of the range of 2.2 to 7.5nm . The analysis of the melting behavior of this liquids confined in applied various porous matrices confirm that an elevation in the melting point is observed for systems where the adsorbate-wall interactions are strong compared to the adsorbate–adsorbate interactions as it was shown also in simulation studies.[7–10]

## 2. Experimental techniques

We report experimental studies of the melting transition of water and nitrobenzene confined in different pores materials such as silica pores (MCM-41, SBA-15, CPG) and carbon nanotubes (CNT) which having different inner diameters of the range of 2.2 to 7.5nm. In our experiments we have used open-tips multi-wall carbon nanotubes with average inner diameter of 3.9nm and 7nm produced and characterized by Nanocyl S.A. The MCM-41 mesoporous silica template was synthesized according to procedure describing in Ref. 13. SBA-15 porous matrices with different inner diameters was synthesized in Department of Chemistry, The University of Hong Kong.[14] Controlled pore glasses (CPG) which we have used in our experiments with an average pore diameter of 7.5 nm and a pore volume of 47 cc/g were made by Corning Co., New Jersey. The pores samples were heated to about 400 K and kept under vacuum ($10^{-3}$ Tr) for a few days to remove the air prior to and during the introduction of the liquids. Differential Scanning Calorimetry (DSC) and Dielectric Relaxation Spectroscopy (DRS) were used to determine the melting temperature of confined liquids. Differential Scanning Calorimeter was applied to determine the melting temperatures of the confinement systems by measuring the heat released in the melting of bulk substances. The DSC used was a DSC Q2000 (TA Instruments) in Departmental Laboratory of Structural Research in Faculty of Physics and DSC 8000 (Perkin Elmer) in NanoBioMedical Centre (AMU). A scanning rates of temperatures of 5-10 K/min in the range between 110 and 330K were used in the experiments. Dielectric Relaxation Spectroscopy was also used to describe melting/freezing behavior of confinement liquids. The dielectric constant is a natural choice of order parameter to study melting of dipolar liquids because of the large change in the orientational polarizability between the solid and the liquid phases.[15] The capacitance C and the tangent loss, of the capacitor filled with the sample was measured at different temperatures T and frequencies $\omega$ using a Solartron 1260 gain impedance analyzer, in the frequency range 1Hz-1MHz. The complex dielectric permittivity $\epsilon = \epsilon' - i\epsilon''$ is related to the measured quantities by $\epsilon' = C/C_0$ and $\epsilon'' = \tan(\delta)/\epsilon'$ (where $\delta$ is the angle by which current leads the voltage). Here C is the electric capacitance and $C_0$ is the capacitance in the absence of the dielectric medium, $C_0 = 66\text{pF}$. For substances confined within nanopores, the sample was introduces between the capacitor plates as a suspension of adsorbate-filled pores in the pure substances. In order to reduce the high conductivity of carbon sample the electrodes were covered with a thin layer of Teflon. The dielectric, orientational relaxation time $\tau$ can be obtained from the experimental data using the Debye dispersion relation which was derived for an isolated dipole rotating under an oscillating electric field in a viscous medium using classical mechanics:

$$\epsilon^* = \epsilon'_\infty + \frac{\epsilon'_s + \epsilon'}{1 + i\omega\tau} \tag{1}$$

In this equation $\omega$ is the frequency of the potential applied and $\tau$ is the orientational relaxation time of a dipolar molecule. The subscript s refers to the static permittivity (the permittivity in the low frequency limit) and subscript $\infty$ refers to the high frequency limit of the permittivity.

## 2.1. *Experimental results*

### 2.1.1. *Water in silica glass*

The melting temperatures of water confined in cylindrical pores MCM-41 with inner diameter of 2.2nm was determined using DSC and DS methods. Result of the measurement of C for this system during the heating process as a function of T and at a frequency of 600kHz is shown in Fig. 1. The sharp increase at 273K corresponds to the bulk melting, while the change of C at 225K is attributed to melting in the pores. The melting temperature of water adsorbed in the pore of MCM-41 is shifted towards lower temperatures by 48K relative to the bulk melting temperature.

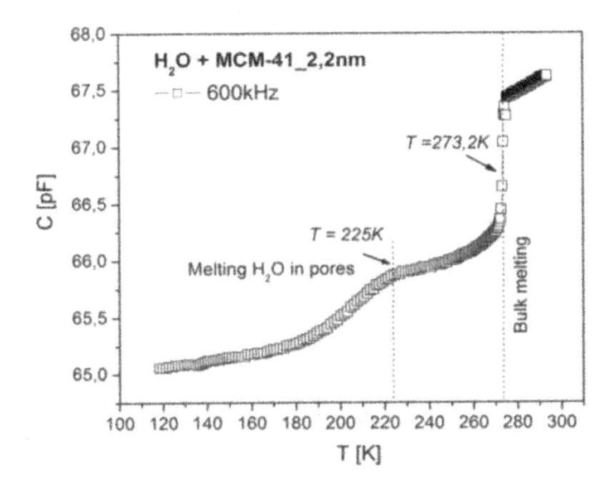


Fig. 1. Capacitance C vs temperature for water in MCM-41 pores of diameter 2.2nm.

Similar result was obtained using the DSC method. The melting temperature was determined from the position of the peaks of the heat flow signals. The DSC scan for $H_2O$ water in MCM-41 is presented in Fig. 2. The large peak at 273K corresponds to the melting of the bulk water and much smaller peak at 225K can corresponds to the melting point of water confined in the pores.

Results of the analysis of the Cole-Cole representation of the complex capacity for water confined in MCM-41 at various temperatures are shown in Figs. 3 and 4. From the plot of $\epsilon'$ and and $\epsilon''$ versus $\log \omega$ the relaxation time can be calculated as the inverse of the frequency $\omega$ corresponding to a saddle point of $\epsilon'$ plot or a maximum of the $\epsilon''$ plot. In Fig. 3a the spectrum plot ($\epsilon$ vs $\omega$) at 189K shows two relaxation mechanisms with re-

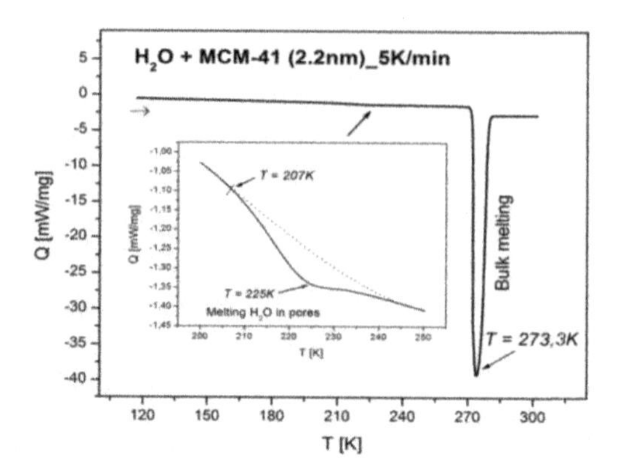


Fig. 2. DSC scan for water confined in MCM-41 silica pores of diameter 2.2nm.

laxation times of the order of $10^{-3}$s and $10^{-6}$s. The value of relaxation time of the order of $10^{-3}$s is typical of the hexagonal ice in this temperature while the relaxation time of the order of $10^{-6}$s is typical of the cubic ice. Each relaxation mechanism is reflected as a semicircle in the Cole-Cole diagram. The corresponding Cole-Cole diagram at 189K is shown in Fig. 3b. The spectrum at 217K (Fig. 4) also represents two different relaxation mechanisms corresponding to double semicircle in the Cole-Cole diagram (Fig. 4b) and is characterized by the relaxation time of the order of $10^{-2}$s, $10^{-4}$s (typical of the hexagonal ice). The longer component of the relaxation is related with the conductivity of the liquid.

The behavior of the relaxation times as a function of temperature for water confinement in MCM-41 is shown in Fig. 5. In the temperature range 160-204K there are two different components of the relaxation times which are characteristic of Ih and Ic ice. The component typical of cubic ice disappears at 204K and for temperatures higher than 204K there are another relaxation mechanism which can be related to the liquid phase of the confined water. This result suggests that for water confined in MCM-41 pores the cubic structure in pores is stable until the melting point (the melting transition of water in MCM-41 is from Ic to liquid). In the case of water confined in larger silica pores diameter[16] the solid-liquid transition occurs rather from hexagonal ice to liquid than from Ic to liquid. In the temperature region between the bulk and the pore melting points that is 204-273K two components of the relaxation times are observed. The shorter component is typical of the Ih in the bulk phase and the longer component, $10^{-2}$s, corresponds to the response of the liquid phase in the pores. For

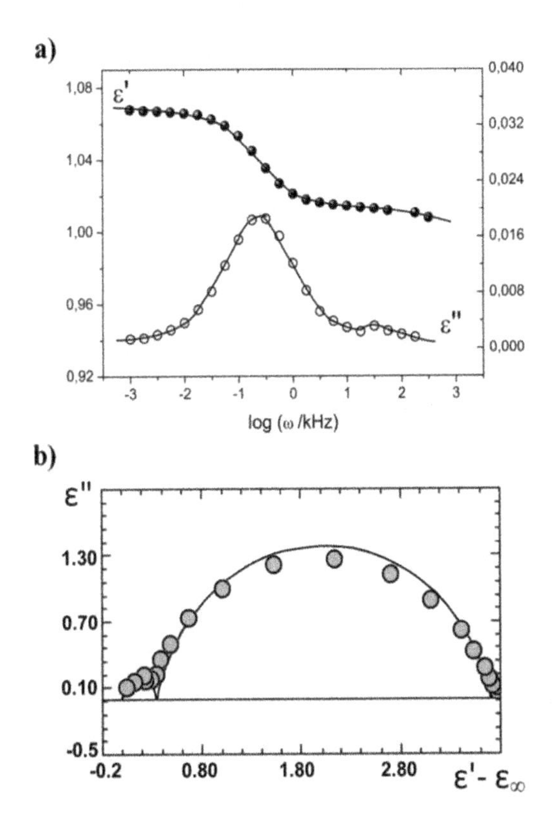


Fig. 3. a) Spectrum plot for $H_2O$ in MCM-41 of 2.2nm diameter at T=189K; b) representation of the spectrum plots in the form of Cole-Cole diagram.

temperatures higher than the bulk melting point one branch of relaxation time is observed which is related to the interfacial Maxwell-Wagner-Sillars polarization. The relaxation mechanism related to the interfacial polarization occurs for a heterogeneous system (porous matrix are suspended in the liquid) when the conducting liquid is enclosed in a material of different conductivity.

The melting temperatures of $H_2O$ and $D_2O$ confined in SBA-15 and CPG pores with different inner diameters of (5.58, 6.63, 7.32 and 7.5) nm were determined using DSC method. Results of DSC measurements for water confined in silica pores are shown in Fig. 6(a–d).

In Fig. 6a the DSC scan for $H_2O$ confined in cylindrical, ordered silica glass SBA-15 of diameter of 5.58 nm is presented. The large endothermic peak at 273.3 K corresponds to the melting of the bulk $H_2O$. In addition,

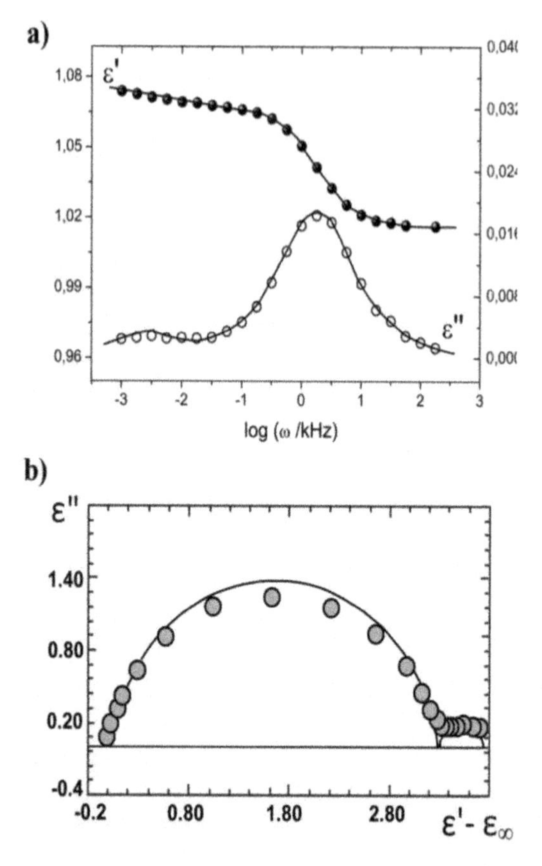


Fig. 4. a) Spectrum plot for $H_2O$ in MCM-41 of 2.2nm diameter at T=217K; b) representation of the spectrum plots in the form of Cole-Cole diagram.

a second peak at 256.3 K is observed, which can correspond to the melting of $H_2O$ in SBA-15. The melting temperature of water adsorbed in SBA-15 of diameter of 5.58nm is shifted toward lower temperatures by about 17 K relative to that of the bulk $H_2O$. Figure 6b presents a DSC scan for $H_2O$ in SBA-15 of 6.63 nm diameter. The large peak corresponds to the melting of the bulk $H_2O$ and the second peak at 258.3 K indicates the melting point of $H_2O$ in the pores. The shift of the melting temperatures $\Delta T$ for water is $\Delta T = 14.98$ K. In Figure 6c the DSC scan for $H_2O$ in SBA-15 of diameter of 7.32 nm is presented. The peak at 273 K indicates the melting of the bulk water, and the smaller peak observed at 259.9 K corresponds to the melting of water in the pores. The shift of melting temperature is

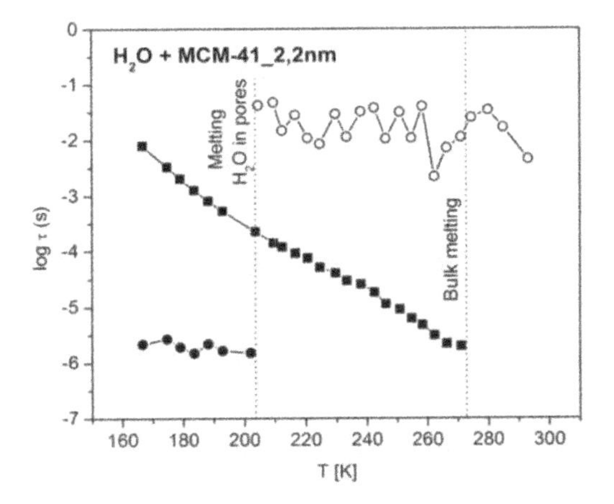


Fig. 5. Dielectric relaxation time vs temperature for water in MCM-41 of diameter 2.2nm

$\Delta T = 13.11$K and is about 4 K lower than for water in SBA-15 of 5.58 nm diameter. Decrease of the melting temperature is also observed for water (D2O) confined in disorder silica glasses CPG with diameter of 7.5nm. In Fig. 6d the DSC scan for this sample is presented; the large peak at 277.8 K corresponds to the melting of the bulk D2O, and a second peak observed at 269.5 K corresponds to the melting of D2O water in the pores of CPG. The melting temperature of D2O adsorbed in the pore of CPG is shifted towards lower temperatures by 8.3 K relative to that of the bulk D2O. For water confined in silica pores we observed decrease of melting temperatures of the adsorbate relative to the bulk melting temperatures. The shift of melting temperatures of water is higher for smaller inner diameter of pores, which is presented in Fig. 7. The value of wetting parameter $\alpha_w$ for water in silica surface is 0.28.[12]

### 2.1.2. *Nitrobenzene in carbon nanopores*

The melting temperature of nitrobenzene confined in carbon nanotubes of pores diameter of 3.9nm was determined using dielectric method. In Fig. 8 the capacity C as a function of temperature for $C_6H_5NO_2$ in CNT is presented. The changing of the value of C versus temperature observed at T=278,8K is associated with the melting point of the bulk nitrobenzene. The increase in the C(T) function observed at T=283,7K is attributed to melting in the pores. The shift of the melting temperature for $C_6H_5NO_2$ in the pores relatively to the bulk melting point is $\Delta T = 4,9$K.

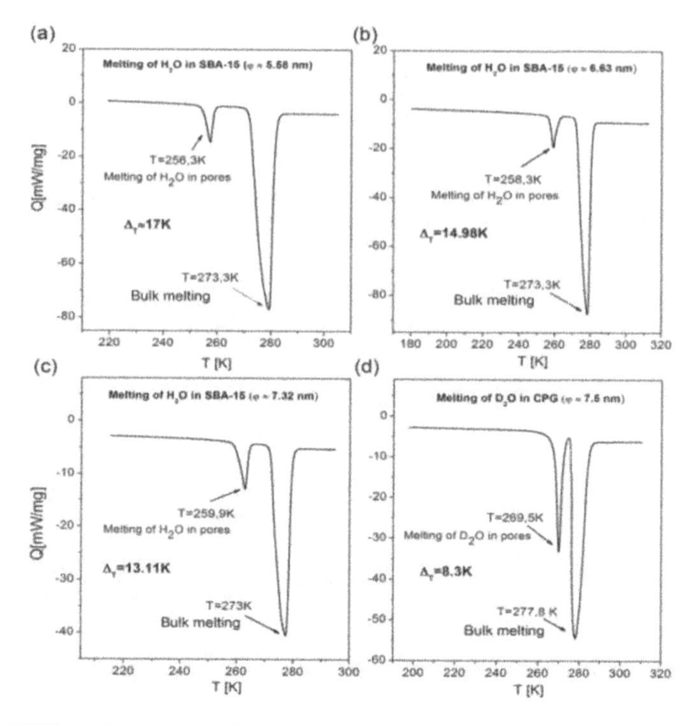


Fig. 6. DSC scan for water confined in SBA-15 silica pores of diameter 5.58nm, 6.63nm, 7.32nm (a,b,c) and CPG silica pores of diameter 7.5nm (d).

Results of the analysis of the Cole-Cole representation of the complex capacity for $C_6H_5NO_2$ confined in CNT of 3.9nm diameter versus temperature are presented in Figs. 9 and 10. In the solid state of $C_6H_5NO_2$ (Fig. 9a) two different relaxation mechanism are observed which are characterized by relaxation times of the order of $10^{-3}$s and $10^{-5}$s. The longer component of the relaxation is related to the crystal phase of the bulk nitrobenzene in the system. The shorter component of the relaxation time is of the order of $10^{-5}$s and can characterize solid phase of $C_6H_5NO_2$ in pores. The corresponding Cole-Cole diagram is shown in Fig. 9b. The spectrum plot at 281K (above the bulk melting point) which is presented in Fig. 11 also represents two kinds of relaxation mechanism corresponding to two semicircles in Cole-Cole diagram. The relaxation time of the order of $10^{-5}$s is related to the solid $C_6H_5NO_2$ in pores and the longer component of the relaxation corresponds to interfacial MWS polarization.

The behavior of the relaxation times as a function of temperature for

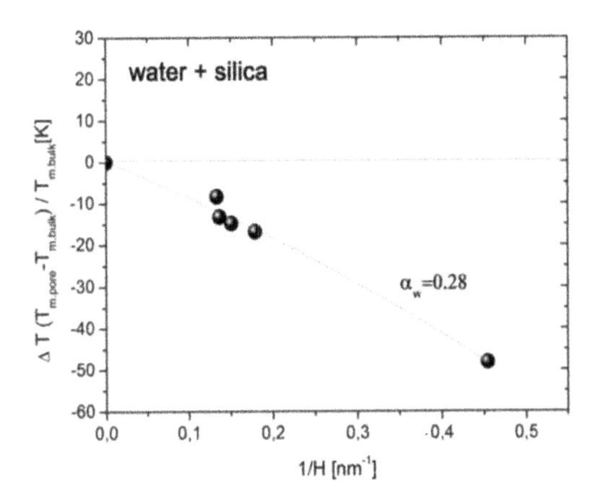


Fig. 7. The shift of melting temperature $\Delta T$ vs pore size for water confined in nanopores.

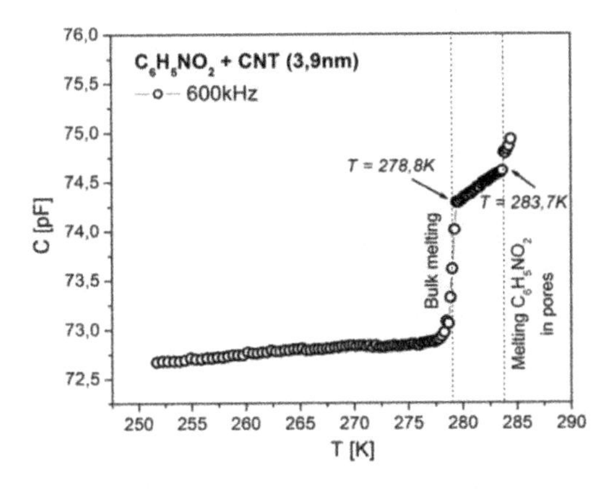


Fig. 8. C(T) for nitrobenzene in CNT-3,9nm.

$C_6H_5NO_2$ in CNTs of diameter of 3.9nm is depicted in Fig. 11. In the temperature range between 170-279K there are two different kind of relaxation. The longer component ($\tau \sim 10^{-3}$) is connected with crystal phase of bulk nitrobenzene. The branch of relaxation time of the order of $10^{-5}$s can be related to the defected or amorphous solid phase confinement in the pores.[17,18] For temperatures higher than 283K (melting point inside the pores) again two branches of relaxation time are observed. The branch of relaxation time of the order of $10^{-3}$s is typical for MWS polarization. The faster component of the relaxation time of the order of $10^{-6} - 10^{-7}$ is too

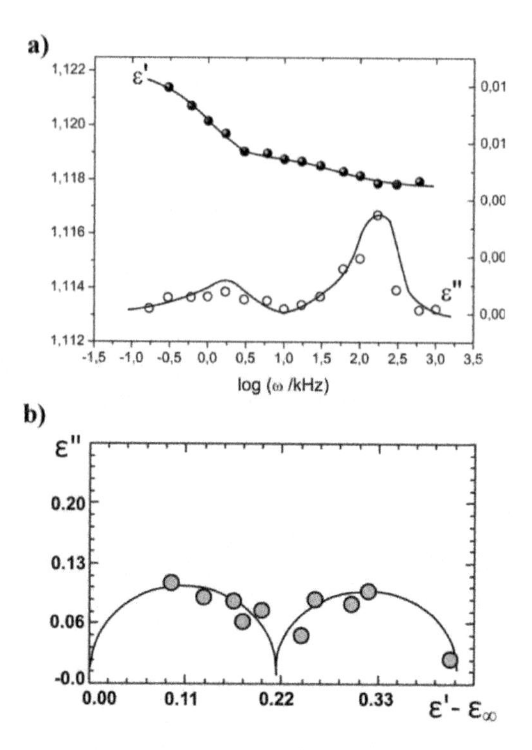


Fig. 9. a) Spectrum plot for $C_6H_5NO_2$ in CNTs of 3.9nm diameter at T=253K; b) representation of the spectrum plots in the form of Cole-Cole diagram.

slow to be relaxation process of liquid phase. Previous results [15] show that contact layers of polar molecules confined in pores reveal much slower dynamics than the dynamic of free substances so this component can describe the dynamic of confinement liquids. Molecular simulation results for simple fluids [15] show same differences in behavior of contact layers and molecules located in the core of pore.

Result of the measurement of C as a function of T for $C_6H_5NO_2$ in CNTs of 7.0 nm diameter is shown in Fig. 12. The large increase in C at 278.9K corresponds to the melting of the bulk $C_6H_5NO_2$. In addition, much smaller change of C(T) function at T=280,4K is observed, which can correspond to the melting of nitrobenzene placed in CNTs. The melting temperature of $C_6H_5NO_2$ adsorbed in CNTs is shifted towards higher temperatures by about 1.5K relative to the bulk melting point. According to these results the elevation of the melting temperature of $C_6H_5NO_2$ in pores decreases with increasing of the diameter of the CNTs.

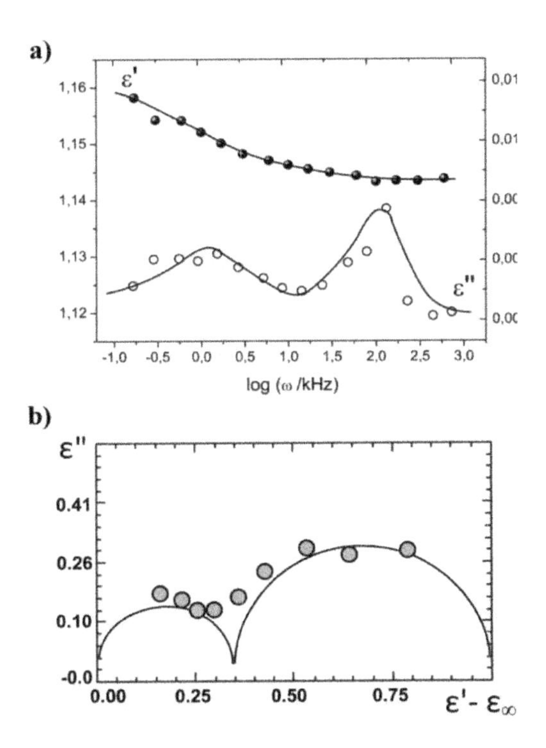


Fig. 10. a) Spectrum plot for $C_6H_5NO_2$ in CNTs of 3.9nm diameter at T=281K; b) representation of the spectrum plots in the form of Cole-Cole diagram.

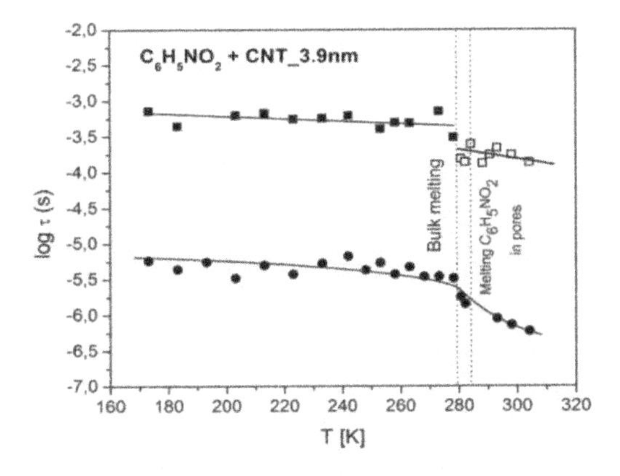


Fig. 11. Dielectric relaxation time $\tau$ vs $T$ for $C_6H_5NO_2$ in CNT-3.9nm.

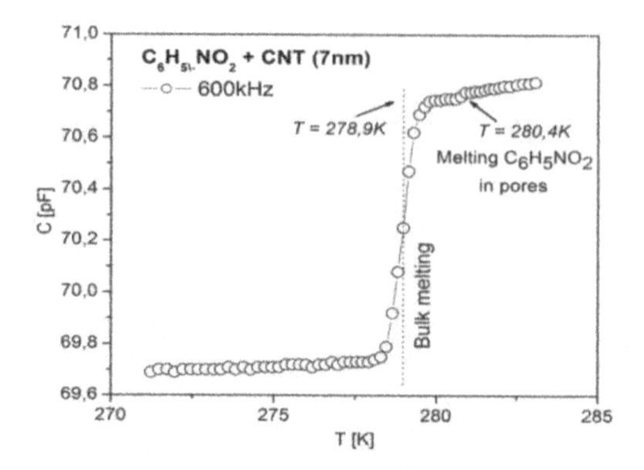


Fig. 12. C(T) for nitrobenzene in CNT-7nm.

For $C_6H_5NO_2$ confined in CNTs of diameter 7.0nm the behavior of relaxation time as a function of temperature is presented in Fig. 13. The distribution of relaxation time has a similar character to that shown in Fig. 11.

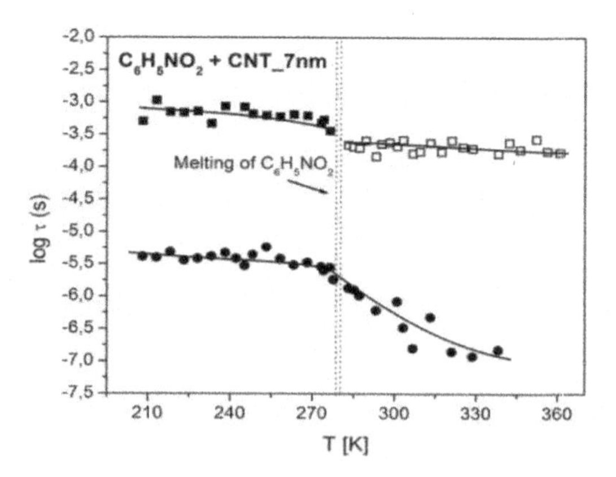


Fig. 13. Dielectric relaxation time $\tau$ vs $T$ for $C_6H_5NO_2$ in CNT-7.0 nm.

The shift of the melting temperatures relative to the bulk for nitrobenzene placed on CNT vs pore size is presented in Fig. 14. We can observe the elevation of the melting temperature of nitrobenzene for this systems. The value of the wetting parameter $\alpha_w$ for nitrobenzene on the carbon surface[12] is $\alpha_w = 1.26$.

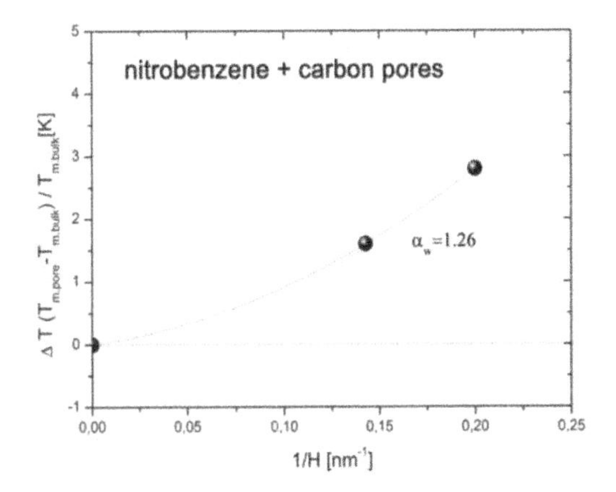


Fig. 14. The shift of melting temperature $\Delta$T vs pore size for nitrobenzene confined in carbon nanopores.

## 3. Discussion and results

We report the melting phenomena in confined phase of water and nitrobenzene in both carbon and silica pores of different inner diameter. We have observed the shift of the melting temperature $T_{mp}$ vs pore size relatively to the bulk for the all systems studied as it is shown in Fig. 15, where $\Lambda T$ vs pore size for the systems with different $\alpha_w$ values are presented. We can observe that the diameter of nanopores H determines the magnitude of shift in the melting temperature of confinement phase; for smaller inner diameter of the pores, the greater shift of the melting temperature of confined phase is observed. The decreasing or elevation in melting point in pores relatively to the bulk for studied substances confined in pores of similar pore diameter can be explained on the basis of the value of wetting parameter $\alpha_w$ calculated for this systems.[12] The wetting parameter, which is the ratio of the fluid-wall to the fluid-fluid attractive interaction is expressed as:

$$\alpha = \frac{\rho_w \epsilon_{fw} \sigma_{fw}^2 \Delta}{\epsilon_{ff}} \tag{2}$$

where $\epsilon_{fw}$ and $\epsilon_{ff}$ are the adsorbate–wall (fluid-wall) and adsorbate–adsorbate (fluid-fluid) attractive Lennard-Jones intermolecular potentials parameters respectively, the density of wall atoms $\rho_w$, the interlayer spacing in the solid $\Delta$, and the fluid–wall diameter parameter - $\sigma_{fw}$.

The value of the $\alpha_w$ parameter determines the direction of change in the melting temperature.[12,19] For $\alpha_w$ higher than 1 we observed increase

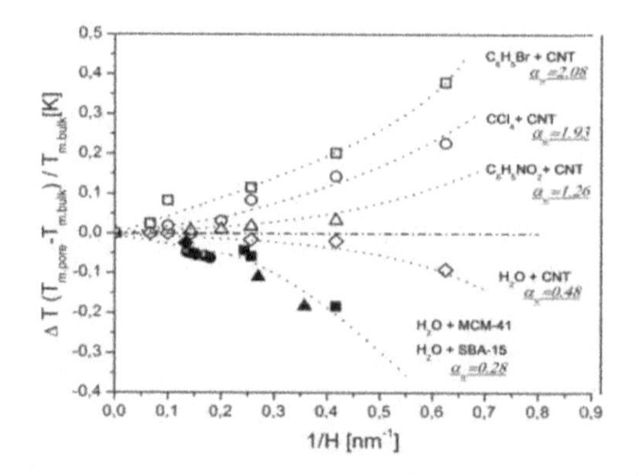


Fig. 15. The shift of melting temperature $\Delta T$ vs pore size for C6H5Br[18], CCl4[18], $C_6H_5NO_2$ and $H_2O/D_2O$ confined in carbon (CNT) and silica pores (MCM-41, SBA-15 and CPG); • adapted from Ref. 18.

in melting temperature of substances confinement in pores, for $\alpha_w$ smaller than 1 our results show depression of the melting point for adsorbate. Thus, the $\alpha_w$ parameter can be used to classify porous materials in studies on freezing and melting of confined fluids; adsorbents with $\alpha_w$ higher than this threshold can be considered as strongly attractive (we can expect an elevation in melting temperature relatively to the bulk ) while those with lower $\alpha_w$ can be considered as weakly attractive (a depression in melting temperature is expected). The experimental results presented in this work confirm this expectations

## Acknowledgments

We thank the National Center of Science for a Grant: DEC 2013/09/B/ST4/03711 in support of this work.

## References

1. D. C. Steytler, J. C. Dore and C. J. Wright, *J. Phys. Chem.* **87**, 2458 (1983).
2. K. Morishige, H. Yasunaga and H. Uematsu, *J. Phys. Chem. C* **113**, 3056 (2009).
3. K. Kaneko, A. Watanabe, T. Ilyama, R. Radhakrishnan and K.E. Gubbins, *J. Phys. Chem. B* **103**, 706 (1999).
4. A. Watanabe and K. Kaneko, *Chem. Phys. Lett.* **305**, 71 (1999).

5. C. Dore, B. Webber and J. H. Strange, *Coll. Surf. Sci. A* **241**, 191 (2004).
6. K. Matsuda, T. Hibi, H. Kadowaki, H. Kataura and Y. Maniwa, *Phys. Rev. B* **74**, 073415 (2006).
7. M. Miyahara, H. Kanda, M. Shibao and K. Higashitani, *J. Chem. Phys.* **112**, 9909 (2000).
8. L. Gelb, R.Radhakrishnan. K.E.Gubbins and M.Sliwinska-Bartkowiak, *Rep. Prog. Phys.* **62** 1573, (1999).
9. M. Sliwinska-Bartkowiak, G. Dudziak, R. Sikorski, R. Gras, R. Radhakrishnan and K. E. Gubbins, *J. Chem. Phys.* **114**, 950 (2001).
10. C. Alba-Simionesco, B. Coasne, G. Dosseh, G. Dudziak, K.E. Gubbins, R. Radhakrishnan and M. Sliwinska-Bartkowiak, *J. Phys. Condens. Matter* **18**, R15 (2006).
11. M. Sliwinska-Bartkowiak, A. Sterczynska, Y. Long, K.E. Gubbins, *Molecular Physics* **112** (17), 2365 (2014).
12. K. E. Gubbins, Y. Long and M. Sliwinska-Bartkowiak, *J. Chemica Thermodynamics* **74**, 169 (2014).
13. Q. Huo and D.I. Margolese, *Nature* **368**, 317 (1994).
14. S. Jun, S. H. Ryoo, R. Kruk, M. Jaroniec, Z. Liu and O. Terasaki, *J. Am. Chem. Soc.* **122**, 10712 (2000).
15. M. Sliwinska-Bartkowiak, G. Dudziak, R. Gras, R. Sikorski, R. Radhakrishnan and K. E. Gubbins, *Colloids and Surface A* **187**, 523 (2000).
16. M. Sliwinska-Bartkowiak, M. Jazdzewska, L. Huang and K.E. Gubbins, *Phys. Chem. Chem. Phys.* **10**, 4909 (2008).
17. M. Jazdzewska, M. Sliwinska-Bartkowiak, A. Beskrovny, S. Vasilowsky, K.Y. Chan, L.L. Huang and K.E. Gubbins, *Phys. Chem. Chem. Phys.* **13**, 4909 (2011).
18. M. Sliwinska-Bartkowiak, M. Jazdzewska, K.E. Gubbins and L. Huang, *J. Chem. Eng. Data* **55** (10), 4183 (2010).
19. Y. Long, J. Palmer, B. Coasne, M. Sliwinska-Bartkowiak, J. Jackson, E. Muller and K.E. Gubbins, *J. Chem. Phys.* **139**, 144701 (2013).

# 3D Topological Dirac Semimetal Based on the HgCdTe*

G. Tomaka, J. Grendysa, M. Marchewka, P. Sliż, C. R. Becker, D. Żak,
A. Stadler and E.M. Sheregii†

*Centre for Microelectronics and Nanotechnology, University of Rzeszow*
*Pigonia St. 1, 35-959 Rzeszow, Poland*
† *E-mail:sheregii@ur.edu.pl*
*www.nanocentrum.univ.rzeszow.pl*

Experimental results of the magneto-transport measurements (longitudinal magneto-transport measurements (longitudinal magneto-resistance $R_{xx}$ and the Hall resistance $R_{xy}$) over a wide interval of temperatures for several samples of $Hg_{1-x}Cd_xTe$ ($x \approx 0.13 - 0.15$) grown by MBE. An amazing temperature stability of the SdH-oscillation period and amplitude is observe in the entire temperature interval of measurements up to 50 K. Moreover, the quantum Hall effect (QHE) behaviour of the Hall resistance was shown in the same temperature interval. These peculiarities of the $R_{xx}$ and $R_{xy}$ for strained thin layers are interpreted using quantum Hall conductivity on topologically protected surface states. The adventures of the HgCdTe alloys as the 3D topological Dirac semimetal analysed.

*Keywords*: Topological insulators; HgCdTe; magnetotransport.

## 1. Introduction

Topological insulator (TI) is a new class of quantum matter with conducting surface states, topologically protected against time-reversal-invariant perturbations and an insulating bulk. The physics of TI links the structures of $d$ dimensions with their boundaries in $d-1$ dimensions: a TI is a state of quantum matter that behaves as an insulator in its interior and as a metal on its boundaries.[1,2] The first topological insulators which were observed via the quantum spin Hall effect concerns a two-dimension electron gas (2DEG) with a spin structure on a 1D edge. They were discussed theoretically,[3,4] predicted to occur in HgTe quantum wells,[5] and then experimentally verified by König et al.[6] Thereafter, Liang Fu et al.[7] proposed

*This work is supported by the authorities of Podkarpackie Voivodship (the Marshal's Office of the Podkarpackie Voivodship of Poland), contract WNDPPK.01.03.00-18-053/12.

the 3D version of topological insulators, predicted to occur in Bi-Sb alloys by Fu and Kane,[8] and experimentally detected with angle-resolved photoemission spectroscopy (ARPES) by Hsieh et al.[9]

Subsequently, the quantum Hall effect (QHE) in the surface layer with states caused by crossing of bands at surface - Topologically Protected Surface States (TPSS) – was observed via transport experiments in strained bulk mercury telluride (HgTe)[10] and semimetal HgCdTe.[11] As predicted by B. A. Bernevig, T. L. Hughes, and S.-C. Zhan,[5] TPSS occur at the surface of HgTe and HgCdTe as massless Dirac points (Dirac cones with a linear dependence of the electron energy on momentum in terms of the band structure) because of the $\Gamma_6$ and $\Gamma_8$ band crossing. Generally, it is caused by strong spin-orbital interaction in these materials which lifts the $\Gamma_8$ band above the $\Gamma_6$ band in the bulk part of the sample.[12]

This kind of TPSS is characterized by a $Z_2$ topological invariant, requiring gapless electronic states to exist on the boundary of the sample, which is a strong topological insulator and is robust in the presence of disorder.[5,11,13,14] Such strong topological insulators have surface states, with the Fermi surface enclosing an odd number of Dirac points and being associated with the Berry phase.[15] This defines a topological metal surface phase, which is predicted to have novel electronic properties.[16–18]

There is some similarity with the half-Heusler alloys. For example, PtMnBi show semi-metallic properties in the $\alpha$ phase, and the effects of the spin-orbit interaction shifts of the valence bands and the indirect semiconducting gap with respect to the spin polarized results.[19,20] Also BiSb alloys, $Bi_2Se_3$, $Bi_2Te_3$ and $Sb_2Te_3$ crystals show the 3D TI properties.[21–24] The next class of bulk topological insulators discovered during the last decade includes $TlSbSe_2$, $TlSbTe_2$, $TlBiSe_2$ and $TlBiTe_2$[25,26] as well as septuple-layer topological insulators.[27]

Recently, well-developed TPSS were reported to occur in the QHE of $BiSbTeSe_2$, an intrinsic TI bulk material, 120 and 160 nm thick.[28] In addition, the dominate contribution of the surface metallic conductivity in the total sample resistance of BiSbTeSe above 120 K, was noticed. Thus, a new feature of the electron transport properties of semiconductors with strong spin-orbital interaction could takes place: the surface metallic conductance. As will be shown below, these features are observed in mercury cadmium telluride solid solutions - arguably the best material for infrared devices.[29–31] The QHE according to Brüne et al.[10] (similar results were presented later in the Ref. 32) was displayed for comparably thin (70 nm) strained samples. The same experimental effect takes place for thin strained layers of semimetal alloys of HgCdTe as well as for no strained layers of these alloys. One of the obvious properties of semiconductors is their strong dependence on external conditions such as temperature, illumination, or pressure. TI has changed our view in this area.

This topic review paper describes experimental results obtained for bulk $Hg_{1-x}Cd_xTe$ (MCT) samples with $x \approx 0.13 - 0.16$ corresponding to a semimetallic type of band-structure, i.e. the $\Gamma_8$ band is higher than $\Gamma_6$ (in normal semi-conductors as CdTe it is contrary: the $\Gamma_6$ band is above the $\Gamma_8$ bands and play role of the conduction band but $\Gamma_8$ bands are valence bands – bands of light and heavy holes), however it is close to the critical point at which the $\Gamma_6$ and $\Gamma_8$ bands cross, that should occur in the temperature region from 0.4 to 150 K depending on the $x$ value. The sections 2 and 3 present the experimental details: the samples grown by MBE technology, the magneto-transport experiment procedure, and experimental data including the SdH oscillations, the QHE behaviour of the Hall resistance observed in a wide temperature region for the samples with different level of doping including comparably thick layers. In the section 4 the origin of the Quantum Hall Conductance behaviour of the Hall resistance for thin strained MCT layers is discussed and for thin not strained layers and comparatively thick ones. Finally, in the section 5 and 6 are summaries these findings and draw conclusions.

## 2. MBE growth of strained and not strained layers

A Riber Compact 21 molecular beam epitaxial (MBE) system was used to grow the $Hg_{1-x}Cd_xTe$ strained and not strained layers. The CdTe and Te fluxes (at constant flux of Hg) were chosen and subsequently adjusted in order to obtain the semimetal $Hg_{1-x}Cd_xTe$ with a Dirac point at approximately 4 – 50 K. Several conditions should be fulfilled during the growth process of the $Hg_{1-x}Cd_xTe$ layer with $0 < x < 0.16$ of high quality – with surface close to ideal and low density of defects (hillocks).

First, the substrate temperature should be controlled with accuracy not worthy then $\pm 0.5^{\circ}$C because its strong influence on composition. The substrate temperature determination need of specific calibration due to lack of physical contact for thermocouple with molybdenum substrate holder. Therefore, this thermocouple is calibrated by using the melting point of In, Sn and PbSn on holder.

Second, the determination and stabilization of the beam equivalent pressure (BEP) of sources (fluxes in others words) is principally important at

MBE growth for each compounds. In case of the $Hg_{1-x}Cd_xTe$ layer growth additional factor is essential: stabilization of the Hg-flux is difficult and long – for two hours long. The composition is determined by the Cd/Te flux ratio whereas, the flux ratio Hg/Te is important for quality of received layers. Stabilization of these two ratio at the growth is crucible: deviation of BEP in the framework of $10^{-8}$ Torr able to change considerably the composition in the case of the Cd/Te ratio as well as to increase the hillock density, electron concentration and decrease the electron mobility, in the case of the Hg/Te ratio.

Two type of substrates were used to perform the experiment: i) (001) oriented GaAs substrates were applied with the MBE grown CdTe buffer that produced $\sim$ 0.3% mismatch with the next thin MBE grown $Hg_{1-x}Cd_xTe$ layer; ii) (112)B orientated CdZnTe substrates were employed and the $Hg_{1-x}Cd_xTe$ was grown directly on the CdZnTe(112)B substrate surface which have a lattice constant practically (with an uncertainty of 0.01%) equal to that of bulk $Hg_{1-x}Cd_xTe$ ($x \approx 0.13$). Consequently, three types of sample were grown and measured, namely: type A – $Hg_{1-x}Cd_xTe$ thin strained layers on the GaAs/CdTe substrates (about 100 nm thick); type AB – $Hg_{1-x}Cd_xTe$ thin not strained layers on the CdZnTe(112)B substrates (about 100 nm thick); type B - $Hg_{1-x}Cd_xTe$ thick not strained layers on the CdZnTe(112)B substrates (samples B4, B6-B9 are about 1 $\mu$m thick, the sample B5 is above 2 $\mu$m thick). The $Hg_{1-x}Cd_xTe$ layers were substantially doped with iodine by means of a $CdI_2$ effusion cell for samples A9, AB9, B8 and B9 and by means of an In effusion cell for samples A4, AB4, B4, B5 and B6. SIMS measurements and the reflection maxima of $E_1$ and $E_1 + \Delta_1$ in the region of fundamental absorption of the $Hg_{1-x}Cd_xTe$ alloys[33] were used to confirm the composition, level of doping and thickness of the grown layers (see Table 1).

Parameters of the resulting layers are shown in Table 1. Fifteen samples with different compositions ($x = 0.13 - 0.16$), thicknesses and levels of doping, were investigated by magnetotransport measurements at low and super-low temperatures.

## 3. Magneto-transport measurements

The magneto-transport measurements were performed using a cryo-magnet system from ICEoxford, which is capable of voltage measurements in the temperature region from 0.25 to 290 K in magnetic fields up to 14 T. The direction of the magnetic field $B$ was perpendicular to the plane of the

Table 1. Parameters of the samples grown by MBE

| Nr | Composition $x$, mol | Substrate | Thickness, nm | Level of the iodine doping $10^{17}$cm$^{-3}$ | Level of the indium doping $10^{17}$cm$^{-3}$ |
|---|---|---|---|---|---|
| A4 | 0.155 | GaAs/CdTe(001) | 100 | | 0.5 |
| A9 | 0.135 | GaAs/CdTe(001) | 100 | 5.0 | |
| AB4 | 0.155 | CdZnTe(112)B | 100 | | 0.5 |
| AB9 | 0.135 | CdZnTe(112)B | 100 | 5.0 | |
| B4 | 0.155 | CdZnTe(112)B | 840 | | 0.5 |
| B5 | 0.135 | CdZnTe(112)B | 2480 | | 0.8 |
| B6 | 0.150 | CdZnTe(112)B | 1280 | | 3.0 |
| B8 | 0.145 | CdZnTe(112)B | 950 | 0.5 | |
| B9 | 0.135 | CdZnTe(112)B | 1100 | 5.0 | |

investigated layer in the samples. Four-terminal longitudinal ($R_{xx}$) and Hall ($R_{xy}$) resistances were measured with standard lock-in techniques at a low-frequency ($< 20$ Hz) and with an excitation current of 0.5 – 1.0 $\mu$A. Four measurements were made, i.e. for the two directions of the magnetic field and for increasing and decreasing magnetic fields, with the subsequent averaging of each measurement of the longitudinal magneto-resistance and Hall resistance at a given temperature.

The results of the magnetotransport measurements obtained for sample A9 – strained thin layer on the GaAs/CdTe substrate – are presented in Fig. 1. The $R_{xx}(B)$ and $R_{xy}(B)$ curves are shown for different temperatures over wide range from 0.4 K to 50 K. The well-defined quantized plateaus in $R_{xy}$ with values $h/(2e^2) = 12.9$ k$\Omega$, accompanied by vanishing $R_{xy}$ is observed at 0.4 K.

It is seen pronounced plateaus in $R_{xy}$ at values equal to about 6.5 k$\Omega$ as well as one less clear at about 4.3 k$\Omega$ The $R_{xx}(B)$ curves exhibit pronounced SdH-oscillations which positions of maxima strongly corresponds to stairs before plateaus on the $R_{xy}(B)$ curve. Described particularities on the $R_{xx}(B)$ and $R_{xy}(B)$ curves of sample A9 explicitly indicate on the Integer Quantum Hall Effect (IQHE) and Shubnikov-de Haas (SdH) oscillations characteristic for 2D electron gas. The quantization in integer multiples of $\sigma_0 = e^2/h$ is evident with the Landau filling factor $\nu$ equals to 2, 4 and 6.

It is necessary to underline that the $R_{xx}(B)$ and $R_{xy}(B)$ curves are reproducible up to 20 K and above this temperature the Integer Quantum Hall Conductivity (IQHC) is observed up to 50 K.

The same magneto-transport measurements are repeated for sample AB9 – no strained thin layer on the CdZnTe(112)B substrate. The remarkable temperature stability of the $R_{xx}(B)$ and $R_{xy}(B)$ curves characteristic for sample A9 is repeated for sample AB9 (see Fig. 2a).

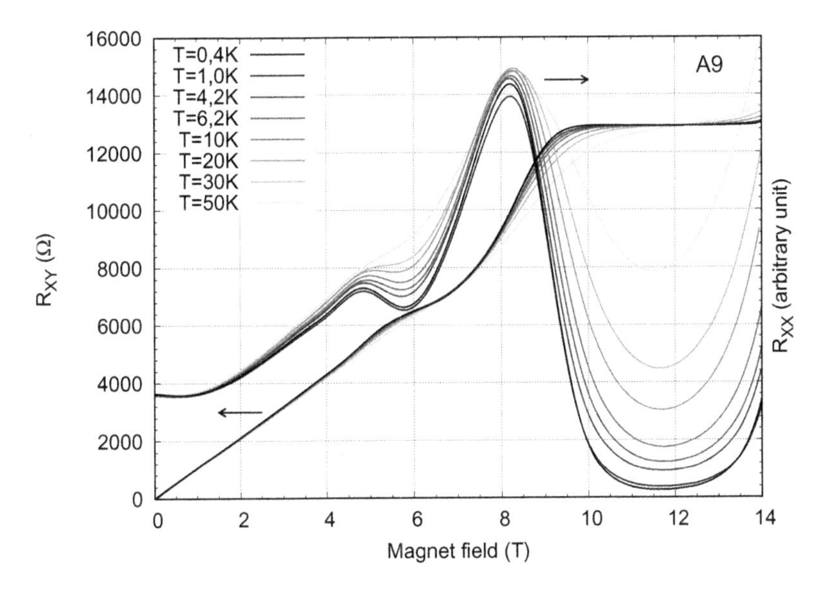


Fig. 1. Magneto-resistances, $R_{xx}$ and $R_{xy}$ vs magnetic field in the temperature region of 0.4 – 50 K for sample A9. Three plateaus are seen on the $R_{xy}$ curves; the resistances in these plateaus in the temperature region of 0.4 – 20 K is equal to values $h/\nu e^2$ when $\nu = 2$,4 and 6.

Similarly, one can see the plateau-like features of the $R_{xy}$ curve and corresponding SdH maxima on the $R_{xx}(B)$ curve, but the quantization in integer multiples of $\sigma_0 = e^2/h$ is not evident. The latter can be attributed to the availability of a parallel conductance channel from the sample interior that decreases the values of the $R_{xy}$ resistance in the plateaus.[34,35] In order to more quantitatively estimate the surface contribution to the total conductance, we fit our data to a simple model used in Refs. 35 and 36, where the total conductance is the parallel sum of the bulk conductance of $Hg_{1-x}Cd_xTe$ ($x = 0.13$) and the 2D surface conductance.

According to this scheme, the values of $R_{xy}$ for the plateaus of QHE decrease proportionally due to the contribution of the bulk part of the sample. The calculation of the $R_{xy}$ resistance (for the whole sample) assumed that the QHE values $R_{xx} = h/\nu e^2$ of the voltage are generated in the surface sheet, and the classic Hall Effect occurs in the bulk part of the sample. The results of calculation and experimental data obtained for the sample AB9 are compared in Fig. 2b. It can be seen that the results of simulation approximate very well the observed plateaus in the case of odd values of the filling factor: $\nu = 3$, 5, 7. Similar experimental data take place for sample

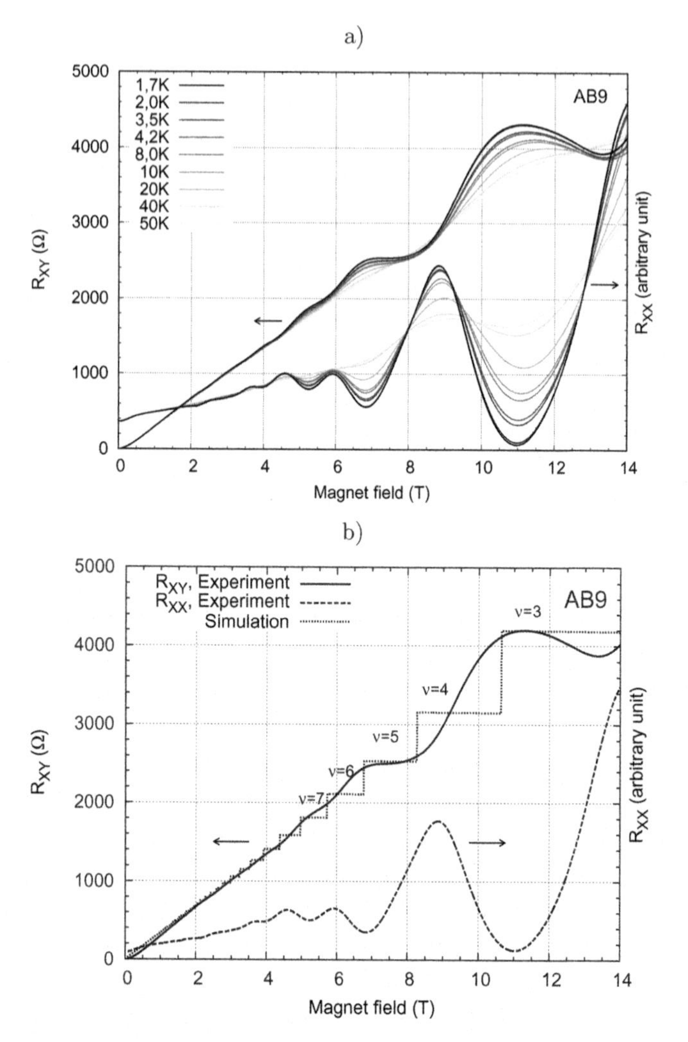


Fig. 2. *a*) Magneto-resistances, $R_{xx}$ and $R_{xy}$ vs magnetic field in the temperature region of 0.4 – 50 K for sample AB9; *b*) Calculations (dotted line) of the $R_{xy}$ resistance for the entire sample AB9 at 1.7 K performed according to a scheme of taking the parallel sum of the QHE values of the voltage generated in the surface sheet and a classic Hall voltage in the bulk part of sample AB9. It can be seen that the odd values of filling factor $\nu = 3$, 5, 7 and 9 correspond to experimentally observed plateaus at temperature 1.7 K.

AB4 also thin layer of semimetal $Hg_{1-x}Cd_xTe$ ($x = 0.155$) grown on the CdZnTe(112)B substrate.

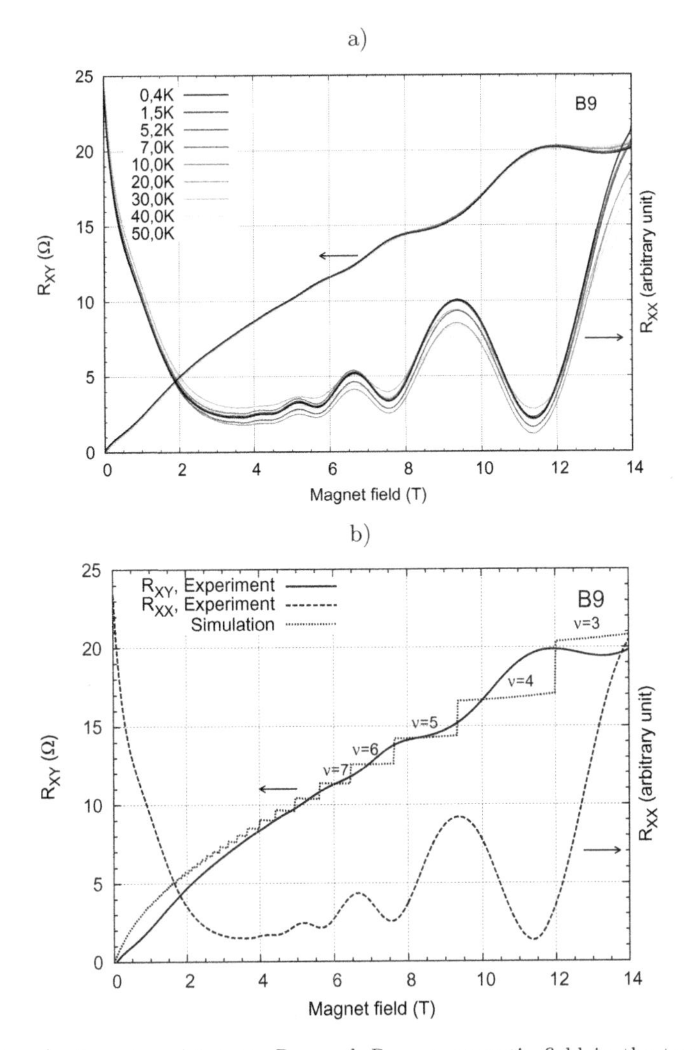


Fig. 3. *a*) Magneto-resistances, $R_{xx}$ and $R_{xy}$ vs magnetic field in the temperature region of 0.4 – 50 K for sample B9; *b*) Calculations (dotted line) of the $R_{xy}$ resistance for the entire sample B9 at 4.2 K performed according to the same scheme as for sample AB9. It can be seen that the odd values of filling factor $\nu$ = 3, 5 and 7 correspond to experimentally observed plateaus.

Figs. 3-4 show the measurements for four samples from those listed in Table 1: B4, B5, B6 and B9 which are thick not strained layers. The $R_{xx}(B)$ and $R_{xy}(B)$ curves recorded for sample B9, are shown in Fig. 3a for different temperatures over a wide range from 0.4 to 50 K. It is noteworthy

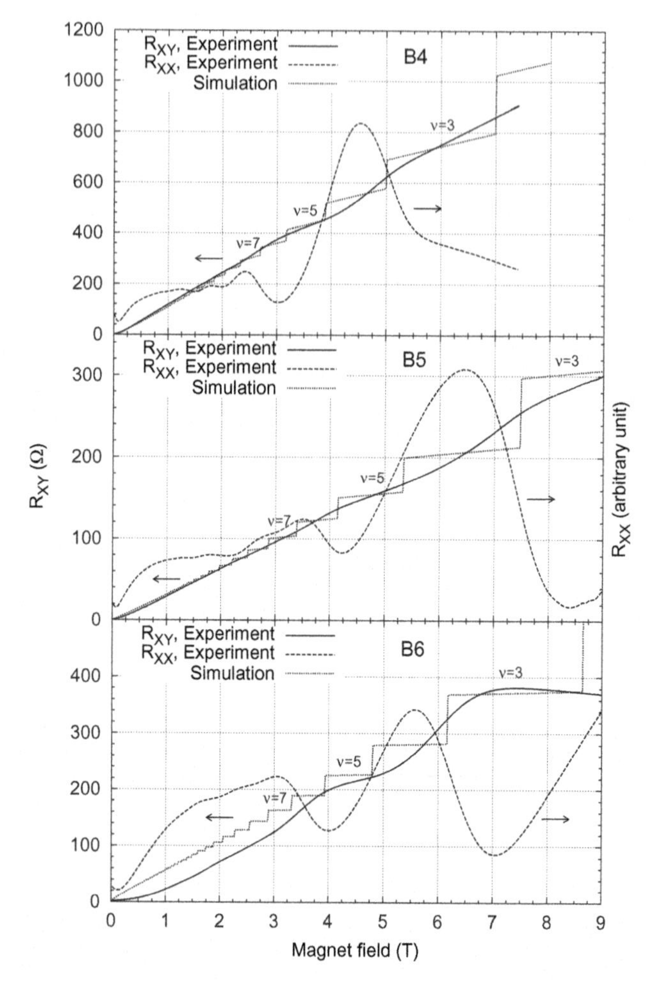


Fig. 4. Magneto-resistances, $R_{xx}$ and $R_{xy}$ vs magnetic field in the temperature region of 0.4 – 2.3 K for sample B4, B5 and B6.

that the results are reproducible at different temperatures up to 50 K that seems unexpected for this thick sample (about 1 $\mu$m, see Table 1). Temperature increase beyond 50 K makes the plateaus less pronounced. The amazing temperature stability also concerns the positions of the observed SdH-oscillation maxima. These positions are reproducible up to 45 K and the amplitude of the SdH-maxima decreases slowly with increasing temperature above 50 K. The plateaus like features in $R_{xy}$ in Figs. 3a and 4 are reminiscent of the IQHE but in the present case the values of the $R_{xy}$ resistance in the plateaus for sample B9 are low, of the order 10 - 20 Ω. From other hand, in case of sample B9 the ratio $R_{xy}^{III}(B)/R_{xy}^{V}(B) = 20.0\Omega/13.0\Omega = 1/3 : 1/5$ and $R_{xy}^{V}(B)/R_{xy}^{VII}(B) = 13.0\Omega/11.0\Omega = 1/5 : 1/7$ where $R_{xy}^{III}$ the is the value of resistance in the plateaus at 11.0 – 14.0 T, $R_{xy}^{V}(B)$ is the value of the resistance in the plateaus at 7.0 – 8.5 T and $R_{xy}^{VII}(B)$ is the resistance at 5.5 – 6.5 T. Thus, these values of the resistance correspond to the magnitude for corresponding values of the filling factor of $\nu = 3$, 5 and 7, but they are smaller due to the parallel resistance in the sample. Similarly to the sample AB9, the calculation of the $R_{xy}$ resistance (for the whole sample) assumed that the QHE values $R_{xy} = h/\nu e^2$ ($\nu$ is an integer) of the voltage are generated in the surface sheet, and the classic Hall Effect occurs in the bulk part of the sample. The results of the calculation are compared with the experimental curve for sample B9 in Fig. 3b. It can be seen that the results of simulation approximate satisfactorily the observed plateaus in the case of odd values of the filling factor: $\nu = 3, 5, 7$.

Similar curves were obtained for all samples of the B series. The results of the magneto-transport measurements obtained for samples B4, B5 and B6 are shown in Fig. 4 at approximately the same temperature (within an interval of 0.42 – 2.25 K). They exhibit the QHE behavior similar to that observed for the sample B9, even though the plateaus values of the resistance are different due to their dependence on the resistance of the bulk part of the samples.

Moreover, the proportions between values of the $R_{xy}$ resistance corresponding to odd values of the $\nu$ factor in different plateaus are retained. For sample B5 the plateaus in the $R_{xy}(B)$ curve are less distinct. The results of calculating the $R_{xy}(B)$ resistance obtained in a similar way as for the samples AB9 and B9, are shown in Fig. 4 also. The stability of the $R_{xy}$ and $R_{xx}$ curves for samples B4 and B6 in the temperature interval of 0.4 – 45 K is similar to that of the sample B9. Other temperature behaviour is observed for sample B5: the plateaus and SdH oscillations disappeared after reaching 30 K.

It is interesting to note that in the case of 2DEG in the HgCdTe /HgTe/ HgCdTe quantum well the resistance at the charge neutrality point was found to be temperature independent at low temperatures.[37]

## 4. Interpretation: Electron transport on topologically protected surface states

### 4.1. *Model of two Dirac cones*

The model proposed for bulk HgTe[10] seems the most appropriate tool to explain the above results of magnetotransport measurements made for the $Hg_{1-x}Cd_xTe$ strained layer A9 (similar ones are for A4). The TPSS exist at the border of the sample surface layer with air as well as at the border with the substrate whose 2D-layers with Dirac fermions contribute to the electron transport of the whole sample with the energy gap in the interior. Based on the studies made by cited above authors[11,30,34] and considering earlier works[38,39] we can understand the observed IQHC as the sum of a half integer QHC from the top surface and another half integer QHC from the bottom surface. In other words, the IQHC is quantized according to

$$\sigma_{xy}^{total} = \nu\frac{e^2}{h} = \sigma_{xy}^{top} + \sigma_{xy}^{bottom} = (\nu_t + \nu_b)\frac{e^2}{h} = (N_t + N_b + 1)\frac{e^2}{h} \qquad (1)$$

with top (bottom) surface QHC

$$\sigma_{xy}^{top(bottom)} = \nu_{t(b)}\frac{e^2}{h} = \left(N_{t(b)} + \frac{1}{2}\right)\frac{e^2}{h}, \qquad (2)$$

where $\nu_{t(b)}$ and $N_{t(b)}$ are the Landau filling factor and Landau level index of top (bottom) surface corresponding to the QH state.

When top and the bottom surface have the same filling factor, and both surfaces have the same density, i.e. $N_t = N_b$ then the resulting expression for the QHC, according to (1), is as follows

$$\sigma_{xy} = (2N + 1)\frac{e^2}{h}. \qquad (3)$$

Hence, the filling factor $\nu$ can only have odd values.

In case of asymmetric conductivity along top and the bottom surface the expression (1) should be applied and the filling factor $\nu$ can be even or half-integer value if the conductivity along the bottom surface disappeared.

### 4.2. *Strained thin layers: A series*

The strong TI exists in the case of the sample A9: the energy gap of about 10 meV is formed in the sample interior due to the tension, and the TPSS are on surfaces due to crossing of the $\Gamma_8$ and $\Gamma_6$ bands, similarly to HgTe strained layer.[11] On the other hand, we should assume that the conditions of electron transport are different on the top and bottom surfaces because the first one is practically ideal (according to the AFM picture), but the bottom surface could be in a different situation at the interface with the CdTe buffer layer due to mismatch. It means that the asymmetric conductivity along the top and bottom surfaces could be produced by latter fact, namely – different density of the fermions.

As was shown in section 2 the $R_{xy}(B)$ curves demonstrate the QHE with the Landau filling factor $\nu$ equals to 2, 4 and 6 in the case of sample A9. So, the filling factor is even, and, according to the expression (1) the next the situation may be realized, for example: in case of $\nu = 2 - N_t = 0$ and $N_b = 1$, for $\nu = 4 - N_t = 1$ and $N_b = 2$ and for $\nu = 6 - N_t = 2$ and $N_b = 3$. The theoretical analyses of this results will be performed in section 5.

### 4.3. *No strained thin layers: AB series*

The sample AB9 was obtained by MBE growth on the CdZnTe(112)B substrate with practically ideal match. Thus, there are no strains, and there is no origin of the energy gap in the sample interior. On the other hand, it is difficult to refute that in semimetallic $Hg_{1-x}Cd_xTe$ ($x < 0.16$), between the nominally designated conduction and valence bands, a small gap can occur due to various reasons (finally, this gap may be generated by a magnetic field). So, it possible to assume that TPSS can exists in case of sample AB9 and AB4 on the background of a small energy gap or on the background of the heavy hole states[40,41] (without gap between of the $\Gamma_8^{1/2}$ and $\Gamma_8^{3/2}$ states, which they could be named as some resonance interface state, see more discussions below). This makes possible the same density of the Dirac fermions on top and bottom surfaces. In this case the IQHC according Eq. (3) could be realized with odd values of the filling factor $\nu$ but the parallel classic Hall effect takes place in the interior of the sample without an energy gap. This assuming is confirmed by the Fig. 2b where the simulation curve approximate very well the observed plateaus on the $R_{xy}$ experimental curve for sample AB9 and odd values of the filling factor: $\nu = 3, 5, 7$. High temperature stability of observed plateaus visible on the $R_{xy}$ curves and of the SdH oscillations on the $R_{xx}$ curves can be explained by this assuming the electron transport on the TPSS (or on the resonance interface state[40,41] (RIS)) dominates the electron transport in the whole sample AB9. This assuming concerns the sample AB4 also.

If the model of the electron transport applied to the sample AB9 is adequate, it becomes important to determine the thickness of a layer in which the Dirac fermions transport dominates. The experimental data obtained for the samples of B series could give an answer to this question.

### 4.4. *No strained thick layers: B series*

It is appropriate to start with the sample B9 which can be considered as a thicker version (1100 nm thick) of the sample AB9. As was shown in section 2, for the sample B9, the plateaus at 11.0 – 14.0 T corresponds to $\nu = 3$ and next one at 7.0 – 8.5 T – to $\nu = 5$, and another one at 5.5 – 6.5 T – to $\nu = 7$ (see Fig. 3b). Thus, the odd integer IQHC is observed as it has been described for the sample AB9, but the plateaus resistances are considerably less because of the substantial contribution from the classical Hall Effect in the sample bulk. High temperature stability of plateaus on the $R_{xy}$ curves as well as of the SdH oscillations on the $R_{xx}$ curves shown in Fig. 3a in wide temperature range, allows advancing the hypothesis concerning a major role which the electron transport on the RIS (or on the TPSS) plays in the electron transport within the whole sample B9 as well.

As to the samples B4 and B6, it should be noted that experimental data confirms the hypothesis given above. The only exception is the sample B5 with the thickness of 2840 nm (Table 1).

### 4.5. *Landau index as function on 1/B*

The above mentioned interpretation of the experimental results concerning the Landau level structure of a Dirac system can be confirmed by plotting the Landau level index as a function of $1/B$.[11,34,39] That concerns mostly the samples of the series AB and B. Fig. 5 is created through taking the magnetic field values corresponding to the Hall plateaus with $\nu = 9$, 7, 5, 3 from the $R_{xy}$ curve in Figs. 2b, 3b and 4 to plot the resulting $N$ as a function of $1/B$ according to Eq. (3). The intercept of these plots for infinite magnetic field gives in case of samples AB9, B9, B4 and B6 a value of $-1/2$, that provides additional evidence for describing the observed IQHC by the two Dirac cones model. For the sample B5 the intercept gives a 0 value that means the peculiarities visible on the $R_{xx}$ curves can be attributed rather to the bulk part of the sample.

It is an unexpected result that the 2D-TPSS conductance (in case of thick samples B9, B4 and B6 we prefer to call: the electron transport on the RSI) contributes to the total conductance in these slab-shaped samples with parallel top and bottom surfaces which surround (together with the side walls) a thick (about 1 $\mu$m) semimetal bulk component (topologically it is the same as a sphere). Surprisingly the Dirac point in the surface layer can exist in conjunction with the (heavy hole) band in the bulk part of the sample (see inset in Fig. 5) without an energy gap (in case of samples of series AB and B) which usually takes place in TI. As was shown in Ref. 40, such TPSS states, designated as interface states, can couple to a heavy hole state and thus could be modified. In the work of Jie Ren et al.[41] the topological phase transition from bulk CdTe to HgTe upon alloying and the massless Dirac-Kane semimetal phase at the critical composition ($x < 0.16$) are illustrated by computations based on a mixed-pseudopotential simulation confirmed by the ARPES experiment: a topological surface state (TSS) band connecting from the $\Gamma_6$ band to the upper $\Gamma_8$ band (above the Fermi level) takes place. An alternative interpretation of above presented results on the AB and B series samples could be the possibility of a surface layer at interface layer-air produced due to increase of the mercury content at surface of the layer. That could forms a narrow quantum well with 2DEG at interface what could generate observed IQHC. Probability of that is negligible due to whatever any increase in the Hg content at surface was registered by SIMS in samples investigated. On the other hand, if such quantum well could exists at surface, the QHE would show all integer values of the Landau filling factor not only odd values and the intercept of the resulting $N$ plotting as function of the $1/B$ would give value 0 not $-1/2$ as it takes place for samples AB9, B9, B4 and B6 (see Fig. 5).

Another interesting feature is that the observed IQHC is more pronounced in samples B6 and B9 which have higher electron densities. Specific screening properties of Dirac systems[42–44] can originate this phenomenon.

## 5. Calculations according to the kp model and the Dirac Hamiltonian

The experimental results on magneto-transport (QHC and SdH) obtained for the strained 100 nm thickness $Hg_{1-x}Cd_xTe$ layer has been interpreted on the basis of the $8 \times 8$ **kp** model.[45] The **kp** model used an eight band description of the band structure including all second-order terms representing the remote-band contributions with the first-order terms attributed to the $\Gamma_8^{3/2}$ – heavy hole, $\Gamma_8^{1/2}$ – light hole bands, and the $\Gamma_6^{1/2}$ – conduction band as well $\Gamma_7^{1/2}$ – cleaved by the spin-orbital interaction. The available band-structure parameters used in the eight band description, are presented in Table 2. The detailed descriptions of the calculation and the results for the mixed $Hg_{1-x}Cd_xTe$ 3D TI for the different thickness of the layers and

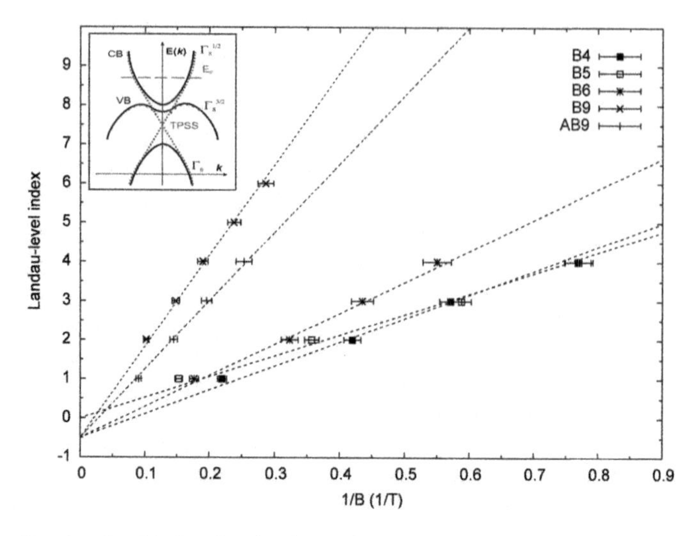


Fig. 5. Landau level index for the data of Fig. 2b, 3b and 4 plotted as a function of inverse magnetic field. The intercept of this plot for infinite magnetic field gives a value of $-1/2$ for samples AB9, B9, B4 and B6, which provides evidence that the observed IQHC can be well described by the two Dirac cones model. In the case of the B5 sample a spin splitted maxima of the SdH-oscillations are visible and the positions corresponded with TPSS is shown.

different tensile strain are presented in Ref. 45. The calculations of the energy spectrum are provided in the framework of the envelope function approach where $z$-axis coincides with the growth direction of the 3D system. The finite difference method with the common central difference form is particularly employed in the discretization procedure. The results of magneto-transport measurements are compared with the calculation of the Landau Level (LL's) energy obtained for A9 sample which was grown on GaAs/CdTe substrate (see Table 1). The LL energy are presented in Fig. 6a. The even filling factor $\nu$ values of 2, 4 and 6 means that two independent surfaces should be taken into account in the fermion transport – on top and bottom of sample. It is interesting to note that the same calculations performed for pure HgTe ($x = 0$) lead to the Fermi level velocity $v_f = 0.43 \cdot 10^6$ m/s what confirms the result obtained by authors of Ref. 11 ($hv_f = 280\text{meV} \cdot \text{nm}$). Using the relation $E_f = hk_fv_f$, it is possible to obtain from the Fig. 6 the Fermi velocity for the electron at the TPSS. For the HgCdTe strained layers the velocity of the charges for the TPSS located on the right side of the structures - at the boundary with the vacuum (see Fig. 6c) - is two times higher than for the pure HgTe, and is equal $v_f = 0.9 \cdot 10^6$ m/s.

Table 2. Band-structure parameters of the $8 \times 8$ kp-model of the CdTe and HgTe applied proportionally to composition x to $Hg_{1-x}Cd_xTe$

| | HgTe | CdTe | | HgTe | CdTe |
|---|---|---|---|---|---|
| $E_g$ | -0.303 eV | 1.606 eV | $C$ | -3.83 | -4.06 |
| $E_v$ | 0 | -570 meV | $a$ | 0 | -0.17 |
| $\Delta$ | 1.08 eV | 0.91 eV | $b$ | -1.5 eV | -1.17 |
| $E_P$ | 18.8 eV | 18.8 eV | $d$ | -2.08 eV | -3.2 |
| $F$ | 0 | -0.09 | $C_{11}$ | 53.6 GPa | 53.6 |
| $\gamma_1$ | 4.1 | 1.47 | $C_{12}$ | 36.6 GPa | -37.0 |
| $\gamma_2$ | 0.5 | -0.28 | $C_{44}$ | 21.2 | -19.9 |
| $\gamma_3$ | 1.3 | 0.03 | | | |
| $k$ | -0.4 | -1.31 | | | |

Such velocity allow to observe the quantized Hall conductance on TPSS even without external gate voltage and on the background of the bulk states. It is interesting to remark that the TPSS are visible in magneto-transport for the wide range of the gate voltage applied into the pure HgTe 3D 70-nm-thick sample due to their remarkable screening properties according to Ref. 44. This hypotheses could be applied to our $Hg_{0.865}Cd_{0.135}Te$ sample also because high electron density. Besides, the high temperature stability of observed TPSS was shown in Ref. 2. It is shown in the Fig. 6 that the LL's fan calculated for TPSS for studied HgCdTe alloy using the $8 \times 8$ **kp** model corresponds to the one calculated by graphene Hamiltonian and characterized by $v_f \approx 0.95 \cdot 10^6$ m/s what is in excellent agreement with experimental IQHE- and SdH-curves. On the other hand, this value of the Fermi velocity is two times greater as for the strained HgTe TI[11] and for others kind of TI – approximately $5 \cdot 10^5$ ms$^{-1}$, for $Bi_2Se_3$[20], or $3.4 \cdot 10^5$ ms$^{-1}$ for the same compound according to Ref. 21.

As it is shown in Fig. 6c and d to get the filling factor $\nu = 2$ it is necessary to have the Landau index $N_b = 1$ (bottom surface) and $N_t = 2$ for top surface. The LL's presented on Fig. 6 were calculated from above mentioned eight band kp model but the experimental data can also be fitted by the LL's fan energy obtained from graphene-like Hamiltonian. In order to obtain a good agreement with the graphene-like LL's energies and experimental curves of $R_{xy}$ and $R_{xx}$ the difference between the positions of the Dirac points for two surfaces should by equal about 40 meV. It is necessary to underline that the LL's fan calculated by the $8 \times 8$ **kp** model with band-structure parameters presented in Table 2, are in excellent agreement with that obtained using the graphene Hamiltonian as it is presented in Fig. 7a but then the Fermi level velocity is equal to about $0.95 \cdot 10^6$ m/s.

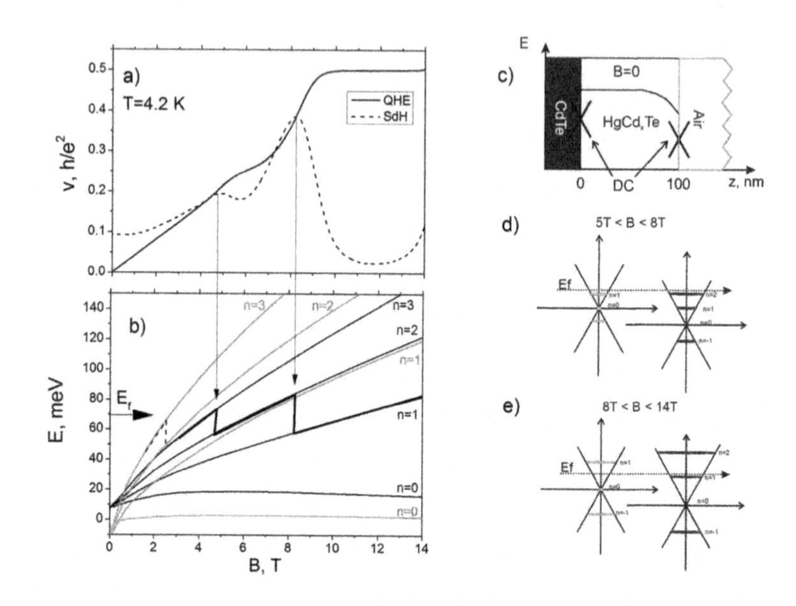


Fig. 6. *a*) The SdH oscillations and the IQHE for 4.2 K for 100 nm wide $Hg_{0.865}Cd_{0.135}Te$ and *b*) the LL's fan calculated for two Dirac cones for two interfaces; *c*) and *d*) represent the LL's under the magnetic field region between 5 and 8 T and 8 and 14 T, respectively.

So, the strained 3D $Hg_{0.865}Cd_{0.135}Te$ is Topological Insulator with high value of the Fermi velocity (approximately the same as for graphene) what means significant advantages for future applications. Besides, the more high value of the energy gap inside sample as in case of strained HgTe, as well as more high position of the Dirac points on the energy scale (all this advantages follow from significantly lower value of the electron momentum on the Fermi level in comparison with pure HgTe: the energy dispersion is closer to linearity in the wider range of the momentum) lead to an increase in the attractiveness of the Topological Insulator based on semimetal HgCdTe alloy for future applications: as massless Weyl fermions for example with addition of non-compensated spins of Mn.[46]

## 6. Conclusions

Presented above review of results concerning the semi-metal MCT enable us to formulate some new properties of this class of material.

First, a quantized Hall conductance in the 3D $Hg_{1-x}Cd_xTe$ samples ($x < 0.16$) with thickness from 100 to 1280 nm and with different levels of n type doping, using either iodine or indium as dopants, was observed[2] over a wide temperature region from 0.4 to 50 K.

Second, the experimental results lead to the hypothesis that quantum Hall conductance on the 2D-TPSS contributes to conductance of entire samples of semi-metallic MCT what explains amazing temperature stability of the electron transport in semimetal HgCdTe.

Third, a simple model with two Dirac cones satisfactorily explains the most salient features of the transport measurement of MBE grown and comparatively thick samples (up to 1280 nm) of semimetal $Hg_{1-x}Cd_xTe$. In another words, the conductance due to the TPSS at the interfaces of the approximately 3 nm thick lower layers with the semimetal bulk part of the investigated sample with thickness from 100 to 1280 nm, can dominates the conductance of the entire sample (see Fig. 6). That are a new conceptions[2] of the electron transport in the semimetal bulk $Hg_{1-x}Cd_xTe$ that could require a reinterpretation of previous experimental results on semimetal $Hg_{1-x}Cd_xTe$ layers (with thickness less as 1300 nm) which have been obtained over the last 40 years.[29–31]

Forth, in comparison with pure HgTe the energy dispersion of semimetallic HgCdTe is closer to linearity in the wider range of the momentum[45] what lead to an increase in the attractiveness of the Topological Insulator based on semimetal HgCdTe alloy for future applications.


## Acknowledgment

We acknowledge support from the Marshals Office of the Podkarpackie Voivodship of Poland, contract WNDPPK.01.03.00-18-053/12.


## References


1. H. C. Manoharan, A romance with many dimentions, *Nature Nanotechnology* **5**, 477 (2010).
2. Shun-Qing Shen, *Topological Insulators. Dirac Equation in Condensed Matter.* Springer Series in Solid State Science, Vol. 174. (Springer, Berlin, 2013).

3. C. L Kane, E. J. Mele, Quantum Spin Hall Effect in Graphene, *Phys. Rev. Lett.* **95**, 226801 (2005).
4. B. A. Bernevig and S.-C. Zhang, Quantum Spin Hall Effect, *Phys. Rev. Lett.* **96**, 106802 (2006).
5. B. A. Bernevig, T. L. Hughes, and S.-C. Zhang, Quantum spin Hall effect and topological phase transition in HgTe quantum wells. *Science* **314**, 1757 (2006).
6. M. König, S. Wiedmann, C. Brüne, A. Roth, H. Buhmann, L. W. Molenkamp, Xiao-Lian Qiand and S.-C. Zhang, Quantum Spin Hall Insulator State in HgTe Quantum Wells, *Science* **318**, 766 (2007).
7. Liang Fu, C. L. Kane, and E. J. Mele, Topological Insulators in Three Dimensions, *Phys. Rev. Lett.* **98**, 106803 (2007).
8. Liang Fu and C. L. Kane, Topological insulators with inversion symmetry, *Phys. Rev. B* **76**, 045302 (2007).
9. D. Hsieh, D. Qian, L. Wray, Y. Xia, Y. S. Hor, R. J. Cava, M. Z. Hasan, A topological Dirac insulator in a quantum spin Hall phase, *Nature* **452**, 970-974 (2008).
10. C. Brüne, C. X. Liu, E. G. Novik, E. M. Hankiewicz, H. Buhmann, Y. L. Chen, X. L. Qi, Z. X. Shen, S.-C. Zhang, and L. W. Molenkamp, Quantum Hall Effect from the Topological Surface States of Strained Bulk HgTe, *Phys. Rev. Lett.* **106**, 126803 (2011).
11. G. Tomaka, J. Grendysa, P. Śliż, C. R. Becker, J. Polit, R. Wojnarowska, A. Stadler, E. M. Sheregii, High-temperature stability of electron transport in semiconductors with strong spin-orbital interaction, *Phys. Rev. B* **93**, 205419 (2016).
12. A. Delin and T. Kluner, Excitation spectra and ground-state properties from density-functional theory for the inverted band-structure systems b-HgS, HgSe, and HgTe, *Phys.Rev. B* **66**, 035117 (2002).
13. M. Z. Hasan and C. L. Kane, Topological Insulators, *Rev. Mod. Phys.* **82**, 3045 (2010).
14. Xiao-Liang Qi and Shou-Cheng Zhang, Topological insulators and superconductors, *Rev. Mod. Phys.* **83**, 1057 (2011).
15. M. Berry, Geometric phase memories, *Nature Physics* **6**, 148 (2010).
16. J. E. Moore and L. Balents, Topological invariants of time-reversal-invariant band structures, *Phys. Rev. B* **75**, 121306 R (2007).
17. Jeffrey C. Y. Teo, Liang Fu, and C. L. Kane, Surface states and topological invariants in three-dimensional topological insulators: Application to $Bi_{1-x}Sb_x$, *Phys Rev. B* **78**, 045426 (2008).
18. Liang Fu and C. L. Kane, Superconducting Proximity Effect and Ma-

jorana Fermions at the Surface of a Topological Insulator, *Phys. Rev. Lett.* **100**, 096407 (2008).

19. I. Galanakis, P. H. Dederichs and N. Papanikolaou, Slater-Pauling behavior and origin of the half-metallicity of the full-Heusler alloys, *Phys. Rev. B* **66**, 174429 (2002).
20. Wenjie Xie, Anke Weidenkaff, Xinfeng Tang, Qingjie Zhang, Joseph Poon and Terry M. Tritt, Recent Advances in Nanostructured Thermoelectric Half-Heusler Compounds, *Nanomaterials* **2**, 379 (2012).
21. Y. Xia, D. Qian, D. Hsieh, L. Wray, A. Pal, H. Lin, A. Bansil, D. Grauer, Y. S. Hor, R. J. Cava, and M. Z. Hasan, Observation of a large-gap topological-insulator class with a single Dirac cone on the Surface, *Nature Physics* **5**, 398 (2009).
22. H. Zhang, C.-X. Liu, X.-L. Qi, X. Dai, Z. Fang, and S.-C.Zhang, Topological insulators in $Bi_2Se_3$, $Bi_2Te_3$ and $Sb_2Te_3$ with a single Dirac cone on the Surface, *Nature Physics* **5**, 438 (2009).
23. Y. L. Chen, J. G. Analytis, J. H. Chu, Z. K. Liu, S. K. Mo, X. L. Qi, H. J. Zhang, D. H. Lu, X. Dai, Z. Fang, S.-C. Zhang, I. R. Fisher, Z. Hussain, and Z. X. Shen, Experimental realization of a three-dimensional topological insulator $Bi_2Te_3$, *Science* **325**, 178 (2009).
24. D. X. Qu, Y. S. Hor, J. Xiong, R. J. Cava and N. P. Ong, Quantum oscillations and hall anomaly of surface states in the topological insulator $Bi_2Te_3$, *Science* **329**, 821 (2010).
25. Tong Zhang, Jeonghoon Ha, Niv Levy, Young Kuk, and Joseph Stroscio, Electric-Field Tuning of the Surface Band Structure of Topological Insulator $Sb_2Te_3$ Thin Films, *Phys. Rev. Lett.*, **111**, 056803 (2013).
26. Bahadur Singh, Ashutosh Sharma, H. Lin, M. Z. Hasan, R. Prasad and A. Bansil, Topological electronic structure and Weyl semimetal in the $TlBiSe_2$ class of semiconductors, *Phys. Rev. B* **86**, 115208 (2012).
27. T. Sato, Kouji Segawa, K. Kosaka, S. Souma, K. Nakayama, K. Eto, T. Minami, Yoichi Ando and T. Takahashi, Unexpected mass acquisition of Dirac fermions at the quantum phase transition of a topological insulator, *Nature Physics* **7**, 840 (2011).
28. Y. Xu, I. Miotkowski, C. Liu, J. Tian, H. Nam, N. Alidoust, J. Hu, C.-K. Shih, M. Z. Hasan, Y. P. Chen, Observation of topological surface state quantum Hall effect in an intrinsic three-dimensional topological insulator, *Nature Physics* **10** 956 (2014).
29. R. Dornhaus and G. Nimtz, *Narrow Gap Semiconductors*, Springer Tracts in Modern Physics, Vol. 98 (Springer, Berlin, 1983).

30. A. Rogalski, History of infrared detectors, *Opto-Electronics Review*, **20**, 279 (2012).
31. A. Rogalski, Recent progress in infrared detector technologies, *Infrared Physics & Technology* **54**, 136 (2011).
32. D.A. Kozlov, Z.D. Kvon, E.B. Olshanetsky, N.N. Mikhailov, S.A. Dvoretsky, and D. Weiss, Transport Properties of a 3D Topological Insulator based on a Strained High-Mobility HgTe Film, *Phys. Rev. Lett.* **112**, 196801 (2014).
33. A. Kisiel, M. Podgórny, A. Rodzik, W. Giriat, Fundamental Reflection of $Cd_xHg_{1-x}Te$ Crystals in the 1.9 to 3.1 eV Energy Range, *Phys. Stat. Sol. B.* **71**, 457 (1975).
34. J.G. Analytis, R.D. McDonald, S.C. Riggs, J.-H. Chu, G.S. Boebinger, and I. R. Fisher, Two-dimensional surface state in the quantum limit of a topological insulator, *Nature Physics* **6**, 960 (2010).
35. B. F. Gao, P. Gehring, M. Burghard and K. Kern, Gate-controlled linear agnetoresistance in thin $Bi_2Se_3$ sheets, *Appl. Phys. Lett.* **100**, 212402 (2012).
36. J. Olea, G. Gonzalez-Daz, D. Pastor, I. Martil,1 A. Mart, E. Antoln, and A. Luque, Two-layer Hall effect model for intermediate band T-implanted silicon, *J. Appl. Phys.* **109**, 063718 (2011).
37. G. M. Gusev, Z. D. Kvon, E. B. Olshanetsky, A. D. Levin, Y. Krupko, J. C. Portal, N. N. Mikhailov and S. A. Dvoretsky, Temperature dependence of the resistance of a two-dimensional topological insulator in a HgTe quantum well, *Phys. Rev. B* **89**, 125305 (2014).
38. J. S. Novoselov, A. K. Geim, S. V. Morozov, D. Jiang, M.I. Katsnelson, I. V. Grigorieva, S. V. Dubonos, and A. A. Firsov, Two-dimensional gas of massless Dirac fermions in graphene, *Nature* **438**, 197 (2005).
39. Y. Zhang, Y. W. Tan, H. L. Stormer, and P. Kim, Experimental observation of the quantum Hall effect and Berry's phase in graphene, *Nature* **438**, 201 (2005).
40. Yia-Chung Chang, J. N. Schulman, G. Bastard, Y. Guldner and M. Voos, Effects of quasi-interface states in HgTe-CdTe superlattices, *Phys.Rev. B* **31**, 2557 (1985).
41. Jie Ren, Guang Bian, Li Fu, Chang Liu, Tao Wang, Gangqiang Zha, Wanqi Jie, Madhab Neupane, T. Miller, M. Z. Hasan, and T.-C. Chiang, Electronic structure of the quantum spin Hall parent compound CdTe and related topological issues, *Phys. Rev. B* **90**, 205211 (2014).
42. Christoph Brüne, Cornelius Thienel, Michael Stuiber, Jan Böttcher, Hartmut Buhmann, Elena G. Novik, Chao-Xing Liu, Ewelina M. Han-

kiewicz, and Laurens W. Molenkamp, Dirac-Screening Stabilized Surface-State Transport in a Topological Insulator, *Phys. Rev. X* **4**, 041045 (2014).

43. E. H. Hwang and S. Das Sarma, Dielectric function, screening, and plasmons in two-dimensional graphene, *Phys. Rev. B* **75**, 205418 (2007).
44. S. Wiedmann, A. Jost, C. Thienel, C. Brüne, P. Leubner, H. Buhmann, L. W. Molenkamp, J. C. Maan, and Uli Zeitler, Temperature-driven transition from a semiconductor to a topological insulator, *Phys. Rev. B* **91**, 205311 (2015).
45. M. Marchewka, Formation of Dirac point and the topological surface states inside the strained gap for mixed 3D $Hg_{1-x}Cd_xTe$, *Physica E* **84**, 407 (2016).
46. D. Bulmash, Chao-Xing Liu, and Xiao-Liang Qi, Prediction of a Weyl semimetal in $Hg_{1-x-y}Cd_xMn_yTe$, *Phys. Rev. B* **89**, 081106R (2014).

# Effects of Anomalous Velocity in Spin-orbit Coupled Systems

Sh. Mardonov

*The Samarkand Agriculture Institute, 140103 Samarkand, Uzbekistan*
*The Samarkand State University, 140104 Samarkand, Uzbekistan*

M. Modugno

*Department of Theoretical Physics and History of Science*
*University of the Basque Country UPV/EHU, 48080 Bilbao, Spain*
*IKERBASQUE Basque Foundation for Science, Bilbao, Spain*

E. Ya. Sherman

*Department of Physical Chemistry*
*University of the Basque Country UPV/EHU, 48080 Bilbao, Spain*
*IKERBASQUE Basque Foundation for Science, Bilbao, Spain*
*E-mail: evgeny.sherman@ehu.eus*

Spin-orbit coupling for itinerant particles is usually represented in solid state- and condensed matter physics as a symmetry-determined sum of products of operators describing the Cartesian components of particle spin and momentum. By general rules of quantum mechanics, this coupling causes a spin-dependent term in the particle velocity operator, the so-called *anomalous velocity* contribution. Here we present and discuss general concept of this anomalous velocity and analyze two examples of its critical effects in condensed matter physics. As a first example we consider collapse of spin-orbit coupled self-attractive Bose-Einstein condensates. We show that the collapse can be inhibited if the anomalous velocity is taken into account. Second, we analyze short-term spin dynamics of Bose-Einstein condensate in a random one-dimensional potential. Here the anomalous velocity strongly influences the spin evolution by producing random mixed spin states and, therefore, leads to a novel spin relaxation mechanism.

*Keywords*: Spin-orbit coupling, anomalous velocity, Bose-Einstein condensate.

# 1. Introduction: Velocity and spin states with spin-orbit coupling

## 1.1. *Momentum and velocity in quantum mechanics*

Non-relativistic single-particle quantum mechanics [1] begins with the Hamiltonian

$$H = \frac{p^2}{2M} + U(\mathbf{r}) \tag{1}$$

that has the same expression as energy in classical mechanics. Here $p^2/2M$ is the particle kinetic energy, $\mathbf{p}$ is the momentum operator, $M$ is the particle mass, and $U(\mathbf{r})$ is the potential the particle moves in. The particle wavefunction $\psi(\mathbf{r},t)$ satisfies the non stationary Schrödinger equation

$$i\hbar\frac{\partial\psi(\mathbf{r},t)}{\partial t} = H\psi(\mathbf{r},t), \tag{2}$$

with $|\psi(\mathbf{r},t)|^2$ being the probability density to find the particle at time $t$ at point $\mathbf{r}$.

In order to find the expression for the momentum operator in terms of the particle's wavefunction, let us consider $\psi(\mathbf{r},t)$ and its variation under a small change in coordinate $\delta\mathbf{r}$ such that

$$\psi(\mathbf{r}+\delta\mathbf{r},t) \equiv \mathcal{T}(\delta\mathbf{r})\,\psi(\mathbf{r},t) = \psi(\mathbf{r},t) + (\delta\mathbf{r}\boldsymbol{\nabla})\,\psi(\mathbf{r},t), \tag{3}$$

where $\boldsymbol{\nabla} \equiv \partial/\partial\mathbf{r}$. In this way we obtain operator of infinitesimal translation $\mathcal{T}(\delta\mathbf{r}) = 1 + (\delta\mathbf{r}\boldsymbol{\nabla})$ expressed using the scalar product $(\delta\mathbf{r}\boldsymbol{\nabla})$. If the system is translationally invariant, its Hamiltonian does not change under such a translation, meaning that operator $\boldsymbol{\nabla}$ corresponds to a conserved quantity, in our case, the momentum. The proportionality coefficient between this operator and momentum can be found from the correspondence between classical and quantum mechanics in the limit of small Planck constant $\hbar$. Let us take the semiclassical wavefunction [1]

$$\psi(\mathbf{r},t) = C\exp\left(i\frac{S}{\hbar}\right), \tag{4}$$

where $C$ is a constant and $S$ is the classical mechanics action defined as:

$$S = \int dt\left(M\frac{v^2}{2} - U(\mathbf{r})\right), \tag{5}$$

and for applying the classical-quantum correspondence we assume $S \gg \hbar$. Then we obtain for $\psi(\mathbf{r},t)$ in Eq.(4):

$$\boldsymbol{\nabla}\psi(\mathbf{r},t) = \frac{i}{\hbar}\psi(\mathbf{r},t)\,\boldsymbol{\nabla}S, \tag{6}$$

and by taking into account that in this limit[1] $\mathbf{p}\psi(\mathbf{r},t) = \psi(\mathbf{r},t)\nabla S$, we arrive at the definition $\mathbf{p} \equiv -i\hbar\nabla$. The eigenfunctions of the momentum operator are plane waves $\psi(\mathbf{r}) = A\exp(i\mathbf{pr}/\hbar)$. As a result, the momentum operator defines how the wavefunction is transformed under spatial translations:

$$\mathcal{T}(\mathbf{a})\,\psi(\mathbf{r},t) = e^{i\mathbf{pa}/\hbar}\psi(\mathbf{r},t)\,. \tag{7}$$

To obtain the velocity operator, that is the time derivative of the particle position, we calculate the evolution of the wavefunction with the non stationary Schrödinger equation such that

$$\psi(\mathbf{r},t+\delta t) = \exp(-iH\delta t/\hbar)\,\psi(\mathbf{r},t) \tag{8}$$

and the change in the position is

$$\langle\mathbf{r}(t+\delta t)\rangle - \langle\mathbf{r}(t)\rangle = \langle\psi(\mathbf{r},t+\delta t)\,|\mathbf{r}|\,\psi(\mathbf{r},t+\delta t)\rangle - \langle\psi(\mathbf{r},t)|\,\mathbf{r}\,|\psi(\mathbf{r},t)\rangle\,. \tag{9}$$

For small $\delta t$ we expand $\exp(-iH\delta t/\hbar) = 1 - iH\delta t/\hbar$ and from Eq.(9) obtain the operator of the time derivative of the displacement in the commutator form:

$$\mathbf{v} = \frac{i}{\hbar}\,[H,\mathbf{r}]\,. \tag{10}$$

As expected, for the Hamiltonian $H$ in Eq.(1) we obtain $\mathbf{p} = M\mathbf{v}$, corresponding to the Galilean invariance of classical mechanics, where momentum conservation occurs in all the inertial frame systems. However, this result as well as the presence of the Galilean invariance is a Hamiltonian-specific property and usually does not hold for other systems. Note that stationary localized eigenstates of the Hamiltonian $H$ have zero expectation value of velocity as they would be propagating otherwise.

The difference between the momentum and the velocity is crucial for many aspects of particle's dynamics in free space and in external electromagnetic fields.[2] Below we consider two systems where this difference, attributed to spin-orbit coupling (SOC), is very important and leads to qualitatively new physics.

### 1.2. *Spin-orbit coupling Hamiltonians: Anomalous velocity*

Spin-orbit coupling, important for semiconductors and cold atomic matter,[3] leads to a rich variety of new physical phenomena. In solids one deals with real spin 1/2 particles and the relativistic effects of electron motion in the crystal field causing this coupling.[4] In cold atomic gases the situation is different. The atoms there can be bosons (e.g. $^{87}$Rb or $^{7}$Li), that is can have zero (or an integer) total spin built by spins of nuclei and electrons. However, if these atoms are located in highly coherent resonant optical fields, the optical fields couple different atomic states. This coupling can be reduced to an effective two-level system, which is equivalent to a *pseudospin* 1/2. For moving atoms, the Doppler effect shifts the optical frequency from the resonance with the optical transitions in the atom, leading to a coupling of the optically produced pseudospin to the atomic momentum (for review see Refs. 5,6). At very low temperatures and sufficient concentrations, the bosonic atoms can undergo Bose-Einstein condensation transition and form the Bose-Einstein condensate (BEC).[7] The physics of the condensate becomes much richer in the presence of this synthetic spin-orbit coupling, that is the optically produced coupling between the atomic pseudospin and atomic momentum. The SOC can be produced in cold atomic matter in various forms[5,6] known in solid state physics. Of course, a similar synthetic spin-orbit coupling can be produced for fermionic atoms (such as $^{40}$K and $^{6}$Li isotopes) as well.[8,9] Since we are interested in the effects common for the real- and pseudospin systems, we will usually use word "spin" rather than "pseudospin" here. We will concentrate on two-dimensional and one-dimensional systems, where the effects of spin-orbit coupling are especially interesting.

Spin 1/2 particles are described by a two-component wave function[1] $\psi = [\psi_\uparrow(\mathbf{r},t), \psi_\downarrow(\mathbf{r},t)]^{\mathrm{T}}$. As a generic example of two-dimensional spin-orbit interaction we consider the Rashba coupling with the Hamiltonian:[4]

$$H_{\mathrm{so}}^{[2D]} = \alpha\left(\sigma_y k_x - \sigma_x k_y\right), \tag{11}$$

and the coupling constant $\alpha$. Here $\sigma_i$ are the Pauli matrices and $k_i$ are the Cartesian components of the operator $\mathbf{k} \equiv \mathbf{p}/\hbar$. The spectrum of the Rashba Hamiltonian taking into account the kinetic energy is given by

$$E(k) = \frac{\hbar^2 k^2}{2M} \pm \alpha k, \tag{12}$$

as shown in Fig. 1.

The eigenstates of $H_{\mathrm{so}}^{[2D]}$ in Eq.(11) have the form of spin-dependent plane waves, namely:

$$\psi_1 = \frac{1}{\sqrt{2}}\begin{bmatrix} e^{-i\phi} \\ i \end{bmatrix} e^{i\mathbf{kr}}, \qquad \psi_2 = \frac{1}{\sqrt{2}}\begin{bmatrix} e^{-i\phi} \\ -i \end{bmatrix} e^{i\mathbf{kr}}. \tag{13}$$

The angle $\phi$ is determined by $\mathbf{k}$ as $\phi = \arctan(k_y/k_x)$ and $\mathbf{r} \equiv (x,y)$. In the absence of an external magnetic field these states are double-degenerate as

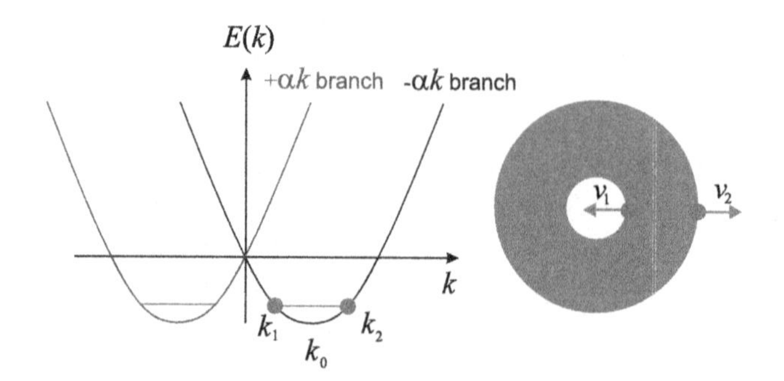


Fig. 1. Two branches of the spectrum of a two-dimensional particle in the presence of the Rashba spin-orbit coupling with $E(k) = \hbar^2 k^2/2M^* \pm \alpha k$. This shape of the spectrum corresponds to the anomalous velocity in Eq.(15). The zero-momentum states have nonzero velocities whereas finite-momentum states at $k_0 = M\alpha/\hbar$ have zero velocities. The opposite velocities $v_1$ and $v_2$ in the right panel correspond to the parallel momenta $k_1$ and $k_2$ in the left panel, respectively.

imposed by the time-reversal symmetry of the spin-orbit coupled Hamiltonian.

Taking into account that the position operator can be presented in the form $\mathbf{r} = i\hbar\partial/\partial\mathbf{k}$, the corresponding spin-dependent terms in Eq.(10) become

$$\frac{i}{\hbar}\left[H_{\rm so}^{[2D]}, x\right] = \frac{\partial H_{\rm so}^{[2D]}}{\hbar\partial k_x} = \alpha\sigma_y, \quad \frac{i}{\hbar}\left[H_{\rm so}^{[2D]}, y\right] = \frac{\partial H_{\rm so}^{[2D]}}{\hbar\partial k_y} = -\alpha\sigma_x \tag{14}$$

resulting in the total velocity operators

$$v_x = \frac{\hbar k_x}{M} + \frac{\alpha}{\hbar}\sigma_y, \qquad v_y = \frac{\hbar k_y}{M} - \frac{\alpha}{\hbar}\sigma_x. \tag{15}$$

As we can see, the velocity in Eq.(15) acquires spin-dependent *anomalous* [10] terms determined by the spin-orbit coupling. A striking feature here is the fact that even at $k_x = k_y = 0$, the particle can move with the velocity proportional to the spin-orbit coupling $\alpha$.

For one-dimensional systems with one spatial and three spin-related degrees of freedom, we consider a typical spin-orbit coupling Hamiltonian as

$$H_{\rm so}^{[1D]} = \alpha k\sigma_z \tag{16}$$

and the corresponding velocity $v = \hbar k/M + \alpha\sigma_z/\hbar$.

The anomalous velocity causes the experimentally observed spin-dipole oscillations [11] and *Zitterbewegung*. [12] In the spin-dipole oscillations of a con-

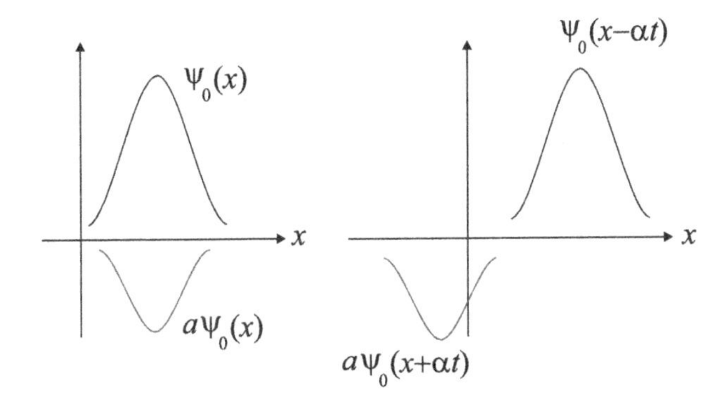


Fig. 2. Spinor components for pure (left) and mixed (right) one-dimensional states. In the pure state we have $\psi_{\downarrow}(x,t) = a\psi_{\uparrow}(x,t)$, where $a$ is a complex constant. In a strongly mixed state $\psi_{\downarrow}(x,t)$ and $\psi_{\uparrow}(x,t)$ have considerably different spatial dependences. For a condensate localized in a harmonic potential, opposite spin components start to oscillate out of phase and form a spin-dipole moment.

densate localized in a one-dimensional harmonic trap, the anomalous velocity drives the condensate motion inside the trap. If the initial spin state is not the eigenstate of the Hamiltonian (16), opposite spin components oscillate out of phase, and the condensate acquires a spin-dipole moment. In the *Zitterbewegung*, that is a trembling motion of a spin-orbit coupled wavepacket, different spin components move with different velocities, causing the trembling in the expectation value of the wave packet position.

### 1.3. *Spin precession, mixed states, and the density flux*

To characterize the spin evolution caused by the spin-orbit coupling, we first find that the spin precession rate corresponding to the spin splitting $2\alpha k$ in Eq.(12) is $2\alpha k/\hbar$ and the resulting precession angle is $\varphi = 2\alpha kt/\hbar$. Taking into account that the velocity associated with momentum is $v = \hbar k/M$ and particle displacement is $L = vt$, we obtain $\varphi = 2L/L_{\mathrm{so}}$. Here we introduced the universal precession length $L_{\mathrm{so}} \equiv \hbar^2/M\alpha$ required for a particle to be displaced to rotate the spin. Thus, $L_{\mathrm{so}}$ describes the spatial scale corresponding to the spin-orbit coupling in two-dimensional and one-dimensional systems.

In our case another consequence of the spin-orbit coupling, strongly influencing the spin states, will be important. Let us first introduce the reduced spin density matrix $\boldsymbol{\rho}$ with the components

$$\boldsymbol{\rho} \equiv \begin{bmatrix} \rho_{11} & \rho_{12} \\ \rho_{21} & \rho_{22} \end{bmatrix} = \int \begin{bmatrix} |\psi_{\uparrow}(\mathbf{r})|^2 & \psi_{\uparrow}^*(\mathbf{r})\psi_{\downarrow}(\mathbf{r}) \\ \psi_{\uparrow}(\mathbf{r})\psi_{\downarrow}^*(\mathbf{r}) & |\psi_{\downarrow}(\mathbf{r})|^2 \end{bmatrix} d^D r, \tag{17}$$

where $D = 1, 2$ is the dimensionality of the system. The spin components are given by the expectation values

$$\langle \sigma_i \rangle = \int \boldsymbol{\psi}^{\dagger}(\mathbf{r}, t)\sigma_i \boldsymbol{\psi}(\mathbf{r}, t) d^D r = \mathrm{tr}\left(\sigma_i \boldsymbol{\rho}\right). \tag{18}$$

To characterize the spin subsystem, we introduce the rescaled purity $P = 2\mathrm{tr}\boldsymbol{\rho}^2 - 1$ with $0 \leq P \leq 1$ given by

$$P = 1 + 4\left[|\rho_{12}|^2 - \rho_{11}\rho_{22}\right]. \tag{19}$$

Here $P = 1$ corresponds to a pure state (Fig. 2, left panel) with $\mathrm{tr}\boldsymbol{\rho}^2 = 1$ while $P = 0$ corresponds to a fully mixed one (Fig. 2, right panel) with $\mathrm{tr}\boldsymbol{\rho}^2 = 1/2$. Note that this purity can be be expressed in terms of the expectation values of the spin components as $P = \sum_i \langle \sigma_i \rangle^2$.

The production of mixed states by the anomalous velocity can be seen in the following simple one-dimensional consideration. Assume that at $t = 0$ we have the state in the form $\boldsymbol{\psi}(x, 0) = \left[\psi_0(x), a\psi_0(x)\right]^{\mathrm{T}}$ and the particle has a very large mass such that on this timescale of interest we can neglect the state broadening determined by the coordinate-momentum uncertainty. Then, at $t \geq 0$, by applying the one-dimensional Hamiltonian in Eq.(16), we obtain:

$$\boldsymbol{\psi}(x, t) = \begin{bmatrix} \psi_0(x - \alpha t) \\ a\psi_0(x + \alpha t) \end{bmatrix}. \tag{20}$$

As a result, spin-up and spin-down components become spatially separated and a mixed spin state with purity $P < 1$ is formed, as it is shown in the right panel of Fig. 2.

The evolution of the probability density $\boldsymbol{\psi}^{\dagger}\boldsymbol{\psi}$ is given by the continuity equation

$$\frac{\partial \boldsymbol{\psi}^{\dagger}\boldsymbol{\psi}}{\partial t} + \boldsymbol{\nabla}\mathbf{J}(\mathbf{r}, t) = 0. \tag{21}$$

In the presence of the anomalous velocity, the Cartesian component of the flux density $J_i(\mathbf{r}, t)$ is presented as the sum

$$J_i(\mathbf{r}, t) = J_i^{\mathrm{nor}}(\mathbf{r}, t) + J_i^{\mathrm{an}}(\mathbf{r}, t), \tag{22}$$

where the normal term is

$$J_i^{\mathrm{nor}}(\mathbf{r}, t) = \frac{i\hbar}{2M}\left[\boldsymbol{\psi}_i^{\dagger}\boldsymbol{\psi} - \boldsymbol{\psi}^{\dagger}\boldsymbol{\psi}_i\right], \tag{23}$$

and the term determined by the anomalous velocity reads

$$J_i^{\mathrm{an}}(\mathbf{r},t) = \boldsymbol{\psi}^{\dagger}\frac{\partial H_{\mathrm{so}}}{\hbar\partial k_i}\boldsymbol{\psi}. \tag{24}$$

Here $\boldsymbol{\psi}_x \equiv \partial\boldsymbol{\psi}/\partial x$ and $\boldsymbol{\psi}_y \equiv \partial\boldsymbol{\psi}/\partial y$. It is important to stress that the spinor wavefunction $\boldsymbol{\psi}$ here is calculated taking into account the spin-orbit coupling.

## 2. Inhibition of collapse of Bose-Einstein condensate by Anomalous velocity

Macroscopic ensembles of interacting particles are one of the most interesting objects in condensed matter physics.[13] If the interaction between particles is attractive and sufficiently strong, the system can experience a collapse, that is shrinking to (almost) zero size after a finite time and then explosion. The collapse, being a fundamental problem in several branches of physics,[14] including the physics of Bose-Einstein condensates, strongly depends on the system dimensionality. The main features of the collapse of a free, not confined by an external potential, Bose-Einstein condensate, are determined by the interplay of its positive quantum kinetic (arising due to the Heisenberg momentum-coordinate uncertainty) and negative attraction energies, both dependent on its characteristic size $a$. For a two-dimensional system both the attraction and kinetic energy scale as $a^{-2}$ and the collapse occurs only at a strong enough interaction, which can overcome the kinetic energy tending to expand the condensate.

One of the advantages for the theory of cold atomic gases in optical potentials is the fact that due to a very large particle wavelength (typically, of the order of 1 micron) compared with the atomic radius of the order of $10^{-8}$ cm, the interatomic interaction can be accurately described by a single parameter. This parameter is the scattering length $a_s$, where $a_s > 0$ ($a_s < 0$) corresponds to the repulsion (attraction) between the atoms. The attraction can be achieved by means of the Feshbach resonance in the interatomic scattering[15] produced by a real (not synthetic) magnetic field a certain range of the system parameters. In this Section we consider the Bose-Einstein condensate with attractive interaction between the atoms and show how its collapse is influenced by the anomalous spin-dependent velocity and can eventually be prevented if the spin-orbit coupling is strong enough.[16]

### 2.1. *Gross-Pitaevskii equation and two-dimensional collapse without spin-orbit coupling*

To have a reference point, we begin with a pancake-shaped condensate of pseudospin 1/2 particles described by spinor $\psi = [\psi_{\uparrow}(\mathbf{r},t), \psi_{\downarrow}(\mathbf{r},t)]^{\mathrm{T}}$, normalized here to the total number of particles $N \gg 1$. Usual number of particles in atomic Bose-Einstein condensates[7] varies between $N \sim 10^3$ and $N \sim 10^6$. Although this is much smaller than the number of particles in conventional condensates such as low-temperature helium, it is sufficient to observe all the relevant physical effects.

We consider an initial state prepared, for example, in a parabolic potential:

$$\psi(\mathbf{r}, t=0) \equiv A(0) \exp\left[-\frac{r^2}{2a^2(0)}\right] \psi(0), \tag{25}$$

where $\psi(0)$ is the coordinate-independent initial spinor, $A(0) = \sqrt{N/\pi}/a(0)$ is a normalization factor, and $a(0)$ is the initial width of the two-dimensional distribution. Here all atoms are condensed in the ground state of the corresponding parabolic potential, in contrast to the conventional Bose-Einstein condensation, which occurs in the momentum- rather than in the coordinate space. At $t = 0$, the optically-produced parabolic potential is switched off, the interatomic interaction is switched on by the Feshbach resonance, and the evolution of the condensate begins.

This evolution is described by the nonlinear Schrödinger equation in the Gross-Pitaevskii form, where interaction between the particles is approximately presented by a density-determined potential $g_2 |\psi|^2$:

$$i\hbar\frac{\partial\psi}{\partial t} = \left[-\frac{\hbar^2}{2M}\Delta + H_{\mathrm{so}}^{[2D]} - g_2 |\psi|^2\right] \psi. \tag{26}$$

The self-attraction coupling constant in Eq.(26) is given by $g_2 = -4\pi\hbar^2 a_s/Ma_z$, which we assume for simplicity to be spin-independent, where $a_z$ is the pancake extension along the $z-$ axis and the interatomic scattering length $a_s$ is negative.[17–19]

We begin with the collapse without spin-dependent effects, that is with $\alpha = 0$. Here the energy of the system is the sum of kinetic and interaction contributions presented as:

$$E = -\frac{1}{2}\int \left[\frac{\hbar^2}{M}\psi^{\dagger}\Delta\psi + g_2 |\psi|^4\right] d^2r. \tag{27}$$

The evolution corresponding to solution of Eq.(26) can be described sufficiently accurately by a two-parametric Gaussian ansatz[18]

$$\psi(\mathbf{r},t) = A(t)\exp\left[-\frac{r^2}{2a_{\rm v}^2(t)}\left(1+ib_{\rm v}(t)\right)\right]\psi(0), \tag{28}$$

with variational parameters $b_{\rm v}(t)$ and $a_{\rm v}(t)$. The time dependence of the width according to this ansatz becomes[18]

$$a_{\rm v}(t) = v_c\sqrt{T_c^2 - t^2} \tag{29}$$

with the collapse time $T_c \equiv Ma^2(0)/\hbar\sqrt{\Lambda}$ and the characteristic velocity $v_c \equiv a(0)/T_c \sim \sqrt{\Lambda}/a(0)$, where $\Lambda \equiv (\widetilde{g}_2 N - 2\pi)/2$ and the dimensionless interaction $\widetilde{g}_2 \equiv 4\pi\,|a_s|\,/a_z$. The collapse in this model occurs if $\widetilde{g}_2 N$ exceeds a threshold value of $2\pi$, corresponding to the condition of a sufficiently strong self-attraction.

## 2.2. *Collapse with spin-orbit coupling: Anomalous velocity against attraction*

Now consider a condensate with the Rashba coupling in Eq.(11), where the spatial scale of SOC is described by the spin-flip distance $L_{\rm so}$. Assume that the condensate is initially prepared in the state fully polarized along the $z-$axis such that the component antiparallel to the $z-$axis is zero. At small $t \ll T_c$ we can obtain from Eq.(26) that this spin-down component begins to grow at distances $r \sim a(0)$ with a rate proportional to $\alpha$ such that the density can be redistributed from the center of the condensate to its periphery. At a sufficiently large $\alpha$ this growth can eventually lead to the collapse prevention, as we will see below.

Let us consider a numerical solution of Eq.(26) for a strong attraction, $\widetilde{g}_2 N \gg 1$ and obtain spinor $\psi(\mathbf{r},t)$ as a result. Figure 3 shows the time-dependent width of the packet defined as the participation ratio for the state $\psi(\mathbf{r},t)$ as

$$a(t) \equiv \frac{N}{\sqrt{2\pi}}\left[\int\left(|\psi_\uparrow(\mathbf{r},t)|^2 + |\psi_\downarrow(\mathbf{r},t)|^2\right)^2 d^2r\right]^{-1/2}. \tag{30}$$

Clearly, the collapse (if occurs) corresponds to the time when $a(t) = 0$. At $\alpha = 0$ the values of $a(t)$ in Eq.(30) as shown by black solid line in Fig.3 and $a_{\rm v}(t)$ in Eq.(29) are very close. When spin-orbit coupling is included, several main features can be seen in the wavepacket evolution.

(i) At small $t \ll T_c$, the attraction-induced velocity, understood here as the dependence of the wavepacket width on time, develops linearly with $t$. This can be seen in Eq.(29), where $a(0) - a_{\rm v}(t) \sim a(0)\left(t/T_c\right)^2$ and, therefore,

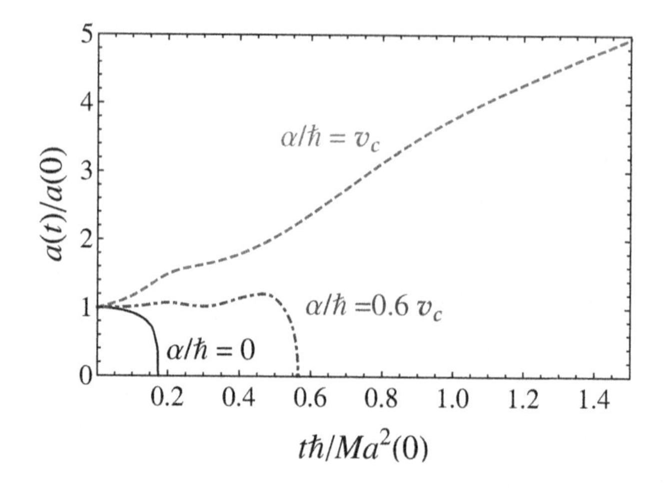


Fig. 3. Packet width as a function of time for different values of spin-orbit coupling. Here $\widetilde{g}_2 N = 16\pi$. We clearly see transition from collapse to stable "no-collapse" regime with the increase in $\alpha$.

the corresponding velocity behaves as $a(0)t/T_c^2$. For the initial spin state of our interest, that is $[1,0]^{\mathrm{T}}$, the velocity components given by Eq.(15) are zero. Therefore, the anomalous velocity is determined by the displacement of the atoms from the initial positions and needs some time to develop as well. One can prove that it increases as $(\alpha/\hbar) \times (t/T_c)^2 \times a(0)/L_{\mathrm{so}}$, being initially much smaller than the velocity due to the self-attraction. As a result, the time-dependence of the packet width corresponding to the compression is universal for small $t \ll T_c$, independent of $\alpha$. Since the spin precession angle at the displacement of $a(0)$ is of the order of $a(0)/L_{\mathrm{so}}$, starting from the fully polarized state, the atoms acquire the anomalous velocity of the order of $(\alpha/\hbar) \times a(0)/L_{\mathrm{so}}$ for the weak SOC, that is $a(0) \lesssim L_{\mathrm{so}}$, or of the order of $\alpha/\hbar$ otherwise. The criterion of a large spin precession in the collapse (again, if it occurs) is $a(0) > L_{\mathrm{so}}$, while the condition of a sufficiently large developed anomalous velocity is $\alpha/\hbar \geq v_c$. If the latter inequality is satisfied, the centrifugal contribution to the density flux $\mathbf{J}(\mathbf{r},t)$ (see Eq.(21)) caused by the SOC can prevent the collapse.

(ii) With the increase in $\alpha$, the packet width initially shows a slight increase with time, reaches a broad plateau, and then falls to zero. Thus, the collapse still can occur. However, it takes a longer, $\alpha-$dependent time $t_c(\alpha) > T_c$.

(iii) There exists a critical value $\alpha_{\mathrm{cr}} \approx 0.7\hbar v_c$, where the counter-flux

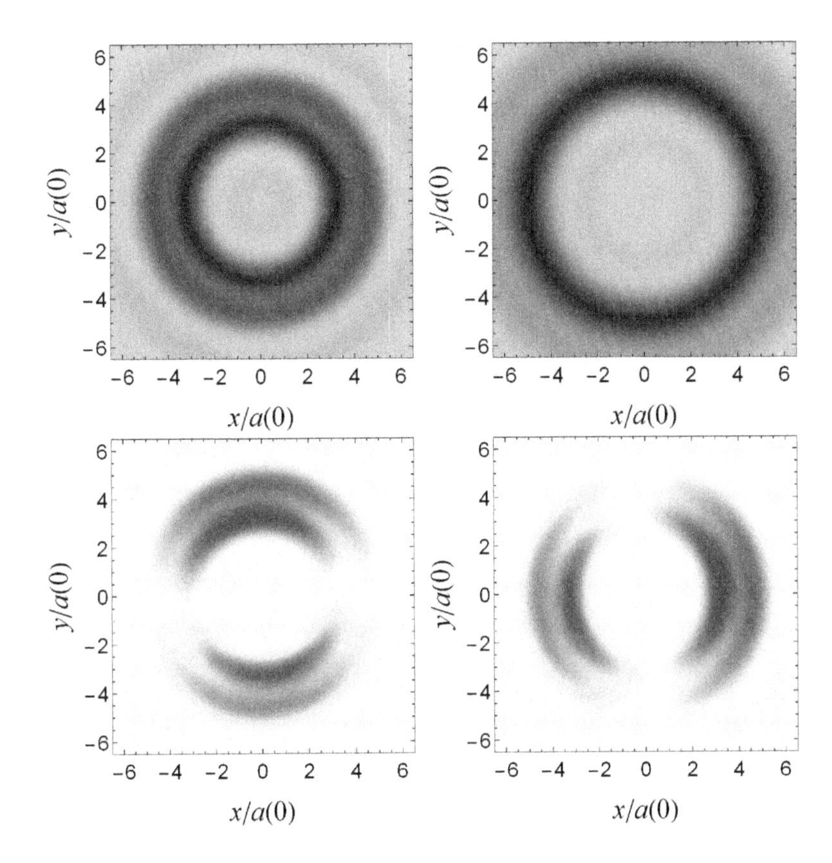


Fig. 4. Upper panel: Density profile (in arbitrary units) $\psi^\dagger(\mathbf{r},t)\psi(\mathbf{r},t)$ for $\widetilde{g}_2 N = 16\pi$ in the no-collapse regime, where the spin-orbit coupling is strong enough to inhibit the collapse. Left: $t = Ma^2(0)/\hbar$, right: $t = 2Ma^2(0)/\hbar$. Lower panel: spin density profile (in arbitrary units) at $t = Ma^2(0)/\hbar$ in the same regime. Left: $\psi^\dagger(\mathbf{r},t)\,\sigma_x\psi(\mathbf{r},t)$, right: $\psi^\dagger(\mathbf{r},t)\,\sigma_y\psi(\mathbf{r},t)$.

produced by the anomalous velocity is sufficient to prevent the collapse at $\alpha > \alpha_{\rm cr}$. The divergent dependence of $t_c(\alpha)$ on the SOC strength near the "collapse - no collapse" transition point can be described as $t_c(\alpha) \sim T_c\alpha_{\rm cr}(\alpha_{\rm cr}-\alpha)^{-1}$.

To get insight into the origin of the SOC effects in the collapse, we depict the mass and spin density profiles in Fig. 4. The resulting density distribution is given by a ring of radius $R(t)$ (produced by the centrifugal contribution to the flux density) and width $w(t)$ with $a(t) \sim \sqrt{R(t)w(t)}$. This ring is responsible for the broad plateau in Fig. 3 at a subcritical spin-orbit coupling $\alpha = 0.6\hbar v_c$. At $R(t) \gg a(t)$ the interatomic interaction

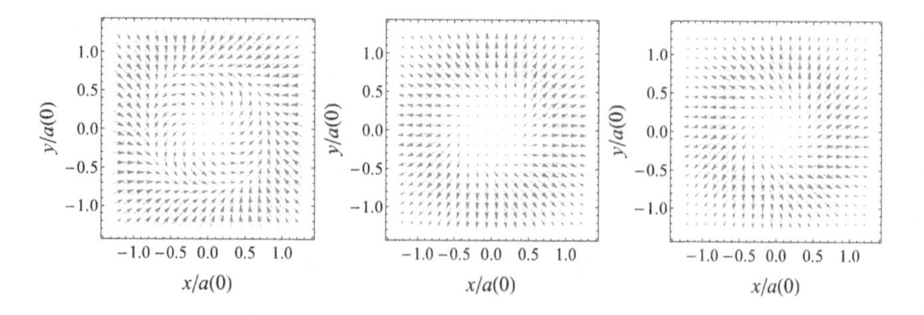


Fig. 5. Density flux (in arbitrary units) for $\tilde{g}_2 N = 16\pi$ at $t = 0.2Ma^2(0)/\hbar$. Here $\alpha/\hbar = v_c$, that is we are in the no-collapse regime. Lengths and directions of the arrows correspond to the absolute values and directions of the flux. Left panel - $\mathbf{J}^{\text{nor}}(\mathbf{r},t)$, middle panel $\mathbf{J}^{\text{an}}(\mathbf{r},t)$, right panel - total flux $\mathbf{J}(\mathbf{r},t)$. Note that the total flux is very small in the vicinity of the origin with $r < 0.5a(0)$, showing that the collapse does not occur.

energy tends to zero as $-1/R(t)w(t)$, and the conserved total energy is the sum of the kinetic and SOC terms. At $\alpha < \alpha_{\text{cr}}$, the attraction is still strong enough to reverse the splitting and to restore the collapse. At $\alpha > \alpha_{\text{cr}}$ (see Fig. 3) the anomalous velocity takes over, the splitting continues, and the collapse does not occur. This process is accompanied by evolution of the condensate spin presented in the lower panel of Fig. 4. To complete the picture, we present in Fig. 5 the density flux components: $\mathbf{J}^{\text{nor}}(\mathbf{r},t), \mathbf{J}^{\text{an}}(\mathbf{r},t)$, and total $\mathbf{J}(\mathbf{r},t)$. As one can see in the Figure, close to the center of the condensate, the flux is very weak, demonstrating that the collapse cannot occur.

To make connections to possible experiments, we estimate the constant $\tilde{g}_2$ as 0.05 for $|a_s| \sim 100a_B \sim 0.05$ $\mu$m and $a_z \sim 1$ $\mu$m. The condition $\tilde{g}_2 N > 2\pi$ can be satisfied even for small condensates with $N \sim 100$ particles. The velocity of the collapse is $v_c \sim \hbar\sqrt{\tilde{g}_2 N}/Ma(0)$. At $a(0) \sim 10$ $\mu$m and $N \sim 10^3$ this estimate yields $v_c \sim 0.03$ cm/s and the corresponding time scale $T_c = a(0)/v_c \sim 0.3$ s. Such a small value of $v_c$ demonstrates that even a relatively weak experimentally achievable spin-orbit coupling[5,6] can prevent the BEC from collapsing.

## 3. Spin relaxation of one-dimensional condensate in a random potential

Out-of- equilibrium spin polarization, produced in a medium, will relax after some time to equilibrium value with the relaxation time strongly dependent on the system properties. For example, in disordered solids, directions of the electron momenta are randomized as a result of electron scattering by impurities. As a result, randomization of directions of spin precession axes determined by the direction of the electron momentum (see Eq.(11)), leads to the famous Dyakonov-Perel mechanism of spin relaxation.[20]

Studies of BEC spin dynamics in a random potential are strongly different from those of carriers in solids since one can access evolution of a single BEC wavepacket and study the effects of the anomalous spin-dependent velocity in different regimes of disorder. Here we investigate these qualitatively new effects in the spin evolution of a one-dimensional noninteracting Bose-Einstein condensate. Usually, one is interested in the long time behavior of the BEC, where Anderson localization takes over[21,22] and prevents the infinite spread of the condensate. In the presence of SOC, this behavior was studied in Ref. 23. However, the evolving spin density is sufficiently large for the experimental observation only at relatively short times. Therefore, we consider here the initial stage of the evolution[24] and study the particle position and spin behavior at this pre-localization stage.

### 3.1. *Hamiltonian: Disorder combined with the spin-orbit coupling*

We consider here a condensate tightly confined in the transverse directions. This confinement forms a quasi one-dimensional system, subject to a random optical field, which can be produced by using coherent light speckles,[25] giving rise to a disorder potential $U_{\rm rnd}(x)$. Initially, the Bose-Einstein condensate is confined, in addition to the disorder, in a parabolic potential and is subject to a synthetic Zeeman field producing its spin polarization. At $t=0$, the optical fields producing the parabolic potential and the Zeeman interaction, are switched off, and the fields producing a spin-orbit coupling are switched on. After this the BEC starts to spread in a random potential in the presence of the spin-orbit coupling.

The effective time-dependent Hamiltonian has the form:

$$\begin{aligned} H(t<0) &= \frac{\hbar^2k^2}{2M} + \frac{M\omega_0^2}{2}x^2 + U_{\rm rnd}(x) + \frac{\Delta_Z}{2}\left(\boldsymbol{\sigma}\mathbf{m}\right), \\ H(t>0) &= \frac{\hbar^2k^2}{2M} + U_{\rm rnd}(x) + \alpha\sigma_z k. \end{aligned} \tag{31}$$

Here the frequency of the trap at $t<0$ is $\omega_0$. The oscillator-related energy quantum $\hbar\omega_0$, the length $a_{\rm ho}=\sqrt{\hbar/M\omega_0}$, and velocity $v_{\rm ho}=\sqrt{\hbar\omega_0/M}$ represent the scales for the system description. For example, for a $^{87}$Rb condensate and $\omega_0=2\pi\times10$ Hz, we obtain $a_{\rm ho}\approx3\times10^{-4}$ cm and

$v_{\rm ho} \approx 0.02$ cm/s.

The Zeeman term $\Delta_Z (\boldsymbol{\sigma}\mathbf{m})/2$, with $\mathbf{m}$ being the synthetic magnetic field direction, prepares the initial spin state. The subsequent dynamics is, thus, a response of the systems to the instantaneous change in the potential, interaction, and spin-orbit coupling. We will look at this dynamics at a relatively short time scale to see how the spread develops and what happens with the spin at this stage.

The disorder is the sum of local Gaussian "impurity" potentials $U_0 \exp\left[-(x-x_j)^2/\xi^2\right]$ at random positions $x_j$ and with the mean concentration $n$ given by:

$$U_{\rm rnd}(x) = U_0 \sum_j s_j \exp\left[-(x-x_j)^2/\xi^2\right]. \tag{32}$$

Here $s_j = \pm 1$ is a random function of $j$ with the mean value $\langle\langle s_j \rangle\rangle = 0$ and, correspondingly $\langle\langle U_{\rm rnd}(x)\rangle\rangle = 0$, where $\langle\langle \ldots \rangle\rangle$ stands for the disorder-averaged quantities. We consider a small width $\xi \ll a_{\rm ho}$ and assume a completely uncorrelated white-noise distribution of positions of impurities, resulting in:[26]

$$\langle\langle U_{\rm rnd}(x_1) U_{\rm rnd}(x_2)\rangle\rangle = \langle\langle U_{\rm rnd}^2\rangle\rangle \exp\left[-(x_1-x_2)^2/2\xi^2\right], \tag{33}$$

with $\langle\langle U_{\rm rnd}^2\rangle\rangle = \sqrt{\pi/2}U_0^2 n\xi$. By using Fermi's golden rule we define a scattering time $\tau = v_{\rm ho}\hbar^2/\pi n U_0^2 \xi^2$, and a free path $\ell = v_{\rm ho}\tau$. We consider a weak disorder with $\ell/a_{\rm ho} \equiv \omega_0\tau \gg 1$, which relatively weakly influences the localization in the parabolic potential. As the initial condition, we take the ground state of $H(t<0)$ in Eq.(31) at sufficiently strong $\Delta_Z > 0$ and $\mathbf{m} = (-1,-1,0)/\sqrt{2}$ :

$$\psi(x, t=0) = \psi_0(x)\,[1,1]^{\rm T}/\sqrt{2}. \tag{34}$$

With the pure initial state in Eq.(34) one obtains: $\rho_{11}(0) = \rho_{22}(0) = 1/2$, $\langle\sigma_z(0)\rangle = 0$, and the purity $P(0) = \langle\sigma_x(0)\rangle^2 + \langle\sigma_y(0)\rangle^2 = 1$.

We characterize the spatial dynamics by three quantities. One is the center of mass position $\langle x(t)\rangle$ and two others are shape-related parameters such as the width $W(t) \equiv \left[\langle x^2(t)\rangle - \langle x(t)\rangle^2\right]^{1/2}$ and the normalized participation ratio $\zeta(t)$:

$$\zeta(t) = \left[\sqrt{2\pi}\int_{-\infty}^{\infty} |\psi(x,t)|^4 \frac{dx}{N^2}\right]^{-1}, \tag{35}$$

where the normalization factor $\sqrt{2\pi}$ is chosen for satisfying condition $W = \zeta$ for Gaussian wavepackets.

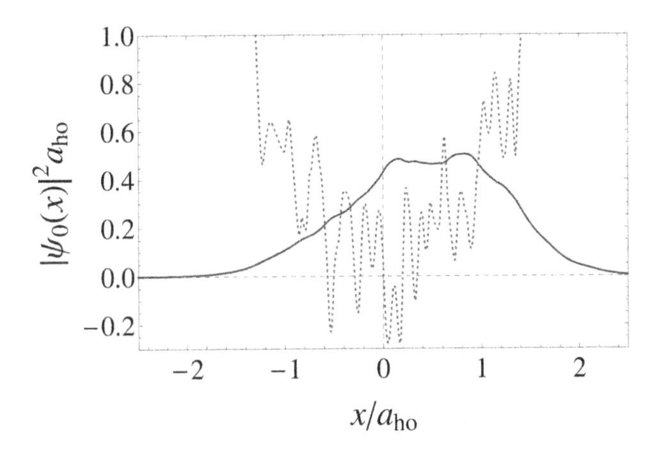


Fig. 6. A realization of the ground state of a BEC (solid line) with $\alpha = 0$ in the total potential (first line of Eq.31 (in $\hbar\omega_0-$units) shown by dashed line. Here and below we use $U_0 = \hbar\omega_0$, $\xi = a_{\rm ho}/32$, and $n = 80/a_{\rm ho}$, corresponding to $\omega_0\tau \approx 4$. This state with $\langle x(0)\rangle > 0$ will be used for calculations of the wavepacket evolution. Note that when the parabolic potential collapses, this state starts to move to the right with the corresponding direction of the spin precession. The total potential is shown only schematically for illustrative reasons.

To study the dynamics, we consider the force $F_\psi$, defined as the derivative of the potential energy with respect to an infinitesimal *virtual* displacement $\delta x$ as $\psi(x,t) \to \psi(x+\delta x,t)$. For a given realization of $U_{\rm rnd}(x)$ one obtains

$$F_\psi = -\frac{1}{N}\int_{-\infty}^{\infty} U_{\rm rnd}(x)\partial_x \left|\psi(x,t)\right|^2 dx, \tag{36}$$

with the disorder-averaged[26] variance:

$$\langle\langle F_\psi^2\rangle\rangle = \frac{4\pi}{N^2}U_0^2 n\,\xi^2 \int \left[\partial_x \left|\psi(x,t)\right|^2\right]^2 dx. \tag{37}$$

This force with $\langle\langle F_\psi^2\rangle\rangle^{1/2} \sim \sqrt{n}\,\xi\, U_0\zeta^{-3/2}(t)$, being determined by the spatial derivative of the density, is sensitive to the local structure of the condensate, resulting in a strong decrease for smooth density distributions. Since at $t \leq 0$ the ground state equilibrium requires the balance of the random force and the force due to the parabolic potential, that is $F_{\psi_0} = M\omega_0^2\langle x(0)\rangle$, the disorder potential causes the displacement of the wavepacket from the center of the trap (cf. Fig. 6) by $\langle x(0)\rangle = F_{\psi_0}/M\omega_0^2 \sim a_{\rm ho}\sqrt{a_{\rm ho}/\ell}$.

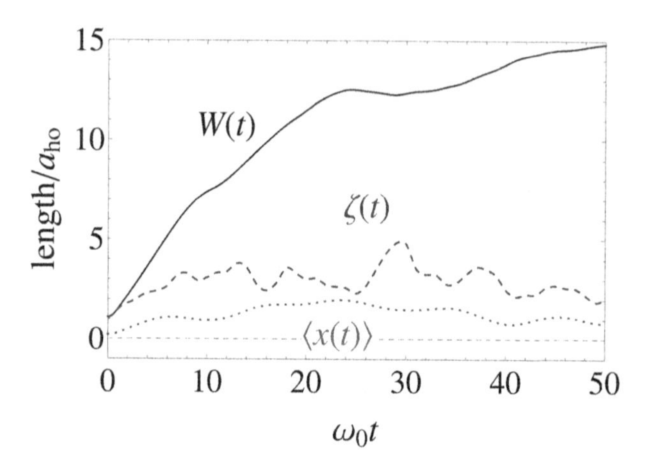


Fig. 7. The wavepacket parameters (as marked near the lines) of a noninteracting BEC with the initial state in Fig. 6. The picture corresponds to the wavepacket fragmentation, where $W(t)$ exceeds $\zeta(t)$. The time dependences do not show qualitative changes with the realization of $U_{\rm rnd}(x)$.

### 3.2. *Spin evolution coupled to particle' propagation*

To understand the effect of the disorder on the spin evolution, one needs to consider two mechanisms, which we will define as *precessional* and *anomalous*. To characterize the precessional mechanism, we use the precession length $L_{\rm so}$, where the spin precession angle $\phi_{12}(t) = 2\left(\langle x(t)\rangle - \langle x(0)\rangle\right)/L_{\rm so}$ is due to the condensate displacement. However, as a result of the anomalous velocity, any initial $\psi(x,0)$, if not an eigenstate of $\sigma_z$, splits into spin-projected components. Therefore, the SOC leads, in addition to the precession, to a reduced off-diagonal component of the spin density matrix $|\rho_{12}(t)|$, decreasing the purity, modifying the spin evolution, and making it dependent on the spin density distribution inside the condensate.

In Fig. 7 we present the evolution of $W(t)$ and $\zeta(t)$ obtained from the numerical solution of the Schrödinger equation with the Hamiltonian (31). The corresponding $P(t)$ (see Eq.(19)) is shown in Fig. 8 and spin evolution is presented in Fig. 9. An irregular behavior of the presented quantities corresponds to propagation of a wavepacket in a random potential with a finite correlation length of the order of $\xi$.

To explain these Figures we note that after releasing the harmonic potential, the condensate starts moving due to the random force (see Eq.(37)) in the direction of $\langle x(0)\rangle$ with the acceleration $F_{\psi_0}/M$. The spin precesses accordingly to the $\langle x(t)\rangle$–displacement, corresponding to a constant acceleration $\omega_0^2\langle x(0)\rangle^2$, with $\langle\sigma_y(t)\rangle = -\omega_0^2 t^2\langle x(0)\rangle/L_{\rm so}$. The evolution of $\langle\sigma_x(t)\rangle = \sqrt{P(t) - \langle\sigma_y(t)\rangle^2}$ has a different origin. In the precessional mechanism with $P(t) = 1$, one expects $\langle\sigma_x(t)\rangle = \sqrt{1 - \langle\sigma_y(t)\rangle^2}$, and, therefore, a quartic $\sim t^4$ initial behavior. However, the initial behavior in Fig. 9 is parabolic since the time dependence of $P(t)$ is important. At small $t$ the wavefunction behaves as

$$\psi(x,t) = \begin{bmatrix} \psi_0(x - \alpha t) \\ \psi_0(x + \alpha t) \end{bmatrix}, \tag{38}$$

and Eq.(17) with $\psi(x,t)$ in Eq.(38) yields $P(t) = 1 - M\alpha^2\omega_0 t^2/4\hbar^3$. In addition, the condensate starts to spread due to the Heisenberg position-momentum quantum uncertainty. As a result of this spread, the force and the acceleration decrease, and $\langle x(t)\rangle$ and $\phi_{12}(t)$ acquire a sub-$t^2$ dependence. Thus, time-dependence of the spin is strongly related to the $U_{\rm rnd}(x)$ where the condensate moves. The behavior of $\langle\sigma_x(t)\rangle$ at this stage is always due to the component separation, demonstrating that two mechanisms of spin evolution are at work simultaneously. On the time scale of the initial wavepacket broadening ($\sim \omega_0^{-1}$), the effect of the precession angle $1 - \cos\phi_{12}(t)$ is of the order of $a_{\rm ho}^2/L_{\rm so}^2 \times a_{\rm ho}/\ell$, while the change in the purity $1 - P(t)$ is of the order of $a_{\rm ho}^2/L_{\rm so}^2 \gg 1 - \cos\phi_{12}(t)$. Therefore, for $\langle\sigma_x(t)\rangle$, the effect of components separation due to the spin-dependent anomalous velocity is larger than the effect of precession by a factor of $\sim \ell/a_{\rm ho}$. Thus, at a weak disorder, the initial evolution of $\langle\sigma_x(t)\rangle$ is due to the purity decrease, while $\langle\sigma_y(t)\rangle$ evolves mainly due to the spin precession.

At the following stage, for $t \geq \tau$, fragmentation of the condensate in the random potential begins, and $\zeta(t)$ becomes smaller than $W(t)$ (Fig. 7). The density distribution becomes relatively sparse and consists of several peaks of different width, in agreement with Ref. 27. At this stage, the purity is related to the details of the wave function components. The evolution of $P(t)$ in Fig. 8 shows a crossover to the oscillating plateau at time satisfying condition $2\alpha t/\hbar \sim \zeta(t)$. The reasons for the change in the purity (Fig. 8) and the corresponding spin dynamics (Fig. 9) can be understood from Fig. 10, showing spinor components $\mathrm{Re}[\psi_{\uparrow,\downarrow}(x,t)]$ and $\mathrm{Im}[\psi_{\uparrow,\downarrow}(x,t)]$ after the fragmentation in a random potential has produced the irregular shape of the wavefunction. In the presence of SOC, the spinor components are considerably different, and their relative oscillations lead to a decrease in $|\rho_{12}(t)|$, resulting in the purity decrease. If $2\alpha\tau/\hbar \leq a_{\rm ho}$, the purity does not fall to zero and the spin length $\sqrt{P(t)}$ remains approximately

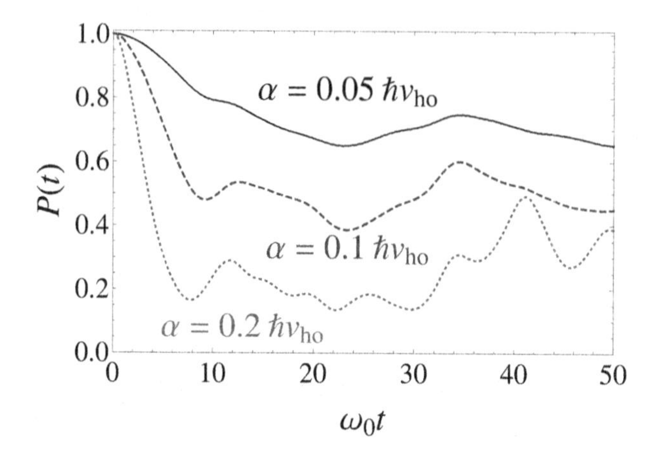


Fig. 8. Purity of the spin state of a noninteracting BEC for different values of SOC. This figure shows the crossover from decreasing $P(t)$ to the randomly oscillating behavior.

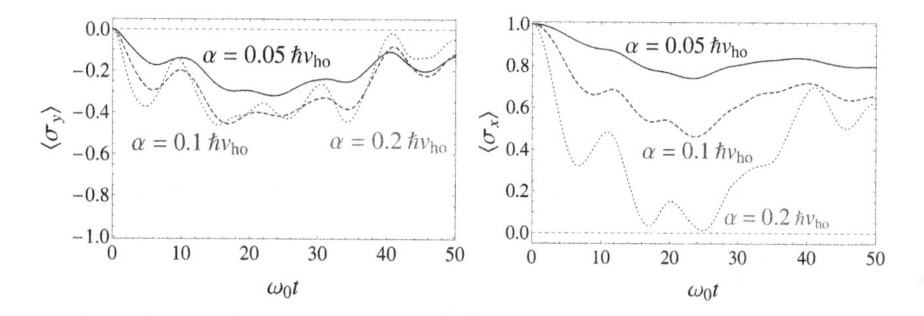


Fig. 9. Time dependence of $\langle\sigma_y(t)\rangle$ and $\langle\sigma_x(t)\rangle$ for different values of $\alpha$. The upper curve in (a) is for the realization of the random potential in Fig. 6 with $\langle x(0)\rangle > 0$. The behavior of $\langle\sigma_y(t)\rangle$ is mainly due to the spin precession.

a constant at this stage. The oscillations correlate with the displacement $\langle x(t)\rangle$ changing on the time scale of the order of $\tau$ due to a nonvanishing force $F_\psi$. The spin precesses with $\langle\sigma_y(t)\rangle \sim \langle x(t)\rangle/L_{\rm so}$ as can be seen in the left panel of Fig. 9. Since $|\langle x(t)\rangle|$ is of the order of $\zeta(t) \sim \ell \ll L_{\rm so}$, at sufficiently weak SOC we have $\langle\sigma_y(t)\rangle \ll 1$.

## 4. Conclusions

We have considered the anomalous velocity[10] produced by the spin-orbit coupling and presented two new effects where it plays a qualitative role. In

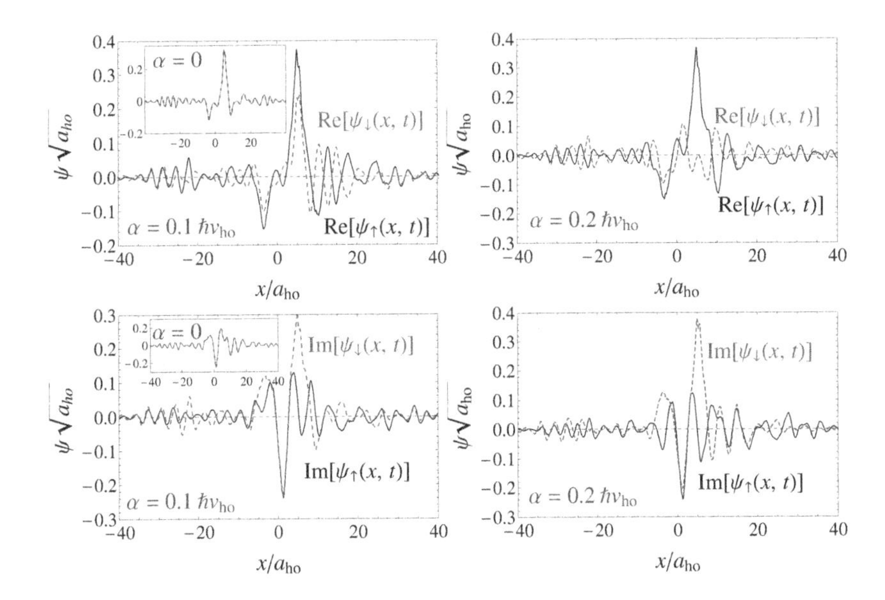


Fig. 10. Components $\psi_\uparrow(x,t)$ and $\psi_\downarrow(x,t)$ (as marked near the plots) at $t = 18\omega_0^{-1}$. The small integral overlap of these functions corresponds to the oscillating plateau in Fig. 7. The insets show the two components at $\alpha = 0$ (as expected, they are identical).

the first effect we have demonstrated that this velocity can prevent collapse of a nonuniform quasi two-dimensional BEC. For Rashba coupling with the spectrum axially symmetric in the momentum space, this velocity leads to a centrifugal component in the two-dimensional density flux. As a result, spin-orbit coupling can prevent collapse of the two-dimensional BEC if this flux is sufficiently strong to overcome the effect of interatomic attraction. In this case, the attraction between the bosons cannot squeeze the initial wavepacket and force it to collapse. These results show that one can gain a control over the BEC collapse by using experimentally available synthetic spin-orbit coupling and, thus, paving the way to study various nontrivial regimes in Bose-Einstein condensates of attracting particles.

As the second effect, we considered the dynamics of a single wavepacket of spin-orbit coupled Bose-Einstein condensates in a random optical potential. In this case, novel experimentally observable features of the spin evolution can emerge. These features include a dependence of the spin dynamics on the initial state and a qualitative difference between the spin components evolution, including a crossover from a decreasing with time to a plateau-like behavior. The striking feature of this process is that two different mechanisms of the spin dynamics, that is a precessional one and that due to the change in the spatial overlap of the spin components, take place simultaneously. The former mechanism is due to the spin precession related to the total condensate displacement, while the latter one, not seen in solids, is caused by the spatial separation of the spin components caused by the anomalous spin-dependent velocity.

## Acknowledgments

This work was supported by the University of Basque Country UPV/EHU under program UFI 11/55, MINECO/FEDER Grant FIS2015-67161-P, and Grupos Consolidados UPV/EHU del Gobierno Vasco (IT-472-10). Sh. Mardonov is supported by the Swiss National Foundation SCOPES project IZ74Z0-160527. We are grateful to A. Rakhimov and J. Li for valuable comments.

## References

1. L.D. Landau and E.M. Lifshitz, *Quantum Mechanics, Non-Relativistic Theory*, Course of Theoretical Physics, Volume 3, (Pergamon Press, Oxford).
2. J. D. Jackson, *Classical Electrodynamics* (John Wiley and Sons, New York).
3. E. I. Rashba, *J. Phys.: Condens. Matter* **28**, 421004 (2016).
4. E. I. Rashba, *Sov. Phys. Solid State* **2**, 1109 (1960).
5. H. Zhai, *Int. J. Mod. Phys. B* **26**, 1230001 (2012).
6. V. Galitski and I. B. Spielman, *Nature* **494**, 49 (2013).
7. W. Ketterle, *Rev. Mod. Phys.* **74**, 1131 (2002).
8. P. Wang, Z.-Q. Yu, Z. Fu, J. Miao, L. Huang, S. Chai, H. Zhai, and J. Zhang, *Phys. Rev. Lett.* **109**, 095301 (2012).
9. L. W. Cheuk, A. T. Sommer, Z. Hadzibabic, T. Yefsah, W. S. Bakr, and M. W. Zwierlein, *Phys. Rev. Lett.* **109**, 095302 (2012).
10. E. N. Adams and E. I. Blount, *J. Phys. Chem. Solids* **10**, 286 (1959).
11. J.-Y. Zhang, S.-C. Ji, Z. Chen, L. Zhang, Z.-D. Du, B. Yan, G.-S. Pan, B. Zhao, Y.-J. Deng, H. Zhai, S. Chen and J.-W. Pan, *Phys. Rev. Let.* **109**, 115301 (2012).
12. Ch. Qu, Ch. Hamner, M. Gong, Ch. Zhang, and P. Engels, *Phys. Rev. A* **88**, 021604 (2013).
13. F. Dalfovo, S. Giorgini, L. P. Pitaevskii and S. Stringari, *Rev. Mod. Phys.* **71**, 463 (1999).

14. C. Sulem and P. L. Sulem, *The Nonlinear Schrödinger Equation: Self-Focusing and Wave Collapse*, Applied Mathematical Sciences, (Springer, New York).
15. C. Chin, R. Grimm, P. Julienne, and E. Tiesinga, *Rev. Mod. Phys.* **82**, 1225 (2010).
16. Sh. Mardonov, E. Ya. Sherman, J. G. Muga, H.-W. Wang, Y. Ban and X. Chen, *Phys. Rev. A* **91**, 043604 (2015).
17. Yu. Kagan, E. L. Surkov, and G. V. Shlyapnikov, *Phys. Rev. Lett.* **79**, 2604 (1997).
18. F. Kh. Abdullaev, J. G. Caputo, R. A. Kraenkel, and B. A. Malomed, *Phys. Rev. A* **67**, 013605 (2003).
19. Y. Kagan, A. E. Muryshev and G. V. Shlyapnikov, *Phys. Rev. Lett.* **81**, 933 (1998).
20. M. I. Dyakonov and V. I. Perel, *Sov. Phys. Solid State* **13**, 3023 (1972).
21. M. Larcher, F. Dalfovo, and M. Modugno, *Phys. Rev. A* **80**, 053606 (2009).
22. I. L. Aleiner, B. L. Altshuler and G. V. Shlyapnikov, *Nature Physics* **6**, 900 (2010).
23. L. Zhou, H. Pu and W. Zhang, *Phys. Rev. A* **87**, 023625 (2013).
24. Sh. Mardonov, M. Modugno and E. Ya. Sherman, *Phys. Rev. Lett.* **115**, 180402 (2015).
25. G. M. Falco, A. A. Fedorenko, J. Giacomelli and M. Modugno, *Phys. Rev. A* **82**, 053405 (2010).
26. For the general theory of disordered systems see A. L. Efros and B. I. Shklovskii, *Electronic Properties of Doped Semiconductors* (Springer, Heidelberg, 1989).
27. Ch. Skokos, D. O. Krimer, S. Komineas and S. Flach, *Phys. Rev. E* **79**, 056211 (2009).

# Fermion Condensation in Strongly Interacting Fermi Liquids

E. V. Kirichenko

*Institute of Mathematics and Informatics, Opole University*
*ul. Oleska 48, 45-052 Opole, Poland*
*E-mail: lenaki@uni.opole.pl*

V. A. Stephanovich

*Institute of Physics, Opole University*
*ul. Oleska 48, 45-052 Opole, Poland*
*E-mail: stef@uni.opole.pl*

This article discusses the construction of a theory which is capable to explain so-called non-Fermi liquid behavior of strongly correlated Fermi systems. We show that such explanation can be done within the framework of a so-called fermion condensation approach. In this approach, as a result of fermion condensation quantum phase transition, ordinary Landau quasiparticles do not decay, but reborn, gaining new properties, as Phoenix from the ashes. The physical reason for that is altering of Fermi surface topology. To be more specific, in contrast to standard Landau paradigm stating that the quasiparticle effective mass does not depend on external stimuli like magnetic field and/or temperature, the effective mass of new quasiparticles strongly depends on them. Our extensive analysis of experimental data gathered on strongly correlated Fermi systems of quite different microscopic nature, shows that the multitude of their non-Fermi liquid properties are due to FC occurrence in them.

*Keywords*: Non-Fermi liquid behavior, fermion condensation, quantum phase transition, heavy-fermion compounds.

## 1. Introduction

The Landau theory of the Fermi liquid has a long history and remarkable results in describing a multitude of physical properties of the electron subsystem in ordinary metals and Fermi liquids of the $^3$He type. The theory is based on the assumption that elementary excitations determine the physics at low temperatures. These excitations behave as quasiparticles of a weakly interacting Fermi gas and thus have a certain effective mass. In this case, the effective mass $M^*$ is independent of the external parameters like temperature and magnetic field strength and is a parameter of the theory.

It is clear, that the above Landau Fermi liquid (LFL) theory fails to explain the experimental results related to the dependence of $M^*$ on the temperature $T$, magnetic field $B$ and other external stimuli. This was the primary reason to conclude that quasiparticles do not survive in some strongly correlated Fermi systems and that the heavy electron does not retain its identity as a quasiparticle excitation.[1–4] The physical phenomena in some strongly correlated fermionic systems (like heavy fermion compounds, high-$T_c$ superconductors, and quasi-two-dimensional Fermi liquids), which cannot be described by LFL theory, have usually been referred to as non-Fermi liquid (NFL) behavior. This behavior is still lacking uniform theoretical explanation.[5–9]

One of the most prominent examples here is so-called heavy - fermion (HF) compounds, where the strong electron - electron correlations generate the quasiparticle effective mass renormalization so that the resulting mass may exceed the ordinary, "bare" electron mass by several orders of magnitude or even become infinitely large. In these compounds, the Landau quasiparticle effective mass depends strongly on temperature, pressure, or applied magnetic field. Such compounds exhibit uncommon power laws in temperature dependence of their low-temperature thermodynamic properties.[1–4] To explain these unusual experimental findings, the ideas that NFL anomalies are related to quantum and thermal fluctuations around LFL state have been suggested.[4,5,7,10–12] The point in the parameters space, where thee anomalies appear, has been called quantum critical point (QCP).

The unusual properties and anomalous behavior observed in high-$T_c$ superconductors and HF compounds are assumed to be determined by various magnetic quantum phase transitions (QPT).[7] Since a QPT occurs at $T = 0$, its control parameters are the composition, the electron (hole) number density $x$, the pressure and the magnetic field strength $B$. A quantum phase transition occurs at QCP, which separates two phases, one of which disappears and the other emerges in this point of external parameters space. It is usually assumed that magnetic (e.g., ferromagnetic and antiferromagnetic) QPTs are responsible for the NFL behavior.

Universal behavior can be expected only if the system under consideration is very close to QCP, i.e. when the correlation length is much longer than the microscopic length scale, and critical quantum and thermal fluctuations determine the anomalous contribution to the thermodynamic functions of a substance. Quantum phase transitions of this type are widespread[6,7] and the essential physics is determined by thermal and quantum fluctuations, destroying quasiparticles. The absence of quasiparticle excitations is considered as the main reason for the NFL behavior of HF metals and high-$T_c$ superconductors.[6,7] However, this approach faces certain difficulties. Namely, the critical behavior in experiments on HF metals containing is observed at temperatures much higher then the effective Fermi temperature $T_k$. For instance, the thermal expansion coefficient $\alpha(T)$, which is a linear function of temperature for the normal LFL, $\alpha(T) \propto T$, demonstrates the temperature dependence of $\sqrt{T}$ type in measurements in $CeNi_2Ge_2$ as the temperature decreases from 6 K to at least 50 mK.[10] Such behavior can hardly be explained in the framework of the above critical point fluctuation theory. Obviously, such situation is possible only as $T \to 0$, when the critical fluctuations make the leading contribution to the entropy and when the correlation length is much longer than the microscopic length scale. At a certain finite temperature this macroscopically large correlation length will obligatory be destroyed by ordinary thermal fluctuations so that the above universal behavior disappears.

The inability to explain the physics of many kinds of strongly interacting Fermi liquids (and HF compounds in particular) within the framework of ordinary QPT approach implies that another important concept introduced by Landau, the order parameter, also ceases to operate (see, e.g., Refs. 2–4). Thus, we are left without the most fundamental principles of many - body quantum physics,[13] and many interesting phenomena associated with the anomalous behavior of strongly correlated Fermi systems remain unexplained.

## 2. Fermi liquid and fermion condensation

### 2.1. *Normal Fermi liquid*

To make the presentation self-contained, here we briefly recapitulate the main ideas of the LFL theory.[13–15] The theory is based on the quasiparticle paradigm, which states that quasiparticles are elementary weakly excited states of Fermi liquids and are therefore specific excitations determining their low-temperature thermodynamic and transport properties. In the case of the electron liquid, the quasiparticles are characterized by the electron quantum numbers and the effective mass $M^*$. The ground state energy of the system is a functional of the quasiparticle occupation numbers (distribution function) $n(\mathbf{p},T)$, and the same is true for the free energy $F(n(\mathbf{p},T))$, the entropy $S(n(\mathbf{p},T))$, and other thermodynamic functions. We can find the distribution function from the minimum condition for the free energy $F = E - TS$ (hereafter we use atomic units, where $k_B = \hbar = 1$)

$$\frac{\delta(F-\mu N)}{\delta n(\mathbf{p},T)} = \varepsilon(\mathbf{p},T) - \mu(T) - T\ln\frac{1-n(\mathbf{p},T)}{n(\mathbf{p},T)} = 0. \tag{1}$$

Here $\mu$ is the chemical potential fixing the number density $x$:

$$x = \int n(\mathbf{p},T)\frac{d\mathbf{p}}{(2\pi)^3} \tag{2}$$

and

$$\varepsilon(\mathbf{p},T) = \frac{\delta E(n(\mathbf{p},T))}{\delta n(\mathbf{p},T)} \tag{3}$$

is the quasiparticle energy spectrum. The quasiparticle spectrum, similar to the ground state energy $E$, is a functional of $n(\mathbf{p},T)$: $\varepsilon = \varepsilon(\mathbf{p},T,n)$. The entropy $S(n(\mathbf{p},T))$ related to quasiparticles is given by the well-known combinatorial expression[13]

$$S(n(\mathbf{p},T)) = -2\int\Big[n(\mathbf{p},T)\ln(n(\mathbf{p},T)) + (1-n(\mathbf{p},T)) \times \ln(1-n(\mathbf{p},T))\Big]\frac{d\mathbf{p}}{(2\pi)^3}. \tag{4}$$

Equation (1) is usually written in the standard form of the Fermi-Dirac distribution,

$$n(\mathbf{p},T) = \left\{1+\exp\left[\frac{(\varepsilon(\mathbf{p},T)-\mu)}{T}\right]\right\}^{-1}. \tag{5}$$

At $T \to 0$, Eqs. (1) and (5) reduce to $n(p,T\to 0) \to \theta(p_F - p)$ if the derivative $\partial\varepsilon(p \simeq p_F)/\partial p$ is finite and positive. Here $p_F$ is the Fermi momentum and $\theta(p_F - p)$ is the unit step function. The single particle energy can be approximated as $\varepsilon(p \simeq p_F) - \mu \simeq p_F(p-p_F)/M^*$, where $M^*$ is the effective mass of Landau quasiparticle,

$$\frac{1}{M^*} = \frac{1}{p}\frac{d\varepsilon(p,T=0)}{dp}\Big|_{p=p_F}. \tag{6}$$

In turn, the effective mass $M^*$ is related to the bare electron mass $m$ by the well-known Landau equation[13–15]

$$\frac{1}{M^*} = \frac{1}{m} + \sum_{\sigma_1}\int\frac{\mathbf{p}_F\mathbf{p}_1}{p_F^3}F_{\sigma,\sigma_1}(\mathbf{p_F},\mathbf{p}_1) \times \frac{\partial n_{\sigma_1}(\mathbf{p}_1,T)}{\partial p_1}\frac{d\mathbf{p}_1}{(2\pi)^3}. \tag{7}$$

where $F_{\sigma,\sigma_1}(\mathbf{p_F}, \mathbf{p}_1)$ is the Landau interaction, which depends on the momenta $\mathbf{p_F}$ and $\mathbf{p}$ and spin indices $\sigma$, $\sigma_1$. For simplicity, we suppress the spin indices in the effective mass as $M^*$ is almost completely spin-independent in the case of a homogeneous liquid and weak magnetic fields. The formal expression for Landau interaction $F$ is given by

$$F_{\sigma,\sigma_1}(\mathbf{p}, \mathbf{p}_1, n) = \frac{\delta^2 E(n)}{\delta n_\sigma(\mathbf{p}) \delta n_{\sigma_1}(\mathbf{p}_1)}. \tag{8}$$

Note that the function $F_{\sigma,\sigma_1}(\mathbf{p}, \mathbf{p}_1, n)$ in LFL theory is not calculated from the expression (8) but, similar to other Landau theories (like theory of phase transitions[13,16]) is supposed to be the series in spherical harmonics with the phenomenological coefficients (so-called Landau amplitudes[13,14]) before these harmonics.

### 2.2. *Pomeranchuk stability conditions*

The notion of Landau interaction as a series over spherical harmonics with phenomenological coefficients permitted Pomeranchuk[17] to formulate the conditions of Fermi liquid stability. Namely, the LFL ground state becomes unstable if at least one Landau amplitude becomes negative and reaches its critical value.[13,15,17]

$$F_L^{a,s} = -(2L+1). \tag{9}$$

Here $F_L^a$ and $F_L^s$ are the dimensionless spin-symmetric and spin-antisymmetric Landau amplitudes, $L$ is the angular momentum related to the corresponding Legendre polynomials $P_L$,

$$F_{\sigma,\sigma_1}(\mathbf{p}, \mathbf{p}_1) = \frac{1}{N} \sum_{L=0}^{\infty} P_L(\Theta) \left[ F_L^a{}_{\sigma,\sigma_1} + F_L^s \right], \tag{10}$$

where $\Theta$ is the angle between momenta $\mathbf{p}$ and $\mathbf{p}_1$ and $N = M^* p_F/(2\pi^2)$ is the density of states. It follows from Eq. (7) that

$$\frac{M^*}{m} = 1 + \frac{F_1^s}{3}, \tag{11}$$

where $F_1^s$ is spin-symmetric Landau amplitude for momentum $L = 1$, i.e. so-called first (or $p$ - wave) Landau harmonic. It follows from Eq. (10) that dimensionless Landau harmonics are related to dimensional ones by multiplication by density of states $N$, which contains the Landau quasiparticle effective mass $M^*$ so that the equation (11) becomes implicit with respect to $M^*$.

To have the effective mass $M^*$ to be positive (which is also in accordance with the Pomeranchuk stability conditions), it is seen from Eq. (11) that it should be $F_1^s > -3$, otherwise $M^* < 0$ leading to unstable and physically irrelevant state.

### 2.3. *LFL ground state energy and the effective mass of Landau quasiparticle*

Good starting point to extend the LFL formalism for fermion condensation consideration is to express the ground state energy $E$ as a functional of the occupations numbers $n(\mathbf{p})$ [18–20]

$$E(n) = \int \frac{p^2}{2m} n(\mathbf{p}) \frac{d\mathbf{p}}{(2\pi)^3} + \frac{1}{2} \int F(\mathbf{p}, \mathbf{p}_1, n)_{|_{n=0}} n(\mathbf{p}) n(\mathbf{p}_1) \frac{d\mathbf{p} d\mathbf{p}_1}{(2\pi)^6}. \tag{12}$$

In this expression, the ground state energy $E$ becomes the functional not only of the occupation numbers but also the function of the number density $x$, $E = E(n(\mathbf{p}), x)$. In this case its variational derivative, Eq. (3), gives the single-particle spectrum. Upon differentiating both sides of Eq. (3) with respect to $\mathbf{p}$ and after some algebra involving integration by parts, we arrive at [9,21]

$$\frac{\partial \varepsilon(\mathbf{p})}{\partial \mathbf{p}} = \frac{\mathbf{p}}{m} + \int F(\mathbf{p}, \mathbf{p}_1, n) \frac{\partial n(\mathbf{p}_1)}{\partial \mathbf{p}_1} \frac{d\mathbf{p}_1}{(2\pi)^3}. \tag{13}$$

Then, substituting Eq. (13) into (6), we obtain well-known Landau equation for quasiparticle effective mass (7). Below we will see that the equation (7) for effective mass, derived from variational derivatives of the functional (12), permit to calculate $M^*$ as a function of temperature $T$, external magnetic field $B$ and other external stimuli. Obviously, $M^*$ acquires latter dependence only in the fermion condensed phase, which, in turn, occurs for some special forms of Landau interaction $F(\mathbf{p}, \mathbf{p}_1)$. [18–20] As we shall see also, it is just this feature of $M^*$ which determines the NFL behavior observed in strongly correlated Fermi systems like HF compounds, $^3$He and others.

Solving the implicit equation (11) for the effective mass, we obtain the explicit dependence of $M^*$ on the dimensional Landau interaction. Introducing the density of states of a free Fermi gas, $N_0 = mp_F/(2\pi^2)$, we obtain[22,23]

$$\frac{M^*}{m} = \frac{1}{1 - F^1/3}, \tag{14}$$

where $F^1 = N_0 f^1$ and $f^1 \equiv f^1(p_F, p_F)$ is the $p$-wave component of the Landau interaction. Since in LFL theory the number density $x$ is related to $p_F$ as $x = p_F^3/3\pi^2$, the Landau interaction can be rewritten as a function of number density $F^1(p_F, p_F) = F^1(x)$. Provided that at a certain critical point $x_{FC}$, the denominator $1 - F^1(x)/3$ equals zero, we can expand it near this point in power series. Namely, at $x \to x_{FC}$ we have $(1 - F^1(x)/3) \propto (x - x_{FC}) + a(x - x_{FC})^2 + ...$, so that substitution to Eq. (14) yields[24]

$$\frac{M^*(x)}{m} \simeq a_1 + \frac{a_2}{x - x_{FC}} \propto \frac{1}{r}, \tag{15}$$

where $a_1$ and $a_2$ are constants and $r = (x - x_{FC})/x_{FC}$ is the "distance" from QCP $x_{FC}$ where $M^*(x \to x_{FC}) \to \infty$. We note that the divergence of the effective mass given by Eq. (15) does preserve the Pomeranchuk stability conditions for $F^1 > -1$, see Eq. (9).

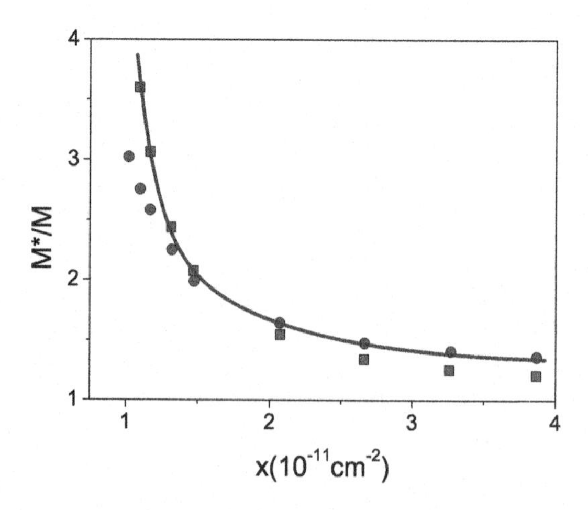


Fig. 1. The ratio $M^*/M$ in a silicon MOSFET as a function of the electron number density $x$. The squares mark the experimental data on the Shubnikov-de Haas oscillations. The data obtained by applying a parallel magnetic field are marked by circles.[25–27] The solid line represents the function (15).

The behavior of $M^*(x)$ described by formula (15) is in good agreement with the experimental results[25–28] and calculations.[29] In the case of electron systems, Eq. (15) holds for $x > x_{FC}$, while for 2D $^3$He we have $x < x_{FC}$ so that always $r > 0$.[28,30] Such behavior of the effective mass is observed in HF metals, which have a fairly flat and narrow conduction band corresponding to a large effective mass, with a strong correlation and the effective Fermi temperature $T_k \sim p_F^2/M^*(x)$ of the order of several dozen kelvins or even lower.[5]

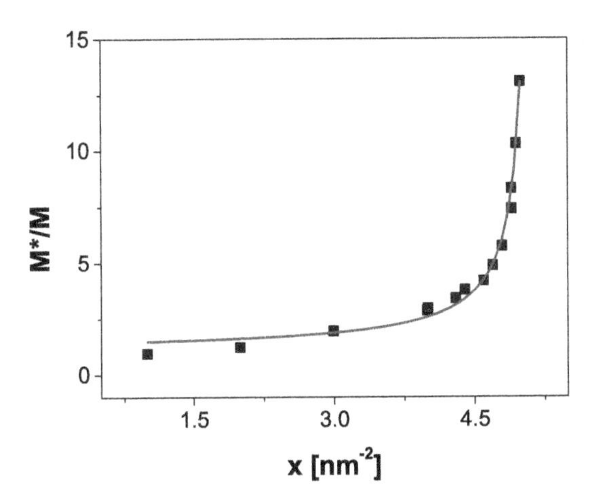


Fig. 2. The ratio $M^*/M$ in 2D $^3$He as a function of the density $x$ of the liquid, obtained from heat capacity and magnetization measurements. The experimental data are marked by black squares,[28] and the solid line represents the function given by Eq. (15), where $a_1$=1.09, $a_2 = 1.68\,\mathrm{nm}^{-2}$, and $x_{FC} = 5.11\ \mathrm{nm}^{-2}$.

The effective mass as a function of the electron density $x$ in a silicon MOSFET (Metal Oxide Semiconductor Field Effect Transistor), approximated by Eq. (15), is shown in Fig. 1. The constants $a_1$, $a_2$ and $x_{FC}$ are taken as fitting parameters. We see that Eq. (15) provides a good description of the experimental results.

The divergence of the effective mass $M^*(x)$ discovered in measurements involving 2D $^3$He[28] is illustrated in Fig. 2. Figures 1 and 2 show that the description provided by Eq. (15) does not depend on the nature of constituting fermions and is in good agreement with the experimental data. This is a reflection of universal features of strongly correlated Fermi system, taking place in the region of their quantum criticality. Below we will see that this universal behavior is intimately related to fermion condensation realization in these substances.

### 2.4. *Fermion condensation*

As it had been shown in Refs. 18,19, the Pomeranchuk stability conditions do not encompass all possible types of instabilities and that at least one related to the divergence of the effective mass given by Eq. (15) was overlooked.[18] This type of instability corresponds to a situation where the effective mass, the most important characteristic of a quasiparticle, can become infinitely large. As a result, the quasiparticle kinetic energy is infinitely small near the Fermi surface and the quasiparticle distribution function $n(\mathbf{p})$ minimizing $E(n(\mathbf{p}))$ (12) is determined by the potential energy. This leads to the formation of a new class of strongly correlated Fermi liquids with fermion condensate (FC),[18,19,31,32] separated from the normal Fermi liquid by fermion condensation quantum phase transition (FCQPT).[33]

It follows from (15) that at $T = 0$ and $r \to 0$ the effective mass diverges, $M^*(r) \to \infty$. Beyond the critical point $x_{FC}$, the distance $r$ becomes negative and, accordingly, so does the effective mass. To avoid an unstable and physically meaningless state with a negative effective mass, the system must undergo a quantum phase transition at the critical point $x = x_{FC}$, which is indeed FCQPT.[33] As the kinetic energy of quasiparticles near the Fermi surface is proportional to the inverse effective mass, their potential energy determines the ground-state one as $x \to x_{FC}$. Hence, a phase transition reduces the energy of the system and transforms the quasiparticle distribution function. Beyond QCP $x = x_{FC}$, the quasiparticle distribution is determined by the ordinary equation for a minimum of the energy functional:[18]

$$\frac{\delta E(n(\mathbf{p}))}{\delta n(\mathbf{p}, T=0)} = \varepsilon(\mathbf{p}) = \mu;\ p_i \leq p \leq p_f. \tag{16}$$

Equation (16) yields the quasiparticle distribution function $n_0(\mathbf{p})$ that minimizes the ground-state energy $E$ (12). This function found from Eq. (16) differs from the step function in the interval from $p_i$ to $p_f$, where $0 < n_0(\mathbf{p}) < 1$, and coincides with the step function outside this interval. In fact, Eq. (16) coincides with Eq. (3) provided that the Fermi surface at $p = p_F$ transforms into the Fermi volume at $p_i \leq p \leq p_f$ suggesting that the single-particle spectrum is absolutely "flat" within this interval. A possible solution $n_0(\mathbf{p})$ of Eq. (16) and the corresponding single-particle spectrum $\varepsilon(\mathbf{p})$ are depicted in Fig. 3. Quasiparticles with momenta within the interval $p_f - p_i$ have the same single-particle energies equal to the chemical potential $\mu$ and form FC, while the distribution $n_0(\mathbf{p})$ describes the new state of the Fermi liquid with FC.[18,19,32] In contrast to the Landau,

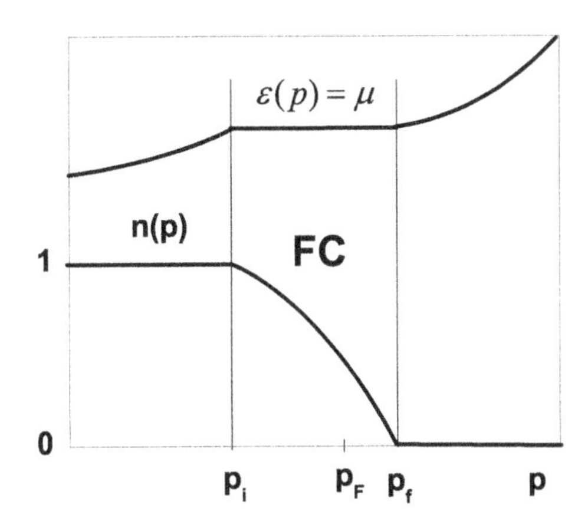


Fig. 3. The single-particle spectrum $\varepsilon(p)$ and the quasiparticle distribution function $n(p)$. As $n(p)$ is a solution of Eq. (16), we have $n(p < p_i) = 1$, $0 < n(p_i < p < p_f) < 1$, and $n(p > p_f) = 0$, while $\varepsilon(p_i < p < p_f) = \mu$. The Fermi momentum $p_F$ satisfies the condition $p_i < p_F < p_f$.

marginal, or Luttinger Fermi liquids,[6,34] which exhibit the same topological structure of the Green's function, in systems with FC, where the Fermi surface spreads into a strip, the Green's function belongs to a different topological class. The topological class of the Fermi liquid is characterized by the invariant[32,35,36]

$$N_t = \mathrm{tr} \oint_C \frac{dl}{2\pi i} G(i\omega, \mathbf{p}) \partial_l G^{-1}(i\omega, \mathbf{p}), \tag{17}$$

where "tr" denotes the trace over the spin indices of the Green's function and the integral is taken along an arbitrary contour $C$ encircling the singularity of the Green's function. The invariant $N_t$ in (17) takes integer values even when the singularity is not of the pole type, cannot vary continuously, and is conserved in a transition from the Landau Fermi liquid to marginal liquids and under small perturbations of the Green's function. As it was shown by Volovik,[32,35,36] the situation is quite different for systems with FC, where the invariant $N_t$ becomes a half-integer and the system with FC transforms into an entirely new class of Fermi liquids with its own topological structure, thus forming a new state of matter.

### 2.5. *Fermion condensate as quantum protectorate*

Like any other phase transition, the FCQPT comprises strong interparticle interaction so that there is no way to describe reliably all its details within first principles approaches. Hence, the only way to confirm or deny the FC existence is to study the model systems, which admit the exact solutions. Such theoretical studies should be augmented by careful examination of experimental data that could be interpreted in favor (or to the detriment) of FC existence. Exactly solvable models unambiguously suggest that Fermi systems with FC exist (see, e.g. Refs. 38–41). Taking the results of topological investigations into account, we can affirm that the new class of Fermi liquids with FC is nonempty, actually exists, and represents an extended family of new states of Fermi systems.[32,35,36]

We note that the solutions $n_0(\mathbf{p})$ of Eq. (16) are new solutions of the well-known equations of the Landau Fermi- liquid theory. Indeed, at $T = 0$, the standard solution given by a step function, $n(\mathbf{p}, T \to 0) \to \theta(p_F - p)$, is not the only possible one. Anomalous solutions $\varepsilon(\mathbf{p}) = \mu$ of Eq. (1) can exist if the logarithmic expression on its right-hand side is finite. This is possible if $0 < n_0(\mathbf{p}) < 1$ for $p_i \leq p \leq p_f$. Then, this logarithmic expression remains finite within this interval as $T \to 0$, the product $T \ln[(1 - n_0(\mathbf{p}))/n_0(\mathbf{p})]_{|T \to 0} \to 0$, and we again arrive at Eq. (16).

Thus, as $T \to 0$, the quasiparticle distribution function $n_0(\mathbf{p})$, which is a solution of Eq. (16), does not tend to the step function $\theta(p_F - p)$ and, correspondingly, in accordance with Eq. (4), the entropy $S(T)$ of this state tends to a finite value $S_0$ as $T \to 0$:

$$S(T \to 0) \to S_0. \tag{18}$$

As the density $x \to x_{FC}$ (or as the interaction force increases), the system reaches QCP where FC is formed. This means that $p_i \to p_f \to p_F$ and that the deviation $\delta n(\mathbf{p})$ from the step function is small. Expanding the function $E(n(\mathbf{p}))$ in Taylor series in $\delta n(\mathbf{p})$ and keeping only the leading terms, we can use Eq. (16) to obtain the following relation that is valid for

$p_i \leq p \leq p_f$:

$$\mu = \varepsilon(\mathbf{p}) = \varepsilon_0(\mathbf{p}) + \int F(\mathbf{p}, \mathbf{p}_1)\delta n(\mathbf{p}_1)\frac{d\mathbf{p}_1}{(2\pi)^2}. \tag{19}$$

Both quantities, the Landau interaction $F(\mathbf{p}, \mathbf{p}_1)$ and the single-particle energy $\varepsilon_0(\mathbf{p})$, are calculated at $n(\mathbf{p}) = \theta(p_F - p)$.

Equation (19) has nontrivial solutions for densities $x \leq x_{FC}$ if the corresponding Landau interaction, which is density-dependent, is positive and sufficiently large for the potential energy to be higher than the kinetic energy. For instance, such a state is realized in a low-density electron liquid. The transformation of the Fermi step function $n(\mathbf{p}) = \theta(p_F - p)$ into a smooth function determined by Eq. (19) then becomes possible.[18,19]

It follows from Eq. (19) that the quasiparticles of fermion condensate form a collective state, since it is determined by the macroscopic number of quasiparticles with the same energy in the momenta interval $p_i < p < p_f$. The shape of the FC single-particle spectrum is independent of the Landau interaction details, which in general is determined by the microscopic properties of the system like chemical composition, structure irregularities and the presence of impurities. The only characteristic determined by the Landau interaction is the length of interval (from $p_i$ to $p_f$) of FC existence. Of course, the interaction must be strong enough for FCQPT to occur. Therefore, we conclude that spectra related to FC have a universal shape. Latter feature can be regarded as characteristic of so-called quantum protectorate, where quantum effects render material properties independent of its microscopic characteristics.[42,43] In our case, the state of matter with FC is also a quantum protectorate.

### 2.6. *The influence of FCQPT at finite temperatures*

According to Eq. (1), the single-particle energy $\varepsilon(\mathbf{p}, T)$ is linear in $T$ for $T \ll T_f$ for $p_f < p < p_i$.[19] Expanding $\ln((1 - n(\mathbf{p}))/n(\mathbf{p}))$ in series in $n(\mathbf{p})$ at $p \simeq p_F$, we arrive at the expression

$$\frac{\varepsilon(\mathbf{p}, T) - \mu(T)}{T} = \ln\frac{1 - n(\mathbf{p})}{n(\mathbf{p})} \simeq \left.\frac{1 - 2n(\mathbf{p})}{n(\mathbf{p})}\right|_{p \simeq p_F}, \tag{20}$$

where $T_f$ is the temperature above which the effect of FC is insignificant:[44]

$$\frac{T_f}{\varepsilon_F} \sim \frac{p_f^2 - p_i^2}{2m\varepsilon_F} \sim \frac{\Omega_{FC}}{\Omega_F}. \tag{21}$$

Here $\Omega_{FC}$ is the volume occupied by FC, $\varepsilon_F$ is the Fermi energy and $\Omega_F$ is the volume of the Fermi sphere. We note that for $T \ll T_f$, the occupation numbers $n(\mathbf{p})$ obtained from Eq. (16) are almost perfectly independent of $T$.[44] At finite temperatures, according to Eq. (20), the dispersionless plateau $\varepsilon(\mathbf{p}) = \mu$ shown in Fig. 3 is slightly rotated counterclockwise relatively to $\mu$. As a result, the plateau is slightly tilted and rounded off at its end points. According to Eqs. (6) and (20), the effective mass $M^*_{FC}$ of the FC quasiparticles is given by

$$M^*_{FC} \simeq p_F \frac{p_f - p_i}{4T}. \tag{22}$$

To derive (22), we approximate $dn(p)/dp \simeq -1/(p_f - p_i)$. Equation (22) shows clearly that for $0 < T \ll T_f$, the electron liquid with FC behaves as if it were placed at a QCP since the electron effective mass diverges as $T \to 0$.

The Eqs. (21) and (22) permit to estimate the effective mass $M^*_{FC}$ in the form

$$\frac{M^*_{FC}}{M} \sim \frac{N(0)}{N_0(0)} \sim \frac{T_f}{T}, \tag{23}$$

where $N_0(0)$ is the density of states of a noninteracting electron gas and $N(0)$ is the density of states at the Fermi surface, see above. Equations (22) and (23) yield the temperature dependence of $M^*_{FC}$.

Multiplying both sides of Eq. (22) by $p_f - p_i$, we obtain an expression for the characteristic energy,

$$E_0 \simeq 4T, \tag{24}$$

which determines the momentum interval $p_f - p_i$ having the low-energy quasiparticles with the energy $|\varepsilon(\mathbf{p}) - \mu| \leq E_0/2$ and the effective mass $M^*_{FC}$. The quasiparticles that do not belong to this momentum interval have an energy $|\varepsilon(\mathbf{p}) - \mu| > E_0/2$ and an effective mass $M^*_L$ that is weakly temperature-dependent.[33] Equation (24) shows that $E_0$ is independent of the condensate volume.

The above discussion shows that a system with FC is characterized by two effective masses, $M^*_{FC}$ and $M^*_L$. This fact manifests itself in the abrupt variation of the quasiparticle dispersion law, which for quasiparticles with energies $\varepsilon(\mathbf{p}) \leq \mu$ can be approximated by two straight lines intersecting at $E_0/2 \simeq 2T$. Figure 3 shows that at $T = 0$, the straight lines intersect at $p = p_i$.

We now estimate the density $x_{FC}$ at which FCQPT occurs. It can be shown that an unlimited increase of the effective mass precedes the appearance of a density wave or a charge density wave formed in electron systems at $r_s = r_0/a_B = r_{cdw}$, where $r_0$ is the average distance between electrons, and $a_B$ is the Bohr radius. Hence, FCQPT certainly occurs at $T = 0$ when $r_s$ reaches its critical value $r_{FC}$ corresponding to $x_{FC}$, with $r_{FC} < r_{cdw}$.[20] We note that the increase of the effective mass at the electron number density decrease has been observed experimentally, see Figs. 1 and 2.

Thus, the formation of FC can be considered as a general property of different strongly correlated systems rather than an exotic phenomenon corresponding to the anomalous solution of Eq. (16). Beyond FCQPT, the condensate volume is proportional to $r_s - r_{FC}$, with $T_f/\varepsilon_F \sim (r_s - r_{FC})/r_{FC}$, at least when $(r_s - r_{FC})/r_{FC} \ll 1$. This implies that[20]

$$\frac{r_s - r_{FC}}{r_{FC}} \sim \frac{p_f - p_i}{p_F} \sim \frac{x_{FC} - x}{x_{FC}}. \tag{25}$$

Since a system state with FC is highly degenerate, FCQPT serves as a stimulator of phase transitions that could lift the degeneracy of the spectrum in the interval $p_f - p_i$. For instance, FC can stimulate the formation of spin density waves, antiferromagnetic and/or ferromagnetic state etc., thus promoting the competition between phase transitions eliminating the degeneracy. The presence of FC facilitates a transition to the superconducting state as both phases have the same order parameter.[20]

### 2.7. *Phase diagram of Fermi system with FCQPT*

At $T = 0$, a quantum phase transition is driven by a nonthermal control parameter like number density $x$. As we have seen, at QCP, $x = x_{FC}$, the effective mass diverges. It follows from Eq. (15) that beyond QCP, the effective mass becomes negative. As such a physically meaningless state cannot be realized, the system undergoes FCQPT leading to the FC formation.

The schematic phase diagrams of the systems which are driven to the FC state by variation of $x$ are reported in Fig. 4. As we have seen, FCQPT occurs as soon as the potential energy of the quasiparticles near the Fermi surface determines the ground-state energy. Therefore, here we have to do with a system composed of particles interacting with each other by forces with strong hardcore repulsion. At elevated densities the potential energy overcomes the kinetic one leading to FC emergence, which is the case for 2D $^3$He films.[28,30]

Figure 4 demonstrates that upon approaching the critical density $x_{FC}$ the system remains in the LFL region at sufficiently low temperatures as it

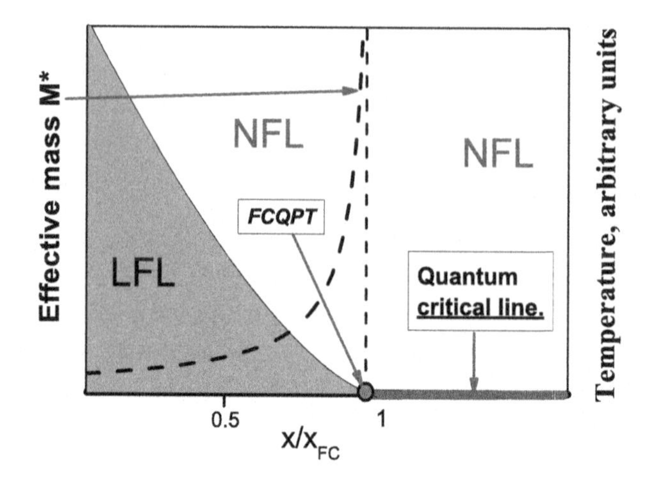


Fig. 4. Schematic phase diagram of the systems with FC. The number density $x$ is taken as the control parameter and depicted as $x/x_{FC}$. The dashed line shows $M^*(x/x_{FC})$ as the system approaches FCQPT, marked by the arrow. The shaded area corresponds to the case $x/x_{FC} < 1$ and sufficiently low temperatures, where the system is in the LFL phase. At $T = 0$ and beyond the FCQPT critical point, the system is at the quantum critical line as shown in the legend. This critical line is characterized by the FC state. At any finite temperature the system undergoes the first order phase transition, possesses finite entropy $S_0$ and exhibits the NFL behavior.

is shown by the shaded area. The temperature range of this area shrinks as the system approaches FCQPT, and $M^*(x/x_{FC})$ diverges as shown by the dashed line and Eq. (15). At FCQPT $x_{FC}$ shown by the arrow in Fig. 4, the system demonstrates the NFL behavior down to the lowest temperatures. Beyond the critical point at finite temperatures the system exhibits NFL properties, which are determined by the temperature-independent entropy $S_0$.[20] In that case at $T \to 0$, the system is approaching a quantum critical line (shown in the panel) rather than a quantum critical point. Upon reaching the quantum critical line from the above at $T \to 0$ the system undergoes the first order quantum phase transition, which is FCQPT taking place at $T = 0$. At the same time, at temperature lowering for $x$ before QCP (i.e. $x < x_{FC}$) the system does not undergo a phase transition and transits smoothly from NFL to LFL phase.

It is seen from Fig. 4 that at finite temperatures there is no boundary (or phase transition) between the states located before or after FC-QPT. Therefore, at elevated temperatures the properties of systems with $x/x_{FC} < 1$ or with $x/x_{FC} > 1$ become indistinguishable. On the other hand, at $T > 0$ the NFL state above the critical line and in the QCP vicinity is strongly degenerate so that the degeneracy stimulates different phase transitions, which finally lift it. The lifting of the degeneracy means that the NFL state can be obscured by the other states like superconducting (for example, in $CeCoIn_5$ [9]) or antiferromagnetic (for example in $YbRh_2Si_2$ [45] or $Cu(C_4H_4N_2)(NO_3)_2$ [46]) etc. The diversity of low-temperature phase transitions is one of the most spectacular features of the HF compounds. The scenario of ordinary quantum phase transitions makes it hard to understand why they are so different and why their critical temperatures are so small. However, such diversity is endemic to systems with a FC.[21]

## 3. Superconducting state with a fermion condensate

### 3.1. *Zero temperature*

To describe superconductivity in the state with fermion condensate we consider new ground state energy functional, which is a sum of normal state energy (12) and superconducting term, proportional to corresponding pairing constant. To be specific, now the ground state energy $E_{gs}[\kappa(\mathbf{p}), n(\mathbf{p})]$ of a 2D electron liquid is a functional of the superconducting order parameter $\kappa(\mathbf{p})$ and the quasiparticle occupation numbers $n(\mathbf{p})$. This energy is determined by the well-known Bardeen-Cooper-Schrieffer (BCS) equations and in the weak-coupling superconductivity theory is given by[47,48]

$$E_{gs}[\kappa(\mathbf{p}), n(\mathbf{p})] = E[n(\mathbf{p})] + \lambda_0 \int V(\mathbf{p}_1, \mathbf{p}_2) \times \kappa(\mathbf{p}_1)\kappa^*(\mathbf{p}_2)\frac{d\mathbf{p}_1 d\mathbf{p}_2}{(2\pi)^4}, \quad (26)$$

where $E[n(\mathbf{p})]$ is determined by Eq. (12) and

$$n(\mathbf{p}) = v^2(\mathbf{p}); \quad \kappa(\mathbf{p}) = v(\mathbf{p})u(\mathbf{p}), \quad (27)$$

where $u(\mathbf{p})$ and $v(\mathbf{p})$ are coefficient of corresponding Bogolyubov transformation (for fermions $v^2(\mathbf{p}) + u^2(\mathbf{p}) = 1$) and $\kappa(\mathbf{p}) = \sqrt{n(\mathbf{p})(1 - n(\mathbf{p}))}$. It is assumed that the constant $\lambda_0$, which determines the magnitude of the pairing interaction $\lambda_0 V(\mathbf{p}_1, \mathbf{p}_2)$, is small. We define the superconducting gap as

$$\Delta(\mathbf{p}) = -\lambda_0 \int V(\mathbf{p}, \mathbf{p}_1)\kappa(\mathbf{p}_1)\frac{d\mathbf{p}_1}{4\pi^2}. \quad (28)$$

Minimizing $E_{gs}$ in $v(\mathbf{p})$ and using (28), we arrive at equations that relate the single-particle energy $\varepsilon(\mathbf{p})$ to $\Delta(\mathbf{p})$ and $E(\mathbf{p})$:

$$\varepsilon(\mathbf{p}) - \mu = \Delta(\mathbf{p})\frac{1 - 2v^2(\mathbf{p})}{2\kappa(\mathbf{p})}, \quad \frac{\Delta(\mathbf{p})}{E(\mathbf{p})} = 2\kappa(\mathbf{p}), \tag{29}$$

where single-particle energy $\varepsilon(\mathbf{p})$ is determined by Landau equation (3):

$$E(\mathbf{p}) = \sqrt{(\varepsilon(\mathbf{p}) - \mu)^2 + \Delta^2(\mathbf{p})}. \tag{30}$$

Substituting the expression for$\kappa(\mathbf{p})$ from (29) in Eq. 28), we obtain the well-known equation of the BCS theory for $\Delta(\mathbf{p})$:

$$\Delta(\mathbf{p}) = -\frac{\lambda_0}{2}\int V(\mathbf{p}, \mathbf{p}_1)\frac{\Delta(\mathbf{p}_1)}{E(\mathbf{p}_1)}\frac{d\mathbf{p}_1}{4\pi^2}. \tag{31}$$

As $\lambda_0 \to 0$, the maximum value $\Delta_1$ of the superconducting gap $\Delta(\mathbf{p})$ tends to zero and Eqs (29) reduce to Eq. (16). Equation (16) shows that for $x < x_{FC}$, the function $n(\mathbf{p})$ is determined from the solution of the standard problem of finding the minimum of the functional $E[n(\mathbf{p})]$.[18,19] We can now study the relations between the state specified by Eq. (16) and the superconducting state.

At $T = 0$, Eq. (16) determines the specific state of a Fermi liquid with fermion condensate, the state for which the absolute value of the order parameter $|\kappa(\mathbf{p})|$ is finite in the momentum interval $p_i \leq p \leq p_f$ as $\Delta_1 \to 0$. Such a state can be considered superconducting with an infinitely small gap $\Delta_1$. Hence, the entropy of this state at $T = 0$ is zero. Obviously, the quantum state with a fermion condensate that emerged as a result of FC-QPT disappears at finite temperatures.[9,19] Any quantum phase transition at $T = 0$ is controlled not by the temperature but by other parameters like pressure, magnetic field strength, or the density $x$ of mobile charge carriers. As we have seen above, in the case of fermion condensate, such parameter is the density $x$, which determines the value of corresponding Landau amplitude.

Solutions $n_0(\mathbf{p})$ of Eqs. (16) constitute a new class of joint BCS and LFL equations solutions. In contrast to the ordinary solutions of the BCS equations,[47] the new solutions are characterized by an infinitely small superconducting gap $\Delta_1 \to 0$, with the order parameter $\kappa(\mathbf{p})$ remaining finite. On the other hand, in contrast to the standard solution of the LFL equations, the new solutions determine the state of an electron liquid with a finite entropy $S_0$ at $T \to 0$, Eq. (18).

We arrive at an important conclusion that the solutions of Eq. (16) can be interpreted as the general solutions of both BCS and LFL equations, while Eq. (16) can be by itself derived either from the BCS or LFL theory. Thus, both states of the system (i.e. superconductivity and FC) coexist at $T \to 0$. As the system passes into a state with the order parameter $\kappa(\mathbf{p})$, the entropy vanishes abruptly so that the system undergoes a first-order transition near which the critical quantum and thermal fluctuations are suppressed. This means that in the vicinity of latter critical point the quasiparticles survive and remain to be well defined excitations. It follows from Eq. (17) that FCQPT is related to altering of the system Green's function topological structure and hence belongs to Lifshits's topological phase transitions, which occur at absolute zero.[32] This fact establishes a relation between FCQPT and quantum phase transitions under which the Fermi sphere splits into a sequence of Fermi layers.[9,20] We note that in the state with finite superconducting order parameter $\kappa(\mathbf{p})$, the system entropy $S = 0$ and the Nernst theorem holds.

If $\lambda_0 \neq 0$, the gap $\Delta_1$ becomes finite, leading to a finite value of the effective mass $M^*_{FC}$, which may be obtained from Eq. (29) by taking the derivative with respect to the momentum $p$ of both sides and using Eq. (6) :[9,19,20]

$$M^*_{FC} \simeq p_F \frac{p_f - p_i}{2\Delta_1}. \tag{32}$$

In this case, the characteristic energy scale is determined by the parameter $E_0$:

$$E_0 = \varepsilon(\mathbf{p}_f) - \varepsilon(\mathbf{p}_i) \simeq p_F \frac{(p_f - p_i)}{M^*_{FC}} \simeq 2\Delta_1. \tag{33}$$

### 3.2. *Green's function of the superconducting state with a Fermi condensate at $T = 0$*

We write Gor'kov equations, which determine the Green's functions $F^+(\mathbf{p},\omega)$ and $G(\mathbf{p},\omega)$ of a superconductor, (see, e.g., Ref. 13):

$$\begin{aligned} F^+ &= \frac{-\lambda_0 \Xi^*}{(\omega - E(\mathbf{p}) + i0)(\omega + E(\mathbf{p}) - i0)}; \\ G &= \frac{u^2(\mathbf{p})}{\omega - E(\mathbf{p}) + i0} + \frac{v^2(\mathbf{p})}{\omega + E(\mathbf{p}) - i0}, \end{aligned} \tag{34}$$

The gap $\Delta$ and the function $\Xi$ are given by

$$\Delta = \lambda_0 |\Xi|, \quad i\Xi = \int \int_{-\infty}^{\infty} F^+(\mathbf{p},\omega) \frac{d\omega d\mathbf{p}}{(2\pi)^4}. \tag{35}$$

We recall that the function $F^+(\mathbf{p},\omega)$ has the meaning of the wave function of Cooper pairs and $\Xi$ is the wave function of the motion of these pairs as a whole and is just a constant in a homogeneous system.[13] It follows from Eqs. (29) and (35) that

$$i\Xi = \int_{-\infty}^{\infty} F_0^+(\mathbf{p},\omega)\frac{d\omega d\mathbf{p}}{(2\pi)^4} = i\int \kappa(\mathbf{p})\frac{d\mathbf{p}}{(2\pi)^3}. \tag{36}$$

Taking Eqs. (35) and (29) into account, we can write Eqs. (34) as[49]

$$\begin{aligned} F^{\,+} &= -\frac{\kappa(\mathbf{p})}{\omega - E(\mathbf{p}) + i0} + \frac{\kappa(\mathbf{p})}{\omega + E(\mathbf{p}) - i0}; \\ G &= \frac{u^2(\mathbf{p})}{\omega - E(\mathbf{p}) + i0} + \frac{v^2(\mathbf{p})}{\omega + E(\mathbf{p}) - i0}. \end{aligned} \tag{37}$$

As $\lambda_0 \to 0$, the gap $\Delta \to 0$, but $\Xi$ and $\kappa(\mathbf{p})$ remain finite if the spectrum becomes flat, $E(\mathbf{p}) = 0$, and Eqs. (37) become

$$\begin{aligned} F^+(\mathbf{p},\omega) &= -\kappa(\mathbf{p})\left[\frac{1}{\omega + i0} - \frac{1}{\omega - i0}\right]; \\ G(\mathbf{p},\omega) &= \frac{u^2(\mathbf{p})}{\omega + i0} + \frac{v^2(\mathbf{p})}{\omega - i0} \end{aligned} \tag{38}$$

in the interval $p_i \leq p \leq p_f$. The parameters $v(\mathbf{p})$ and $u(\mathbf{p})$ are determined by the condition that the spectrum is flat: $\varepsilon(\mathbf{p}) = \mu$.

We construct the functions $F^+(\mathbf{p},\omega)$ and $G(\mathbf{p},\omega)$ in the case where the constant $\lambda_0$ is finite but small, such that $v(\mathbf{p})$ and $\kappa(\mathbf{p})$ can be found on the basis of the Fermi-condensate solutions of Eq. (16). Then $\Xi$, $E(\mathbf{p})$ and $\Delta$ are given by Eqs. (36), (35), and (29). Substituting the functions constructed in this manner into (37), we obtain $F^+(\mathbf{p},\omega)$ and $G(\mathbf{p},\omega)$.[49] We note that Eqs. (35) imply that the gap $\Delta$ is a linear function of $\lambda_0$ under the adopted conditions.

### 3.3. *The superconducting state at finite temperatures*

We assume that the region occupied by the Fermi condensate is small: $(p_f - p_i)/p_F \ll 1$ and $\Delta_1 \ll T_f$. Then, the order parameter $\kappa(\mathbf{p})$ is determined primarily by the Fermi condensate, i.e., the distribution function $n_0(\mathbf{p})$.[20] To be able to solve Eq. (31) analytically, we adopt the BCS approximation for the interaction:[47] $\lambda_0 V(\mathbf{p},\mathbf{p}_1) = -\lambda_0$ if $|\varepsilon(\mathbf{p}) - \mu| \leq \omega_D$ ($\omega_D$ is Debye frequency) and the interaction is zero outside this region. As a result, the superconducting gap depends only on temperature, $\Delta(\mathbf{p}) \equiv \Delta_1(T)$, and

Eq. (31) assumes the form

$$N_{FC}\lambda_0 \int\limits_0^{E_0/2} \frac{d\xi}{\sqrt{\xi^2+\Delta_1^2(0)}} + N_L\lambda_0 \int\limits_{E_0/2}^{\omega_D} \frac{d\xi}{\sqrt{\xi^2+\Delta_1^2(0)}} = 1, \qquad (39)$$

where $E_0$ is determined by expression (33) and we introduce the notation $\xi = \varepsilon(\mathbf{p}) - \mu$ and the density of states $N_{FC}$ in the interval $p_f - p_i$. It follows from Eq. (32) that $N_{FC} = p_F(p_f - p_F)/(2\pi\Delta_1)$. Within the energy interval $\omega_D - E_0/2$, the density of states $N_L$ has the standard form $N_L = M_L^*/2\pi$. As $E_0 \to 0$, Eq. (39) becomes the BCS equation for $\Delta_1(0)$ and hence the superconducting phase transition temperature $T_c$. In pure FC phase we assume that $E_0 \leq 2\omega_D$ and discard the second integral in the right-hand side of Eq. (39) to obtain

$$\begin{aligned} \Delta_1(0) &= \frac{\lambda_0 p_F(p_f - p_F)}{2\pi} \ln\left(1+\sqrt{2}\right) \\ &= 2\beta\varepsilon_F \frac{p_f - p_F}{p_F} \ln\left(1+\sqrt{2}\right), \end{aligned} \qquad (40)$$

where $\varepsilon_F = p_F^2/2M_L^*$ is the Fermi energy and $\beta = \lambda_0 M_L^*/2\pi$ is the dimensionless coupling constant. Using the standard value of $\beta$ for ordinary superconductors, e.g., $\beta \simeq 0.3$, and assuming that $(p_f - p_F)/p_F \simeq 0.2$, we obtain a large value $\Delta_1(0) \sim 0.1\varepsilon_F$ from Eq. (40); for ordinary superconductors, this gap has a much smaller value: $\Delta_1(0) \sim 10^{-3}\varepsilon_F$. With the account for above discarded second integral, we find that

$$\begin{aligned} \Delta_1(0) &\simeq 2\beta\varepsilon_F \frac{p_f - p_F}{p_F} \ln\left(1+\sqrt{2}\right) \\ &+ \Delta_1(0)\beta \ln\left(\frac{2\omega_D}{\Delta_1(0)}\right). \end{aligned} \qquad (41)$$

The expression (41) can be regarded as an equation for $\Delta_1(0)$. As $E_0 \to 0$ and $p_f \to p_F$, the first terms in the right-hand sides of Eqns (39), (41) are zero and we obtain the ordinary BCS result. The isotopic effect is small in this case, as $T_c$ depends on $\omega_D$ logarithmically, but the effect is restored as $E_0 \to 0$.

At $T \simeq T_c$, Eqs. (32) and (33) are replaced by Eqs. (22) and (24), which also hold for $T_c \leq T \ll T_f$:

$$M_{FC}^* \simeq p_F \frac{p_f - p_i}{4T_c}, \quad E_0 \simeq 4T_c, \; T \simeq T_c, \qquad (42)$$

$$M_{FC}^* \simeq p_F \frac{p_f - p_i}{4T}, \quad E_0 \simeq 4T, \; T \geq T_c. \qquad (43)$$

To obtain the temperature dependence of superconducting gap, the equation (39) is replaced by its standard finite temperature generalization

$$1 = N_{FC}\lambda_0 \int\limits_0^{E_0/2} \frac{d\xi}{\sqrt{\xi^2+\Delta_1^2}} \tanh\frac{\sqrt{\xi^2+\Delta_1^2}}{2T} + N_L\lambda_0 \int\limits_{E_0/2}^{\omega_D} \frac{d\xi}{\sqrt{\xi^2+\Delta_1^2}} \tanh\frac{\sqrt{\xi^2+\Delta_1^2}}{2T}. \qquad (44)$$

Since $\Delta_1(T \to T_c) \to 0$, Eq. (44) implies a relation that closely resembles the BCS result,[8]

$$2T_c \simeq \Delta_1(0), \qquad (45)$$

where $\Delta_1(0) \equiv \Delta_1(T=0)$ is found from Eq. (41). Comparing (32) and (33) with (42) and (43), we see that both $M^*_{FC}$ and $E_0$ are temperature independent for $T \leq T_c$.

Note that similarly to LFL theory, the theory of high-$T_c$ superconductivity based on the FCQPT notion, deals with quasiparticles that are elementary low-energy excitations. The theory provides a qualitative general description of the superconducting and normal states of a superconductor. Of course, with phenomenological parameters (e.g., the superconductive pairing constant) chosen, we can obtain a quantitative description of superconductivity in this case similar to LFL approach to a normal Fermi liquid like $^3$He. Hence, any theory capable of describing a fermion condensate and compatible with the BCS theory gives the same qualitative picture of the superconducting and normal states as our FC approach.

## 4. Summary and conclusions

The electronic correlations, which cannot be described perturbatively starting from ideal Fermi gas as zeroth approach, generate a fascinating state known as strongly correlated Fermi systems. Almost three decades have passed since appearance of the ideas that led to the development of fermion condensation (extended Landau quasiparticle) paradigm[18,19,32,36] in latter systems. During this time many theoretical and experimental papers appeared using this paradigm to study numerous strongly correlated electron systems, ranging from HF compounds and high-$T_c$ superconductors to the systems like 2D liquid $^3$He and even astronomical objects like black holes[50] and baryons in the early Universe.[20] The basic FC mechanism is a new type of instability of ordinary Fermi liquid relatively the infinite growth of Landau quasiparticle effective mass. To avoid such unphysical situation, the system rearranges its energy spectrum. In other words, the Fermi surface at $p = p_F$ transforms into the Fermi volume at $p_i \leq p \leq p_f$ suggesting that a single - particle spectrum is absolutely "flat" within this interval, i.e. the corresponding energies equal to the chemical potential $\mu$. This state, where all quasiparticles with momenta $p_i \leq p \leq p_f$, have the same energy, resembles a lot the Bose-condensed state. That is why the above state had been called Fermi condensate.[18] It had been shown by Volovik,[32] that the FC state possesses different (than that of ordinary Fermi liquid) topology of the Fermi surface and thus cannot be reached from LFL without the phase transition. This phase transition had been called the Fermion condensation quantum phase transition. The main physical effect of the FC theory is above extended quasiparticle paradigm. Namely, during FCQPT, the quasiparticles survive but they become different from those in ordinary Fermi liquid. More precisely, if in ordinary LFL the quasiparticle effective mass $M^*$ is a constant, in FC state this quantity starts to depend on the external stimuli like temperature $T$ and external magnetic field $B$. This dependence is a clue for explanation of mysterious NFL behavior of many strongly correlated fermionic systems.

Note that many existing approaches to describe the NFL behavior presuppose the absence of quasiparticles at the QCP of HF compounds. Arguments that quasiparticles in strongly correlated electron systems "get heavy and die" at the QCP commonly employ the assumption that the quasiparticle weight factor $z$ vanishes at the point of an associated second-order phase transition.[3,4] Numerous experimental results have been discussed in terms of such an approach, but they were not able to explain quantitatively all subtleties of the HF compounds physics.[9,20] Extensive studies have shown that above discussed FC approach, which preserves quasiparticles while being intimately related to the unlimited growth of $M^*$, delivers an adequate theoretical explanation of vast majority of experimental results in different HF metals. The essential point is that – as before – well-defined quasiparticles determine the thermodynamic and transport properties of strongly correlated Fermi systems, while the dependence of the effective mass $M^*$ on $T$ and $B$ gives rise to the observed NFL behavior. The most fruitful strategy for exploring and revealing the nature of the QCP is to focus on those properties that exhibit the most spectacular deviations from LFL behavior in the zero-temperature limit.

It turns out that the FC occurs in many compounds, generating the

NFL behavior by forming flat bands. This means that different materials with strongly correlated fermions can unexpectedly have a similar behavior despite their microscopic (like lattice symmetry and structure, magnetic ground state, dimensionality etc.) diversity. Thus, the quantum critical physics of different systems with strongly interacting fermions is universal and emerges regardless of the underlying microscopic details of the substances. This identical behavior, induced by the universal quantum critical physics, allows us to view it as the main characteristic of the new state of matter.[32,36,51] Moreover, the presence of fermion condensate can be considered as the universal reason for the NFL behavior observed in various HF metals, liquids, insulators with quantum spin liquids, and quasicrystals.

The FC state represents the topologically protected new state of matter. In the case of Bose system the equation $\delta E/\delta n(p) = \mu$ describes a common instance. In the case of Fermi systems such an equation, generally speaking, is not correct. Thus, it is the FC state, taking place behind FCQPT, that makes this equation applicable to Fermi systems. This means, that Landau quasiparticles with Fermi statistics can behave as those with Bose ones, occupying the same energy level $\varepsilon = \mu$. This state is viewed as the state possessing the supersymmetry (SUSY) that interchanges bosons and fermions eliminating the difference between them. The FC state accompanied by SUSY violates the time invariance symmetry, while emerging SUSY violates the baryon symmetry of the Universe. Thus, restoring one important symmetry, the FC state violates another one. In the future, the domain of problems should be broadened and certain efforts should be made to describe the other macroscopic features of FCQPT that could strongly modify the thermoelectric effects like Seebeck, Peltier etc. due to their relation to the entropy. We recollect here that in FC state the entropy at $T = 0$ is finite thus violating the Nernst theorem.

In addition to the already known materials whose properties obviously manifest the FC presence, there are other materials of enormous interest which could serve as possible objects for studying the FCQPT. To name a few, these objects are neutron stars, black holes, atomic clusters and fullerenes as well as trapped ultracold gases, nuclei, and quark plasma. Another possible area of research is related to the structure of the nucleon, in which the entire "sea" of non-valence quarks may be in FC state. The combination of quarks and gluons that hold them together is especially interesting because gluons, quite possibly, can be in the gluon-condensate phase, which could be qualitatively similar to the pion condensate proposed by Migdal long ago.[52] We believe that FC can be observed in traps, where there is a possibility to control the emergence of a quantum phase transition accompanied by the FC formation by altering the particle number density.

Entirely, the ideas associated with new physics in one area of research stimulates intensive studies of its possible manifestation in other areas. This has happened in the case of metal superconductivity, whose ideas were fruitful in description of atomic nuclei and in a possible explanation of the origin of the mass of elementary particles. This, quite possibly, could be the case for FCQPT.

## References


1. J. Custers, P. Gegenwart, H. Wilhelm, K. Neumaier, Y. Tokiwa, O. Trovarelli, C. Geibel, F. Steglich, C. Pépin and P. Coleman, The break-up of heavy electrons at a quantum critical point, *Nature* **424**, 524 (2003).
2. T. Senthil, M. Vojta, S. Sachdev, Weak magnetism and non-Fermi liquids near heavy-fermion critical points, *Phys. Rev. B* **69**, 035111 (2004).
3. P. Coleman, *Lectures on the Physics of Highly Correlated Electron Systems VI,* ed. F Mancini (New York: American Institute of Physics, 2002).
4. P. Coleman and A. J. Schofield, Quantum criticality, *Nature* **433**, 226 (2005).
5. G. R. Stewart, Non-Fermi-liquid behavior in $d-$ and $f-$ electron metals, *Rev. Mod. Phys.* **73**, 797 (2001).
6. C.M. Varma, Z. Nussionov, W. van Saarloos, Singular or non-Fermi liquids, *Phys. Rep.* **361**, 267 (2002).
7. H. v. Löhneysen, A. Rosch, M. Vojta, P. Wölfle, Fermi-liquid instabilities at magnetic quantum phase transitions, *Rev. Mod. Phys.* **79**, 1015 (2007).
8. V. I. Belyavsky, Yu. V. Kopaev, Superconductivity of repulsive particles, *Physics–Uspekhi* **49**, 441 (2006).
9. V. R. Shaginyan, M. Ya. Amusia and K. G. Popov, Universal behavior of strongly correlated Fermi-systems. *Physics–Uspekhi* **50**, 563 (2007).
10. R. Küchler, N. Oeschler, P. Gegenwart, T. Cichorek, K. Neumaier, O. Tegus, C. Geibel, J. A. Mydosh, F. Steglich, L. Zhu, and Q. Si, Divergence of the Grüneisen Ratio at Quantum Critical Points in Heavy Fermion Metals, *Phys. Rev. Lett.* **91**, 066405 (2003).
11. N. E. Hussey, Strongly correlated electrons: Landau theory takes a pounding, *Nature Phys.* **3**, 445 (2007).

12. P. Gegenwart, T. Westerkamp, C. Krellner, Y. Tokiwa, S. Paschen, C. Geibel, F. Steglich, E. Abrahams, Q. Si, Multiple Energy Scales at a Quantum Critical Point, *Science* **315**, 969 (2007).
13. E. M. Lifshitz, L. P. Pitaevskii, *Statistical Physics, Part 2* (Butterworth-Heinemann, Oxford, 1999).
14. L. D. Landau, The Theory of a Fermi Liquid, *Sov. Phys. JETP* **3**, 920 (1956).
15. D. Pines, Ph. Noziérres *Theory of Quantum Liquids* (Benjamin, New York, Amstredam, 1966).
16. L. D. Landau and E. M. Lifshits, *Electrodynamics of Continuous Media* (Butterworth-Heinemann, Oxford, 2004).
17. I. Ya. Pomeranchuk, On the Stability of a Fermi Liquid, *Sov. Phys. JETP* **8**, 361 (1959).
18. V. A. Khodel, V. R. Shaginyan, Superfluidity in system with fermion condensate, *JETP Lett.* **51**, 553 (1990).
19. V. A. Khodel, V. R. Shaginyan, V.V. Khodel, New approach in the microscopic Fermi systems theory, *Phys. Rep.* **249**, 1 (1994).
20. M. Ya. Amusia, K. G. Popov, V. R. Shaginyan and V. A. Stephanovich, *Theory of Heavy-Fermion Compounds*, (Springer Series in Solid-State Sciences 182, 2014).
21. V. A. Khodel, J. W. Clark, M. V. Zverev, Topology of the Fermi surface beyond the quantum critical point, *Phys. Rev. B* **78**, 075120 (2008).
22. D. Vollhardt, Normal $^3$He: an almost localized Fermi liquid, *Rev. Mod. Phys.* **56**, 99 (1984).
23. D. Vollhardt, P. Wölfle, P. W. Anderson, Gutzwiller-Hubbard lattice-gas model with variable density: Application to normal liquid $^3$He, *Phys. Rev. B* **35**, 6703 (1987).
24. V. A. Khodel, M. V. Zverev and V. M. Yakovenko, Curie Law, Entropy Excess, and Superconductivity in Heavy Fermion Metals and Other Strongly Interacting Fermi Liquids, *Phys. Rev. Lett.* **95**, 236402 (2005).
25. A. A. Shashkin, S. V. Kravchenko, V. T. Dolgopolov, T. M. Klapwijk, Sharp increase of the effective mass near the critical density in a metallic two-dimensional electron system, *Phys. Rev. B* **66**, 073303 (2002).
26. A. A. Shashkin, M. Rahimi, S. Anissimova, S.V. Kravchenko, V. T. Dolgopolov, T. M. Klapwijk, Spin-Independent Origin of the Strongly Enhanced Effective Mass in a Dilute 2D Electron System, *Phys. Rev. Lett.* **91**, 046403 (2003).

27. S.V. Kravchenko, M.P. Sarachik, Metal?insulator transition in two-dimensional electron systems, *Rep. Prog. Phys.* **67**, 1 (2004).
28. A. Casey, H. Patel, J. Nyéki, J. Cowan, B. P. Saunders, Evidence for a Mott-Hubbard Transition in a Two-Dimensional $^3$He Fluid Monolayer, *Phys. Rev. Lett.* **90**, 115301 (2003).
29. Y. Zhang, V. M. Yakovenko, S. Das Sarma, Dispersion instability in strongly interacting electron liquids, *Phys. Rev. B* **71**, 115105 (2005).
30. V. R. Shaginyan, A. Z. Msezane, K. G. Popov, and V. A. Stephanovich, Universal Behavior of Two-Dimensional $^3$He at Low Temperatures, *Phys. Rev. Lett.* **100**, 096406 (2008).
31. V. A. Khodel, J. W. Clark, H. Li, M. V. Zverev, Merging of Single-Particle Levels and Non-Fermi-Liquid Behavior of Finite Fermi Systems, *Phys. Rev. Lett.* **98**, 216404 (2007).
32. G. E. Volovik, A new class of normal Fermi liquids, *JETP Lett.* **53**, 222 (1991).
33. M. Y. Amusia, V. R. Shaginyan, Quasiparticle picture of high-temperature superconductors in the frame of a Fermi liquid with the fermion condensate, *Phys. Rev. B* **63**, 224507 (2001).
34. C.M. Varma, P.B. Littlewood, S. Schmitt-Rink, E. Abrahams, A.E. Ruckenstein, Phenomenology of the normal state of Cu-O high-temperature superconductors, *Phys. Rev. Lett.* **63**, 1996 (1989).
35. G.E. Volovik, Topology of Momentum Space and Quantum Phase transitions, *Acta Phys. Slov.* **56**, 49 (2006)
36. G.E. Volovik, in *Quantum Analogues: From Phase Transitions to Black Holes and Cosmology*, eds. W.G. Unruh, R. Schutzhold, Springer Lecture Notes in Physics, 718 (Springer, Orlando, 2007), p. 31.
37. L. N. Oliveira, E. K.U. Gross, W. Kohn, Density-Functional Theory for Superconductors, *Phys. Rev. Lett.* **60**, 2430 (1988).
38. D.V. Khveshchenko, R. Hlubina, T.M. Rice, Non-Fermi-liquid behavior in two dimensions due to long-ranged current-current interactions, *Phys. Rev. B* **48**, 10766 (1993).
39. I.E. Dzyaloshinskii, Extended Van-Hove Singularity and Related Non-Fermi Liquids, *J. Phys. I (France)* **6**, 119 (1996).
40. D. Lidsky, J. Shiraishi, Y. Hatsugai, M. Kohmoto, Simple exactly solvable models of non-Fermi-liquids, *Phys. Rev. B* **57**, 1340 (1998).
41. V.Y. Irkhin, A.A. Katanin, M.I. Katsnelson, Robustness of the Van Hove Scenario for High-$T_c$ Superconductors, *Phys. Rev. Lett.* **89**, 076401 (2002).

42. R.B. Laughlin, D. Pines, The Theory of Everything, *Proc. Natl. Acad. Sci. USA* **97**, 28 (2000).
43. P. W. Anderson, Sources of Quantum Protection in High-$T_c$ Superconductivity, *Science* **288**, 480 (2000).
44. J. Dukelsky, V. Khodel, P. Schuck, V. Shaginyan, Fermion condensation and non Fermi liquid behavior in a model with long range forces, *Z. Phys. B* **102**, 245 (1997).
45. V.R. Shaginyan, M.Y. Amusia, K.G. Popov, Strongly correlated Fermi-systems: Non-Fermi liquid behavior, quasiparticle effective mass and their interplay, *Phys. Lett. A* **373**, 2281 (2009).
46. V. R. Shaginyan, V. A. Stephanovich, K. G. Popov, E. V. Kirichenko, and S. A. Artamonov, Magnetic quantum criticality in quasi-one-dimensional Heisenberg antiferromagnet $Cu(C_4H_4N_2)(NO_3)_2$, *Ann. Phys. (Berlin)* **528**, 483 (2016).
47. J. Bardeen, L.N. Cooper and J. R. Schrieffer, Theory of Superconductivity, *Phys. Rev.* **108**, 1175 (1957).
48. D.R. Tilley, J. Tilley *Superfluidity and Superconductivity*, (Bristol: Hilger, 1985).
49. V. R. Shaginyan, A. Z. Msezane, V. A. Stephanovich and E. V. Kirichenko, Quasiparticles and quantum phase transition in universal low-temperature properties of heavy-fermion metals, *Europhys. Lett.* **76**, 898 (2006).
50. S. S. Lee Non-Fermi liquid from a charged black hole: A critical Fermi ball, *Phys. Rev. D* **79**, 086006 (2009).
51. D. Yudin, D. Hirschmeier, H. Hafermann, O. Eriksson, A. I. Lichtenstein, M. I. Katsnelson, Fermi Condensation Near van Hove Singularities Within the Hubbard Model on the Triangular Lattice, *Phys. Rev. Lett.* **112**, 070403 (2014).
52. A.B. Migdal, *Theory of Finite Fermi Systems and Applications to Atomic Nuclei* (Wiley, New York, 1967).

# Theory of Electron Transport and Magnetization Dynamics in Metallic Ferromagnets

Gen Tatara

*RIKEN Center for Emergent Matter Science (CEMS)*
*2-1 Hirosawa, Wako, Saitama 351-0198, Japan*
*E-mail: gen.tatara@riken.jp*

Magnetic electric effects in ferromagnetic metals are discussed from the viewpoint of effective spin electromagnetic field that couples to conduction electron spin. The effective field in the adiabatic limit is the spin Berry's phase in space and time, and it leads to spin motive force (voltage generated by magnetization dynamics) and topological Hall effect due to spin chirality. Its gauge coupling to spin current describes the spin transfer effect, where magnetization structure is driven by an applied spin current. The idea of effective gauge field can be extended to include spin relaxation and Rashba spin-orbit interaction. Voltage generation by the inverse Edelstein effect in junctions is interpreted as due to the electric component of Rashba-induced spin gauge field. The spin gauge field arising from the Rashba interaction turns out to coincides with toroidal moment, and causes asymmetric light propagation (directional dichroism) as a result of the Doppler shift. Rashba conductor without magnetization is shown to be natural metamaterial exhibiting negative refraction.

*Keywords*: Spintronics, spin-charge conversion, gauge field, Rashba spin-orbit interaction.

## 1. Introduction

Our technology is based on various electromagnetic phenomena. For designing electronics devices, the Maxwell's equation is therefore of essential importance. The mathematical structure of the electromagnetic field is governed by a U(1) gauge symmetry, i.e., an invariance of physical laws under phase transformations. The gauge symmetry is equivalent to the conservation of the electric charge, and was established when a symmetry breaking of unified force occurred immediately after the big bang. The beautiful mathematical structure of charge electromagnetism was therefore determined when our universe started, and there is no way to modify its laws.

Fig. 1. The spin of a conduction electron is rotated by a strong *sd* interaction with magnetization as it moves in the presence of a magnetization texture, resulting in a spin gauge field. Magnetization texture is therefore equivalent to an effective electromagnetic field for conduction electron spin.

Interestingly, charge electromagnetism is not the only electromagnetism allowed in the nature. In fact, electromagnetism arises whenever there is a U(1) gauge symmetry associated with conservation of some effective charge. In solids, there are several systems which have the U(1) gauge symmetry as a good approximation. Solids could thus display several types of effective electromagnetic fields. A typical example is a ferromagnetic metal. In ferromagnetic metals, conduction electron spin (mostly *s* electron) is coupled to the magnetization (or localized spins of *d* electrons) by an interaction called the *sd* interaction, which tends to align the electron spin parallel (or anti-parallel) to the localized spin. This interaction is strong in most 3*d* ferromagnetic metals, and as a result, conduction electron's spin originally consisting of three components, reduces to a single component along the localized spin direction. The remaining component is invariant under a phase transformation, i.e., has a U(1) gauge symmetry just like the electric charge does. A spin electromagnetic field thus emerges that couples to conduction electron's spin.

The subject of the present paper is this spin electromagnetic field. Spin electromagnetic field drives electron's spin, and thus plays essential roles in spintronics. There is a gauge field for the spin electromagnetic field, a spin gauge field, which couples to spin current of the conduction electron. The gauge coupling describes the effects of spin current on the localized spin dynamics. As we shall see, when a spin-polarized electric current is applied, the adiabatic spin gauge field leads to spin-transfer torque and moves the magnetization structure (Sec. 6). The world of spin electromagnetic field is richer than that of electric charge, since the electron's spin in solids is under influence of various interactions such as spin-orbit interaction. We shall show that even magnetic monopoles can emerge (Sec. 4)

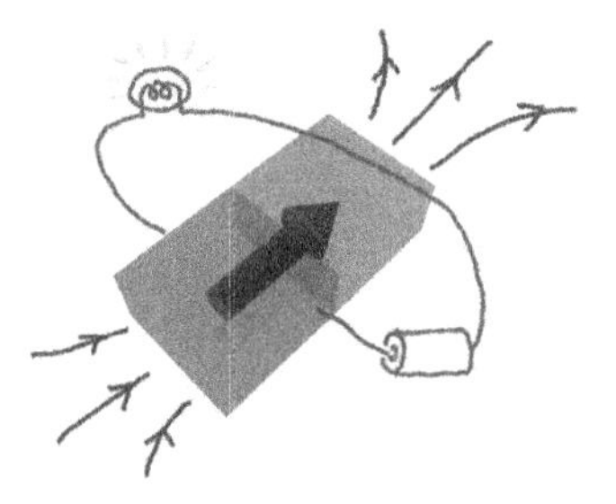

Fig. 2. Ferromagnetic metals have magnetization and conduct electricity, indicating existence of localized spins and conduction electrons.

A spin electromagnetic field was first discussed in the context of a voltage generated by a canting of a driven domain wall by L. Berger,[1] and mathematically rigorous formulation was given by G. Volovik.[2] The idea of effective gauge field was shown to be extended to the cases with spin relaxation,[3] and Rashba interaction.[4–7]

Some of the phenomena discussed in this paper overlaps those in the paper by R. Raimondi in this lecture series, studied base on the Boltzmann equation approach.[8]

## 2. Ferromagnetic metal

Let us start with a brief introduction of ferromagnetic metals (Fig. 2). Ferromagnets have magnetization, namely, an ensemble of localized spins. Denoting the localized spin as $\boldsymbol{S}$, the magnetization is $\boldsymbol{M} = -\frac{\hbar\gamma}{a^3}\boldsymbol{S}$, where $\gamma(>0)$ and $a$ are gyromagnetic ratio and lattice constant, respectively. As the electron has negative charge, the localized spin and magnetization points opposite direction. In $3d$ transition metals, localized spins are aligned spins of $3d$ electrons. Ferromagnetic metals have finite conductivity, indicating that there are conduction electrons, mainly $4s$ electrons. The conduction electrons and $d$ electrons are coupled via $sd$ mixing. As a result, there arises an exchange interaction between conduction electron spin, $\boldsymbol{s}$, and localized spin, which reads

$$H_{sd} = -J_{sd}\boldsymbol{S} \cdot \boldsymbol{s}, \tag{1}$$

where $J_{sd}$ represents the strength. In this article, the localized spin is treated as classical variable, neglecting the conduction of $d$ electrons.

The dynamics of localized spin is described by the Landau-Lifshiz-Gilbert (LLG) equation,

$$\dot{\boldsymbol{n}} = \gamma \boldsymbol{B} \times \boldsymbol{n} + \alpha \boldsymbol{n} \times \dot{\boldsymbol{n}}, \tag{2}$$

where $\boldsymbol{n} \equiv \boldsymbol{S}/S$ is a unit vector representing the direction of localized spin, $\boldsymbol{B}$ is the total magnetic field acting on the spin. The last term of the right hand side represents the relaxation (damping) of localized spin, called the Gilbert damping effect and $\alpha$ is the Gilbert damping constant. The Gilbert damping constant in most metallic ferromagnets are of the order of $10^{-2}$.

We shall now start studying phenomena arising from the exchange interaction, Eq. (1), between localized spin and conduction electron.

Fig. 3. Schematic figures showing conduction electron injected to a domain wall. (a): In the adiabatic limit, i.e., for a large domain wall width, the electron goes through the wall with a spin flip (left). (b): Non adiabaticity due to finite domain wall width leads to reflection and electric resistance (right).

## 3. Electron transport through magnetic domain wall: Phenomenology

We consider a ferromagnetic domain wall, which is a structure where localized spins (or magnetization) rotate spatially (Fig. 3). Its thickness, $\lambda$, in typical ferromagnets is $\lambda = 10 - 100$nm. Let us consider here what happens when a conduction electron goes through a domain wall. The wall is a macroscopic object for electrons, since thickness is much larger than the typical length scale of electron, the Fermi wavelength, $1/k_F$, which is atomic scale in metals. The electron is interacting with localized spin via the *sd* exchange coupling, Eq. (1). We consider the case of positive $J_{sd}$, but the sign does not change the scenario. The *sd* interaction tends to align parallel the localized spin and conduction electron spin. If localized spin is spatially uniform, therefore, the conduction electron is also uniformly polarized, and electron transport and magnetism are somewhat decoupled.

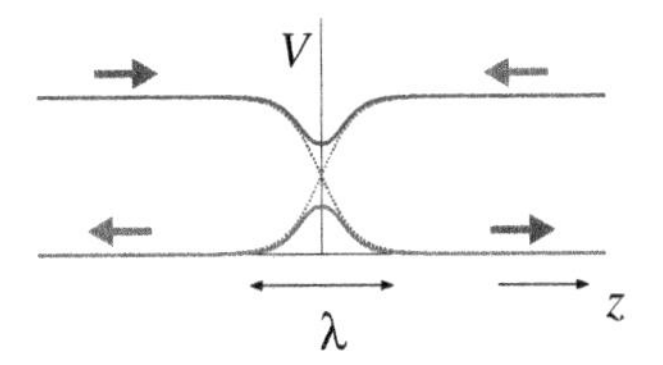


Fig. 4. Potential energy $V(z)$ for conduction electron with spin $\rightarrow$ and $\leftarrow$ as a result of *sd* exchange interaction. Dotted lines are the cases neglecting spin flip inside the wall, while solid lines are with spin flip.

Interesting effects arise if the localized spins are spatially varying like the case of a domain wall. We choose the $z$ axis along the direction localized spins change. The lowest energy direction (magnetic easy axis) for localized spins is chosen as along $z$ axis. (The mutual direction between the localized spin and direction of spin change is irrelevant in the case without spin-orbit interaction.) The wall in this case is with localized spins inside the wall changing within the plane of localized spin, and such wall is called the Nèel wall. At $z = \infty$ the localized spin is $S_z = S$, and is $S_z = -S$ at $z = -\infty$, and those states are represented a $\rightarrow$ and $\leftarrow$, respectively. For $\leftarrow$ electron, the potential in the left regime is low because of *sd* exchange interaction, while that in the right region is high (dotted lines in Fig. 4). That is, the localized spin structure due to a domain wall acts as a spatially varying magnetic field, resulting in potential barriers, $V_{\rightarrow}(z) = -J_{sd}S_z(z)$ and $V_{\leftarrow}(z) = J_{sd}S_z(z)$. Considering the domain wall centered at $x = 0$ having profile of

$$S_z(z) = S \tanh \frac{z}{\lambda}, \quad S_x(z) = \frac{S}{\cosh \frac{z}{\lambda}}, \quad S_y = 0, \tag{3}$$

conduction electron's Schrödinger equation with energy $E$ reads

$$\left[ -\frac{\hbar^2}{2m} \frac{d^2}{dz^2} - J_{sd} S \left( \sigma_z \tanh \frac{z}{\lambda} + \sigma_x \frac{1}{\cosh \frac{z}{\lambda}} \right) \right] \Psi = E \Psi, \tag{4}$$

$\Psi(z) = (\Psi_{\rightarrow}(z), \Psi_{\leftarrow}(z))$ begin the two-component wave function. If the spin direction of the conduction electron is fixed along the $z$ axis, the potential barrier represented by the term proportional to $\sigma_z$ leads to reflection of electron, but in reality, the electron spin can rotate inside the wall as a result of the term proportional to $\sigma_x$ in Eq. (4). The mixing of $\leftarrow$ and $\rightarrow$

electron leads to the smooth potential barrier plotted as solid lines in Fig. 4.

Let us consider an incident $\leftarrow$ electron from the left. If the electron is slow, the electron spin can keep the lowest energy state by gradually rotating its direction inside the wall. This is the adiabatic limit. As there is no potential barrier for the electron in this limit, no reflection arises from the domain wall, resulting in a vanishing resistance (Fig. 3(a)) In contrast, if the electron is fast, the electron spin cannot follow the rotation of the localized spin, resulting in a reflection and finite resistance (Fig. 3(b)). The condition for slow and fast is determined by the relation between the time for the electron to pass the wall and the time for electron spin rotation. The former is $\lambda/v_F$ for electron with Fermi velocity $v_F(=\hbar k_F/m)$ (spin-dependence of the Fermi wave vector is neglected and $m$ is the electron mass). The latter time is $\hbar/J_{sd}S$, as the electron spin is rotated by the $sd$ exchange interaction in the wall. Therefore, if

$$\frac{\lambda}{v_F} \gg \frac{\hbar}{J_{sd}S}, \tag{5}$$

is satisfied, the electron is in the adiabatic limit.[9] The condition of adiabatic limit here is the case of clean metal (long mean free path); In dirty metals, it is modified.[10,11]

The transmission of electron through a domain wall was calculated by G. G. Cabrera and L. M. Falicov,[12] and its physical aspects were discussed by L. Berger.[1,13] Linear response formulation and scattering approach were presented in Refs. 14–16. The adiabaticity condition was discussed by X. Waintal and M. Viret.[9]

### 3.1. *Spin-transfer effect*

As we discussed above, in the adiabatic limit, the electron spin gets rotated after passing through the wall (Fig. 3(a)). The change of spin angular momentum, $2 \times \frac{\hbar}{2} = \hbar$, must be absorbed by the localized spins. (Angular momentum dissipation as a result of spin relaxation is slow compared to the exchange of the angular momentum via the $sd$ exchange interaction.) To absorb the spin change of $\hbar$, the domain wall must shift to the right, resulting in an increase of the spins $\leftarrow$. We consider for simplicity the case of cubic lattice with lattice constant $a$. The distance of the wall shift $\Delta X$ necessary to absorb the electron's spin angular momentum of $\hbar$ is then $[\hbar/(2\hbar S)]a$ (Fig. 5)). If we apply a spin-polarized current through the wall with the density $j_s$ (spin current density is defined to have the same unit of

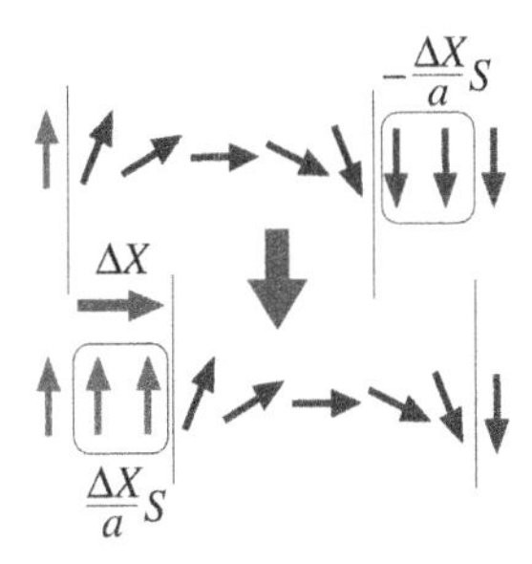


Fig. 5. The shift of the domain wall by a distance $\Delta X$ results in a change of the spin of the localized spins $\frac{\Delta X}{a}S - \left(-\frac{\Delta X}{a}S\right) = 2S\frac{\Delta X}{a}$. The angular momentum change is therefore $\hbar$ if $\Delta X = \frac{a}{2S}$.

$A/m^2$ as the electric current density.) The rate of the angular momentum change of the conduction electron per unit time and area is $\hbar j_s/e$. As the number of the localized spins in the unit area is $1/a^2$, the wall must keep moving a distance of $(j_s/e)(a^3/2S)$ per unit time. Namely, when a spin current density is applied, the wall moves with the speed of

$$v_s \equiv \frac{a^3}{2eS} j_s. \tag{6}$$

This effect was pointed out by L. Berger [1] in 1986, and is now called the spin-transfer effect after the papers by J. Slonczewski [17].

From the above considerations in the adiabatic limit, we have found that a domain wall is driven by spin-polarized current, while the electrons do not get reflected and no resistance arises from the wall. These two facts naively seem inconsistent, but are direct consequence of the fact that a domain wall is a composite structure having both linear momentum and angular momentum. The adiabatic limit is the limit where angular momentum is transferred between the electron and the wall, while no linear momentum is transferred

## 4. Adiabatic phase of electron spin

Transport of conduction electrons in the adiabatic (strong *sd*) limit is theoretically studied by calculating the quantum mechanical phase attached to the wave function of electron spin. We here consider a conduction electron hopping from a site $\boldsymbol{r}$ to a neighboring site at $\boldsymbol{r}' \equiv \boldsymbol{r} + \boldsymbol{a}$ ($\boldsymbol{a}$ is a vector connecting neighboring sites)(Fig. 6). The localized spin direction at those

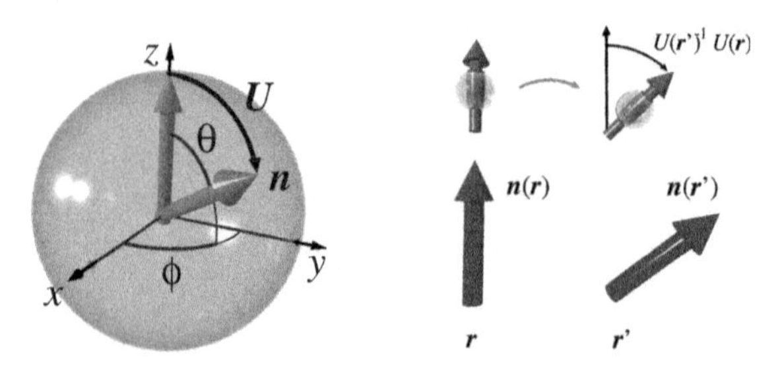


Fig. 6. Left: A Unitary transformation $U(\theta,\phi)$ relates the two spin configurations $|\uparrow\rangle$ and $|\boldsymbol{n}\rangle$ as $|\boldsymbol{n}\rangle = U|\uparrow\rangle$. Right: The overlap of the wave functions at sites $\boldsymbol{r}$ and $\boldsymbol{r}'$ is $\langle \boldsymbol{n}(\boldsymbol{r})|\boldsymbol{n}(\boldsymbol{r}')\rangle = \langle\uparrow|U(\boldsymbol{r}')^{-1}U(\boldsymbol{r})|\uparrow\rangle$.

sites are $\boldsymbol{n}(\boldsymbol{r}) \equiv \boldsymbol{n}$ and $\boldsymbol{n}(\boldsymbol{r}+\boldsymbol{a}) \equiv \boldsymbol{n}'$, respectively, and the electron's wave function at the two sites are

$$|\boldsymbol{n}\rangle = \cos\frac{\theta}{2}|\uparrow\rangle + \sin\frac{\theta}{2}e^{i\phi}|\downarrow\rangle$$
$$|\boldsymbol{n}'\rangle = \cos\frac{\theta'}{2}|\uparrow\rangle + \sin\frac{\theta'}{2}e^{i\phi'}|\downarrow\rangle, \tag{7}$$

where $\theta$, $\phi$ and $\theta'$, $\phi'$ are the polar angle of $\boldsymbol{n}(\boldsymbol{r})$ and $\boldsymbol{n}(\boldsymbol{r}')$, respectively (Fig. 6). The wave functions are concisely written by use of matrices, $U(\boldsymbol{r})$ and $U(\boldsymbol{r}')$, which rotates the spin state $|\uparrow\rangle$ to $|\boldsymbol{n}\rangle$ (Fig. 6), as $|\boldsymbol{n}\rangle = U(\boldsymbol{r})|\uparrow\rangle$ and $|\boldsymbol{n}'\rangle = U(\boldsymbol{r}')|\uparrow\rangle$. The rotation matrix is given by [18] (neglecting irrelevant phase factors)

$$U(\boldsymbol{r}) = e^{\frac{i}{2}(\phi-\pi)\sigma_z}e^{\frac{i}{2}\theta\sigma_y}e^{-\frac{i}{2}(\phi-\pi)\sigma_z} = \begin{pmatrix} \cos\frac{\theta}{2} & \sin\frac{\theta}{2}e^{i\phi} \\ -\sin\frac{\theta}{2}e^{-i\phi} & \cos\frac{\theta}{2} \end{pmatrix}. \tag{8}$$

The overlap of the electron wave functions at the two sites is thus $\langle \boldsymbol{n}'|\boldsymbol{n}\rangle = \langle\uparrow|U(\boldsymbol{r}')^{-1}U(\boldsymbol{r})|\uparrow\rangle$. When localized spin texture is slowly varying, we can expand the matrix product with respect to $\boldsymbol{a}$ as $U(\boldsymbol{r}')^{-1}U(\boldsymbol{r}) = 1 - U(\boldsymbol{r})^{-1}(\boldsymbol{a}\cdot\nabla)U(\boldsymbol{r}) + O(a^2)$ to obtain

$$\langle \boldsymbol{n}'|\boldsymbol{n}\rangle \simeq 1 - \langle\uparrow|U(\boldsymbol{r})^{-1}(\boldsymbol{a}\cdot\nabla)U(\boldsymbol{r})|\uparrow\rangle \simeq e^{i\varphi}, \tag{9}$$

where

$$\varphi \equiv i\boldsymbol{a}\cdot\langle\uparrow|U(\boldsymbol{r})^{-1}\nabla U(\boldsymbol{r})|\uparrow\rangle \equiv \boldsymbol{a}\cdot\boldsymbol{A}_{\rm s}. \tag{10}$$

Since $(U^{-1}\nabla U)^\dagger = -U^{-1}\nabla U$, $\varphi$ is real. A vector $\boldsymbol{A}_{\rm s}$ here plays a role of a gauge field, similarly to that of the electromagnetism, and it is called (adiabatic) spin gauge field. By use of Eq. (8), this gauge field reads (the factor of $\frac{1}{2}$ represents the magnitude of electron spin)

$$\boldsymbol{A}_{\rm s} = \frac{\hbar}{2e}(1 - \cos\theta)\nabla\phi. \tag{11}$$

For a general path $C$, the phase is written as an integral along $C$ as

$$\varphi = \frac{e}{\hbar}\int_C d\boldsymbol{r}\cdot\boldsymbol{A}_{\rm s}. \tag{12}$$

Existence of path-dependent phase means that there is an effective magnetic field, $\boldsymbol{B}_{\rm s}$, as seen by rewriting the integral over a closed path by use of the Stokes theorem as

$$\varphi = \frac{e}{\hbar}\int_S d\boldsymbol{S}\cdot\boldsymbol{B}_{\rm s}, \tag{13}$$

where

$$\boldsymbol{B}_{\rm s} \equiv \nabla\times\boldsymbol{A}_{\rm s}, \tag{14}$$

represents the curvature or effective magnetic field. This phase $\varphi$, arising from strong *sd* interaction, couples to electron spin, and is called the spin Berry's phase. Time-derivative of phase is equivalent to a voltage, and thus we have effective electric field defined by

$$\dot{\varphi} = -\frac{e}{\hbar}\int_C d\boldsymbol{r}\cdot\boldsymbol{E}_{\rm s}, \tag{15}$$

where

$$\boldsymbol{E}_{\rm s} \equiv -\dot{\boldsymbol{A}}_{\rm s}, \tag{16}$$

(For a gauge invariant expression of $\boldsymbol{E}_{\rm s}$, we need to include the time component of the gauge field, $A_{{\rm s},0}$.[19]) In terms of vector $\boldsymbol{n}$ the effective fields read

$$\begin{aligned} \boldsymbol{E}_{{\rm s},i} &= -\frac{\hbar}{2e}\boldsymbol{n}\cdot(\dot{\boldsymbol{n}}\times\nabla_i\boldsymbol{n}) \\ \boldsymbol{B}_{{\rm s},i} &= \frac{\hbar}{4e}\sum\nolimits_{jk}\epsilon_{ijk}\boldsymbol{n}\cdot(\nabla_j\boldsymbol{n}\times\nabla_k\boldsymbol{n}). \end{aligned} \tag{17}$$

These two fields couple to the electron spin and are called spin electromagnetic fields ($\boldsymbol{A}_{\rm s}$ is spin gauge field). They satisfy the Faraday's law,

$$\nabla\times\boldsymbol{E}_{\rm s} + \dot{\boldsymbol{B}}_{\rm s} = 0, \tag{18}$$

as a trivial result of their definitions. Defining the spin magnetic charge as

$$\nabla\cdot\boldsymbol{B}_{\rm s} \equiv \rho_{\rm m}, \tag{19}$$

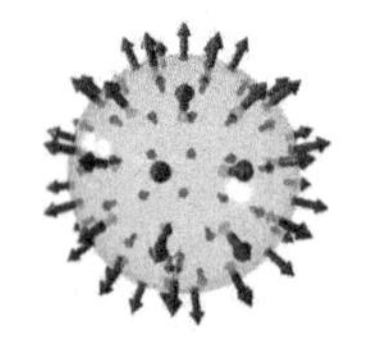
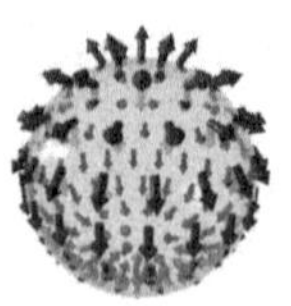

Fig. 7. Magnetization structures, $\boldsymbol{n}(\boldsymbol{r})$, of a hedgehog monopole having a monopole charge of $Q_{\rm m} = 1$ and the one with $Q_{\rm m} = 2$ . At the center, $\boldsymbol{n}(\boldsymbol{r})$ has a singularity and this gives rise to a finite monopole charge.

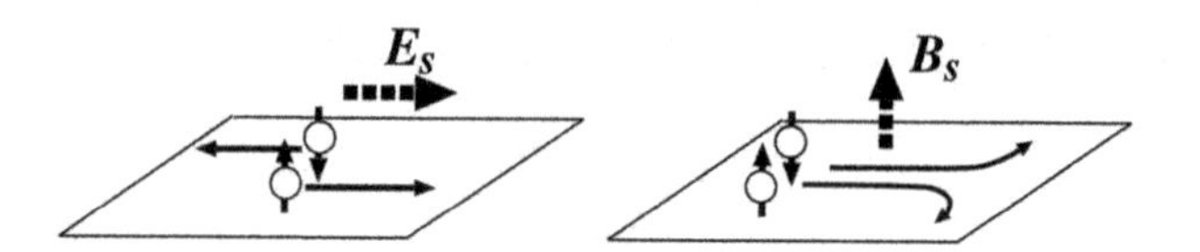


Fig. 8. Spin electric field $\boldsymbol{E}_{\rm s}$ and spin magnetic field $\boldsymbol{B}_{\rm s}$ act oppositely for electrons with opposite spin, and thus are useful for generation of spin current.

we see that $\rho_{\rm m} = 0$ as a local identity, since spin vector with fixed length has only two independent variables, and therefore $\sum_{ijk}\epsilon_{ijk}(\nabla_i \boldsymbol{n})\cdot(\nabla_j \boldsymbol{n} \times \nabla_k \boldsymbol{n}) = 0$. However, there is a possibility that the volume integral, $Q_{\rm m} \equiv \int d^3 r \rho_{\rm m}$, is finite; In fact, using the Gauss's law we can write ($\int d\boldsymbol{S}$ represents a surface integral)

$$Q_{\rm m} = \frac{h}{4\pi e}\int d\boldsymbol{S}\cdot\boldsymbol{\Omega}, \tag{20}$$

and it follows that $Q_{\rm m} = \frac{h}{e}\times$integer since $\frac{1}{4\pi}\int d\boldsymbol{S}\cdot\boldsymbol{\Omega}$ is a winding number, an integer, of a mapping from a sphere in the coordinate space to a sphere in spin space. If the mapping is topologically non-trivial as a result of a singularity, the monopole charge is finite. Typical nontrivial structures of $\boldsymbol{n}$ are shown in Fig. 7. The singular structure with a single monopole charge is called the hedgehog monopole.

The Faraday's law similarly reads $(\nabla \times \boldsymbol{E}_{\rm s})_i + \dot{\boldsymbol{B}}_{{\rm s}i} = \frac{\hbar}{4e}\sum_{ijk}\epsilon_{ijk}\dot{\boldsymbol{n}}\cdot(\nabla_j \boldsymbol{n} \times \nabla_k \boldsymbol{n}) \equiv \boldsymbol{j}_{\rm m}$, which vanishes locally but is finite when integrated, indicating that topological monopole current $\boldsymbol{j}_{\rm m}$ exists.

The other two Maxwell's equations describing $\nabla\cdot\boldsymbol{E}_{\rm s}$ and $\nabla\times\boldsymbol{B}_{\rm s}$ are derived by evaluating the induced spin density and spin current based on linear response theory.[4,19]

## 5. Detection of spin electromagnetic fields

The spin electromagnetic fields are real fields detectable in transport measurements. They couples to the spin polarization of the electrons (Fig. 8), and because spin density and spin current in ferromagnetic metals is always accompanied with electric charge and current, respectively, the effects of the spin magnetic fields are observable in electric measurements. The electric component $\boldsymbol{E}_{\rm s}$ is directly observable as a voltage generation from magnetization dynamics, and the voltage signals of $\mu$V order have been observed for the motion of domain walls and vortices.[20,21] The spin magnetic field causes an anomalous Hall effect of spin, i.e., the spin Hall effect called the topological Hall effect. The spin electric field arises if magnetization structure carrying spin magnetic field becomes dynamical due to the Lorentz force from $\boldsymbol{B}_{\rm s}$ according to $\boldsymbol{E}_{\rm s} = \boldsymbol{v} \times \boldsymbol{B}_{\rm s}$, where $\boldsymbol{v}$ denotes the electron spin's velocity. The topological Hall effect due to skyrmion lattice turned out to induce Hall resistivity of 4 n$\Omega$cm.[22,23] Although those signals are not large, existence of spin electromagnetic fields is thus confirmed experimentally. It was recently shown theoretically that spin magnetic field couples to helicity of circularly polarized light (topological inverse Faraday effect),[24] and an optical detection is thus possible.

## 6. Effects of spin gauge field on magnetization dynamics

As discussed in the previous section, the spin gauge field are measured by transport experiments. Here we study the opposite effect, the effects of spin gauge field on magnetization dynamics when spin current is applied. The spin gauge field is expected to couple to the spin current of the electron, $j_{\rm s}$, via the minimal coupling,

$$H_{A_{\rm s}} = \int d^3r \left[ -\frac{\hbar}{e} \boldsymbol{A}_{\rm s} \cdot \boldsymbol{j}_{\rm s} + \frac{n\hbar^2}{2m} (\boldsymbol{A}_{\rm s})^2 - 2\hbar A_{{\rm s},0} \rho_{\rm s} \right], \tag{21}$$

where $n$ is the electron density, and $\rho_{\rm s} = \frac{1}{2}(n_+ - n_-)$ is the electron spin density, $n_\sigma$ ($\sigma = \pm$) representing the density of electron with spin $\sigma$. The field $A_{{\rm s},0}$ is the time component of spin gauge field (Eq. (11) with spatial derivative replaced by time derivative). (For rigorous derivation of the coupling, see Eqs. (38)(39).) As the spin gauge field is written in terms of localized spin variables, $\theta$ and $\phi$, as a result of Eq. (11), this interaction describes how the spin current and electron density affects the magnetization dynamics. Here we study the adiabatic limit, where the contribution second order in $\boldsymbol{A}_{\rm s}$ (the second term of the right hand side of Eq. (21) is neglected. Including the gauge interaction, the Lagrangian for the localized spin reads

$$L_S = \int d^3r \left[\frac{2}{a^3} A_{s,0} \left(S + \rho_s a^3\right) - \frac{\hbar}{e} \boldsymbol{A}_s \cdot \boldsymbol{j}_s\right] - H_S, \qquad (22)$$

where $H_S$ is the Hamiltonian. We see that the magnitude of localized spin is modified to be the effective one $\overline{S} \equiv S + \rho_s a^3$ including the spin polarization of the conduction electron. Writing the gauge field terms explicitly, we have

$$L_S = \int d^3r \left[\overline{S}(1 - \cos\theta) \left(\frac{\partial}{\partial t} - \boldsymbol{v}_s \cdot \nabla\right) \phi\right] - H_S, \qquad (23)$$

where $\boldsymbol{v}_s \equiv \frac{a^3}{2e\overline{S}} \boldsymbol{j}_s$. The velocity $\boldsymbol{v}_s$ here agrees with the phenomenological one, Eq. (6), if electron spin polarization is neglected (i.e., if $\overline{S} = S$). In the adiabatic limit, therefore, the time-derivative of the localized spin in the equation of motion is replaced by the Galilean invariant form with a moving velocity of $\boldsymbol{v}_s$ when a spin current is present. The equation of motion derived from the Lagrangian (23) reads

$$\left(\frac{\partial}{\partial t} - \boldsymbol{v}_s \cdot \nabla\right) \boldsymbol{S} = -\gamma \boldsymbol{B}_S \times \boldsymbol{S}, \qquad (24)$$

where $\boldsymbol{B}_S$ is the effective magnetic field due to $H_S$. From Eq. (24), it is obvious that the magnetization structure flows with velocity $\boldsymbol{v}_s$, and this effect is in fact the spin-transfer effect discussed phenomenologically in Sec. 3.1. It should be noted that the effect is mathematically represented by a simple gauge interaction of Eq. (22). The equation of motion (24) is the Landau-Lifshiz-Gilbert (LLG) equation including adiabatic spin-transfer effect. It was theoretically demonstrated that the spin-transfer torque induces a red shift of spin wave, resulting in instability of uniform ferromagnetic state under spin-polarized current.[25]

In reality, there is nonadiabatic contribution described by spin-flip interactions. Such contribution leads to a mixing of the electron spins resulting in a scattering of the conduction electron and a finite resistance due to the magnetization structure.[15,16] This scattering gives rise to a force on the magnetization structure as a counter action.[26]

As we have seen, the concept of adiabatic spin gauge field is useful to give a unified description of both electron transport properties in the presence of magnetization structure and the magnetization dynamics in the presence of spin-polarized current.

## 7. Field-theoretic description

So far we discussed that an effective spin gauge field emerges by looking into the quantum mechanical phase factor attached to conduction electron in the presence of magnetization structures. Existence of effective gauge field is straightforwardly seen in field-theoretic description.

A field-theoretical description is based on the Lagrangian of the system,

$$\hat{L} = i\hbar \int d^3r \sum_{\sigma} \hat{c}_{\sigma}^{\dagger} \dot{\hat{c}}_{\sigma} - \hat{H}, \tag{25}$$

where $\hat{H} = \hat{K} + \hat{H}_{sd}$ is the field Hamiltonian. Here

$$\hat{K} = \int d^3r \sum_{\sigma} \hat{c}_{\sigma}^{\dagger} \left(-\frac{\hbar^2}{2m}\nabla^2\right) \hat{c}_{\sigma} = \frac{\hbar^2}{2m} \sum_{\sigma} \int d^3r (\nabla \hat{c}_{\sigma}^{\dagger})(\nabla \hat{c}_{\sigma}) \tag{26}$$

describes the free electron part in terms of field operators for conduction electron, $\hat{c}_{\sigma}$ and $\hat{c}_{\sigma}^{\dagger}$, where $\sigma = \pm$ denotes spin. The *sd* exchange interaction is represented by

$$\hat{H}_{sd} = -\frac{J_{sd}S}{2} \int d^3r \hat{c}^{\dagger} (\boldsymbol{n} \cdot \boldsymbol{\sigma}) \hat{c}. \tag{27}$$

We are interested in the case where $\boldsymbol{n}(\boldsymbol{r}, t)$ changes in space and time slowly compared to the electron's momentum and energy scales. How the electron 'feels' when flowing through such slowly varying structure is described by introducing a rotating frame where the *sd* exchange interaction is locally diagonalized. In Sec. 4, we introduced a unitary matrix $U(\boldsymbol{r}, t)$, and this matrix is used here to introduce a new electron operator as

$$\hat{a}(\boldsymbol{r}, t) = U(\boldsymbol{r}, t)\hat{c}(\boldsymbol{r}, t). \tag{28}$$

The new operator $\hat{a}$ describes the low energy dynamics for the case of strong *sd* exchange interaction. In fact, the *sd* exchange interaction for this electron is diagonalized to be

$$\hat{H}_{sd} = -M \int d^3r \hat{a}^{\dagger} \sigma_z \hat{a}, \tag{29}$$

where $M \equiv \frac{J_{sd}S}{2}$. Instead, the kinetic term for the new electron is modified, because derivative of the electron field is modified as

$$\nabla \hat{c} = U(\nabla + i\mathcal{A}_{\rm s})\hat{a}, \tag{30}$$

where

$$\mathcal{A}_{\rm s} \equiv -iU^{\dagger}\nabla U. \tag{31}$$

Here $\mathcal{A}_{\mathrm{s}}$ is a $2\times 2$ matrix, whose components are represented by using Pauli matrices as

$$\mathcal{A}_{\mathrm{s},i} = \sum_{\alpha=x,y,z} \mathcal{A}_{\mathrm{s},i}^{\alpha}\sigma_{\alpha}. \tag{32}$$

Equation (30) indicates that the new electron field $\hat{a}$ is interacting with an effective gauge field, $\mathcal{A}_{\mathrm{s}}$. This gauge field has three components, is non-commutative and is called the SU(2) gauge field. The three components explicitly read

$$\begin{pmatrix} \mathcal{A}_{\mathrm{s},\mu}^{x} \\ \mathcal{A}_{\mathrm{s},\mu}^{y} \\ \mathcal{A}_{\mathrm{s},\mu}^{z} \end{pmatrix} = \frac{1}{2}\begin{pmatrix} -\partial_\mu\theta\sin\phi - \sin\theta\cos\phi\partial_\mu\phi \\ \partial_\mu\theta\cos\phi - \sin\theta\sin\phi\partial_\mu\phi \\ (1-\cos\theta)\partial_\mu\phi \end{pmatrix}. \tag{33}$$

Due to Eq. (30), the kinetic term $\hat{K}$ is written in terms of $\hat{a}$ electron as

$$\hat{K} = \frac{\hbar^2}{2m}\int d^3r[(\nabla - i\mathcal{A}_{\mathrm{s}})\hat{a}^\dagger][(\nabla + i\mathcal{A}_{\mathrm{s}})\hat{a}]. \tag{34}$$

Similarly, time-component of the gauge field

$$\mathcal{A}_{\mathrm{s},0} \equiv -iU^\dagger\partial_t U, \tag{35}$$

arises from the time-derivative term $(i\hbar\hat{c}_\sigma^\dagger\dot{\hat{c}}_\sigma)$ of the Lagrangian (25). The Lagrangian in terms of $\hat{a}$ electron therefore reads

$$\begin{aligned}\hat{L} = \int d^3r \Bigg[ & i\hbar\hat{a}^\dagger\dot{\hat{a}} - \frac{\hbar^2}{2m}|\nabla\hat{a}|^2 + \epsilon_F\hat{a}^\dagger\hat{a} + M\hat{a}^\dagger\sigma_z\hat{a} \\ & + i\frac{\hbar^2}{2m}\sum_i(\hat{a}^\dagger\mathcal{A}_{\mathrm{s},i}\nabla_i\hat{a} - (\nabla_i\hat{a}^\dagger)\mathcal{A}_{\mathrm{s},i}\hat{a}) - \frac{\hbar^2}{2m}\mathcal{A}_{\mathrm{s}}^2\hat{a}^\dagger\hat{a} - \hbar\hat{a}^\dagger\mathcal{A}_{\mathrm{s},0}\hat{a}\Bigg].\end{aligned} \tag{36}$$

If we introduce electron density operator, $\hat{n} \equiv \hat{a}^\dagger\hat{a}$, and operators for spin density and spin current density as

$$\hat{\rho}_{\mathrm{s}\alpha} \equiv \frac{1}{2}\hat{a}^\dagger\sigma_\alpha\hat{a},\quad \hat{j}_{\mathrm{s},i}^{\alpha} \equiv \frac{-i}{2m}\hat{a}^\dagger\overleftrightarrow{\nabla}_i\sigma_\alpha\hat{a} \equiv \frac{-i}{2m}\left[\hat{a}^\dagger\sigma_\alpha(\nabla_i\hat{a}) - (\nabla_i\hat{a}^\dagger)\sigma_\alpha\hat{a}\right], \tag{37}$$

it reads

$$\hat{L} = \int d^3r\left[i\hbar\hat{a}^\dagger\dot{\hat{a}} - \frac{\hbar^2}{2m}|\nabla\hat{a}|^2 + \epsilon_F\hat{a}^\dagger\hat{a} + M\hat{a}^\dagger\sigma_z\hat{a} - \hat{j}_{\mathrm{s},i}^{\alpha}\mathcal{A}_{\mathrm{s},i}^{\alpha} - \frac{\hbar^2}{2m}\mathcal{A}_{\mathrm{s}}^2\hat{n} - \hat{\rho}_{\mathrm{s}}{}^{\alpha}\mathcal{A}_{\mathrm{s},0}^{\alpha}\right]. \tag{38}$$

In the case of $M/\epsilon_F \gg 1$ (large $J_{sd}$), the electron with spin $\downarrow$ has high energy because of strong spin splitting, Eq. (29), and is neglected. In this case, only the $z$ component of the gauge field, $\mathcal{A}^z_{s,i}$, survives. This component is thus essentially a U(1) gauge field, which coincides with the U(1) gauge field we have obtained from the argument of electron's phase factor, namely, $\boldsymbol{A}_s = \boldsymbol{\mathcal{A}}^z_s$. The total Hamiltonian in the limit of large $J_{sd}$ therefore reduces to the one for a charged particle in the presence of a U(1) gauge field $\boldsymbol{A}_s$;

$$\hat{H} = \int d^3r \left[ \frac{\hbar^2}{2m} [(\nabla - i\boldsymbol{A}_s)\hat{a}^\dagger_\uparrow][(\nabla + i\boldsymbol{A}_s)\hat{a}_\uparrow] - \frac{J_{sd}S}{2} \hat{a}^\dagger_\uparrow \hat{a}_\uparrow \right]. \quad (39)$$

The field-theoretic method present here is highly useful, as it leads to a conclusion of the existence of an effective gauge field for spin simply by carrying out a unitary transformation to diagonalize strong $sd$ exchange interaction.

## 8. Non-adiabaticity and spin relaxation

In reality, there is a deviation from the adiabatic limit we have considered so far. One origin is the fact that the magnetization structure is not in the slowly varying limit, but has a finite length scale of spatial modulation. This effect, we call the non adiabaticity, leads to reflection of conduction electron by magnetization structures as in Fig. 3(b), resulting in a force on the magnetization structure when an electric current is applied.[13,26] In terms of torque, the effect of the force due to reflection is represented by a non-local torque, as it arises from finite momentum transfer.[27] Another effect we need to take into account is the relaxation (damping) of spin schematically shown in the Fig. 9. In metallic ferromagnets, the damping mostly arises from the spin-orbit interaction, as seen from the fact that the Gilbert damping parameter $\alpha$ and the $g$ value has a correlation of $\alpha \propto (g-2)^2$ as shown in Ref. 28. Spin relaxation generates a torque perpendicular to the motion of the spin, resulting in a canting of the precession axis. Similarly, when a spin current $\boldsymbol{j}_s$ is applied, the spin relaxation thus was argued to induce a torque perpendicular to the spin-transfer torque, i.e.,

$$\boldsymbol{\tau}_\beta \equiv \beta \frac{a^3}{2e} \boldsymbol{n} \times (\boldsymbol{j}_s \cdot \nabla)\boldsymbol{n}, \quad (40)$$

where $\beta$ is a coefficient representing the effect of spin relaxation.[29,30]

Those effects of non adiabaticity and spin relaxation can be calculated from a microscopic viewpoint.[27,31] Let us go back to the LLG equation for localized spin interacting with conduction electron spin via the $sd$ exchange interaction. The total Hamiltonian is $H_S - M\sum_{\boldsymbol{r}} \boldsymbol{n}(\boldsymbol{r}) \cdot \boldsymbol{\sigma} - H_e$, where $H_S$

Fig. 9. Spin relaxation induces a torque perpendicular to the spin motion and let the spin relax to the stable direction along the external magnetic field.

and $H_e$ are the Hamiltonian for localized spin and conduction electron, respectively. The equation of motion for localized spin is given by

$$\dot{\boldsymbol{n}} = \gamma \boldsymbol{B}_S \times \boldsymbol{n} + \gamma \boldsymbol{B}_e \times \boldsymbol{n}, \tag{41}$$

where $\gamma \boldsymbol{B}_S \equiv \frac{1}{\hbar}\frac{\delta H_S}{\delta \boldsymbol{n}}$ and

$$\gamma \boldsymbol{B}_e \equiv \frac{1}{\hbar}\frac{\delta H_e}{\delta \boldsymbol{n}} = -\frac{M}{\hbar}\langle \boldsymbol{\sigma} \rangle, \tag{42}$$

are the effective magnetic field arising from the localized spin and conduction electron, respectively. The field $\boldsymbol{B}_e$ is represented by the expectation value of electron spin density, $\langle \boldsymbol{\sigma} \rangle$, and all the effects from the conduction electron is included in this field; Equation (41) is exact if $\langle \boldsymbol{\sigma} \rangle$ is evaluated exactly. Field theoretic approach is suitable for a systematic evaluation of the electron spin density. We move to a rotated frame where the electron spin is described choosing the local $z$ axis along the localized spin. In the case we are interested, namely, when the effect of non adiabaticity and damping are weak, these effects are treated perturbatively.

The spin density in the laboratory frame is written in terms of the spin in the rotated frame $\tilde{\boldsymbol{s}}$ as $s_i = R_{ij}\tilde{s}_j$, where

$$R_{ij} \equiv 2m_i m_j - \delta_{ij}, \tag{43}$$

is a rotation matrix, $\boldsymbol{m} \equiv (\sin\frac{\theta}{2}\cos\phi, \sin\frac{\theta}{2}\sin\phi, \cos\frac{\theta}{2})$ being the vector which define the unitary rotation. The perpendicular components (denoted by $\perp$) of electron spin density in the rotated frame are calculated as[11]

$$\tilde{s}^{\perp} = -\frac{2\rho_s}{M}\mathcal{A}_{s,0}^{\perp} - \frac{a^3}{eM}\boldsymbol{j}_s \cdot \mathcal{A}_s^{\perp} - \frac{\alpha_{sr}}{M}(\hat{\boldsymbol{z}} \times \mathcal{A}_{s,0}^{\perp}) - \frac{\beta_{sr}}{eM}(\hat{\boldsymbol{z}} \times (\boldsymbol{j}_s \cdot \mathcal{A}_s^{\perp})). \tag{44}$$

The effect of spin relaxation is included in $\alpha_{sr}$ and $\beta_{sr} = \hbar/(2M\tau_s)$, both proportional to the spin relaxation time, $\tau_s$.[31] The first term of Eq. (44) represents the renormalization of the localized spin as a result of electron spin polarization and the second term, induced in the presence of applied spin current, describes the adiabatic spin-transfer torque. Using the identity

$$R_{ij}(\mathcal{A}_{\text{s},\mu})_j^{\perp} = -\frac{1}{2}(\boldsymbol{n} \times \partial_\mu \boldsymbol{n})_i, \qquad R_{ij}(\hat{\boldsymbol{z}} \times \mathcal{A}_{\text{s},\mu}^{\perp})_j = \frac{1}{2}\partial_\mu \boldsymbol{n}_i, \tag{45}$$

we see that Eq. (44) leads to

$$(1 + \rho_{\text{s}} a^3)\dot{\boldsymbol{n}} = \alpha \boldsymbol{n} \times \dot{\boldsymbol{n}} - \frac{a^3}{2e}(\boldsymbol{j}_{\text{s}} \cdot \nabla)\boldsymbol{n} - \frac{\beta a^3}{2e}[\boldsymbol{n} \times (\boldsymbol{j}_{\text{s}} \cdot \nabla)\boldsymbol{n}] + \gamma \boldsymbol{B}_S \times \boldsymbol{n}, \tag{46}$$

which is the LLG equation taking into account the torque due to electrons. Here $\alpha \equiv \alpha_{\text{sr}}$ and $\beta \equiv \beta_{\text{sr}}$, neglecting other origins for Gilbert damping and nonadiabatic torque.

## 9. Current-driven domain wall motion

Let us briefly discuss dynamics of a domain wall based on the LLG equation (46) including the current-induced torques. We consider an one-dimensional and rigid wall, neglecting deformation. For a domain wall to be created, the system must have an easy axis magnetic anisotropy energy. We also include the hard-axis anisotropy energy, which turns out to govern the domain wall motion. Choosing the easy and the hard axises along the $z$ and the $y$ directions, respectively, the anisotropy energy is represented by the Hamiltonian

$$H_K \equiv \int \frac{d^3r}{a^3}\left[-\frac{KS^2}{2}\cos^2\theta + \frac{K_\perp S^2}{2}\sin^2\theta \sin^2\phi\right], \tag{47}$$

where $K$ and $K_\perp$ are the easy- and hard-axis anisotropy energies (both are positive). We need to take into account of course the exchange coupling, which is essential for ferromagnetism, which in the continuum expression reads

$$H_J \equiv \int d^3r \frac{JS^2a^2}{2}(\nabla \boldsymbol{n})^2. \tag{48}$$

The domain wall solution obtained by minimizing $H_K$ and $H_J$ is Eq. (3) with $\lambda = \sqrt{J/K}$. Considering a rigid wall, we assume that $K \gg K_\perp$. The low energy dynamics of the wall is then described by two variables (called the collective coordinates), the center coordinate of the wall, $X(t)$, and the angle $\phi(t)$ out-of the easy plane.[11,32] The wall profile including the collective coordinates is

$$n_z(z,t) = \tanh\frac{z-X(t)}{\lambda}, \quad n_\pm(z,t) \equiv n_x \pm in_y = \frac{e^{\pm i\phi(t)}}{\cosh\frac{z-X(t)}{\lambda}}. \tag{49}$$

The equation of motion for domain wall is obtained by putting the wall profile (49) in Eq. (46) and integrating over spatial coordinate as

$$\begin{aligned}\dot{\phi} + \alpha\frac{\dot{X}}{\lambda} =& P\frac{\beta}{\lambda}\tilde{j} \\ \dot{X} - \alpha\lambda\dot{\phi} =& -v_c \sin 2\phi + P\tilde{j},\end{aligned} \tag{50}$$

where $P \equiv j_s/j$ is spin polarization of the current, and both $v_c \equiv \frac{K_\perp \lambda S}{2\hbar}$ and $\tilde{j} \equiv \frac{a^3}{2eS}j$ have dimension of velocity.

When $\beta = 0$, the wall velocity when a constant $\tilde{j}$ is applied is easily obtained as[26]

$$\overline{\dot{X}} = \begin{cases} 0 & (\tilde{j} < \tilde{j}_c^i) \\ \frac{|P|}{1+\alpha^2}\sqrt{\tilde{j}^2 - (\tilde{j}_c^i)^2} & (\tilde{j} \geq \tilde{j}_c^i) \end{cases} \tag{51}$$

and $\tilde{j}_c^i \equiv \frac{v_c}{P}$ is the intrinsic threshold current density.[26] Namely, the wall cannot move if the applied current is lower than the threshold value. This is because the torque supplied by the current is totally absorbed by the wall by tilting the out of plane angle to be $\sin 2\phi = P\tilde{j}/v_c$ when the current is weak ($|P\tilde{j}/v_c| \leq 1$) and thus the wall cannot move. This effect is called the intrinsic pinning effect.[11] For larger current density, the torque carried by the current induces an oscillation of the angle similar to the Walker's breakdown in an applied magnetic field, and the wall speed also becomes an oscillating function of time.

When nonadiabaticity parameter $\beta$ is finite, the behavior changes greatly and intrinsic pinning effect is removed and the wall can move with infinitesimal applied current as long as there is no extrinsic pinning. In fact, when the applied current density is $\tilde{j} > \tilde{j}_a$, where

$$\tilde{j}_a \equiv \frac{v_c}{P - \frac{\beta}{\alpha}}, \tag{52}$$

the solution of Eq. (50) is an oscillating function given by[11]

$$\dot{X} = \frac{\beta}{\alpha}\tilde{j} + \frac{v_c}{1+\alpha^2}\frac{\left(\frac{\tilde{j}}{\tilde{j}_a}\right)^2 - 1}{\frac{\tilde{j}}{\tilde{j}_a} - \sin(2\omega t - \vartheta)}, \tag{53}$$

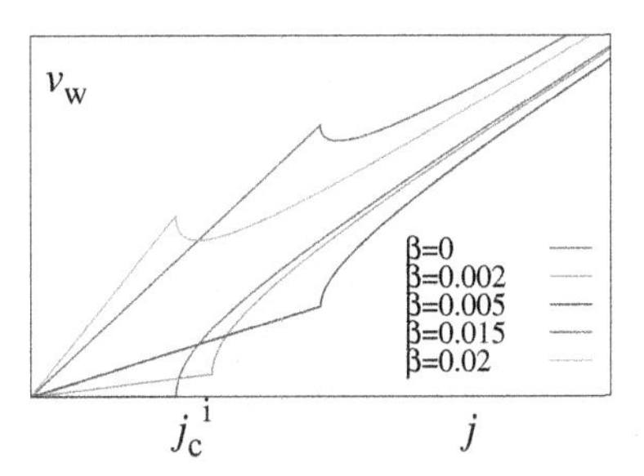


Fig. 10. Time averaged wall velocity $v_w$ as function of applied spin-polarized current $j$ for $\alpha = 0.01$. Intrinsic pinning threshold $j_c^i$ exists only for $\beta = 0$. The current density where derivative of $v_w$ is discontinuous corresponds to $\tilde{j}_a$.

where

$$\omega \equiv \frac{v_c}{\lambda}\frac{\alpha}{1+\alpha^2}\sqrt{\left(\frac{\tilde{j}}{\tilde{j}_a}\right)^2 - 1}, \qquad \sin\vartheta \equiv \frac{v_c}{(\frac{\beta}{\alpha} - P)\tilde{j}}. \tag{54}$$

The time-average of the wall speed is

$$\overline{\dot{X}} = \frac{\beta}{\alpha}\tilde{j} + \frac{v_c}{1+\alpha^2}\frac{1}{\tilde{j}_a}\sqrt{\tilde{j}^2 - \tilde{j}_a^2}. \tag{55}$$

For current density satisfying $\tilde{j} < \tilde{j}_a$, the oscillation in Eq. (53) is replaced by an exponential decay in time, and the wall velocity reaches a terminal value of

$$\dot{X} \to \frac{\beta}{\alpha}\tilde{j}. \tag{56}$$

The angle of the wall also reaches a terminal value determined by

$$\sin 2\phi \to \left(\frac{\beta}{\alpha} - P\right)\frac{\tilde{j}}{v_c}. \tag{57}$$

The averaged wall speed (Eq. (55)) is plotted in Fig. 10.

The intrinsic pinning is a unique feature of current-driven domain wall, as the wall cannot move even in the absence of pinning center. In the unit of $A/m^2$, the intrinsic pinning threshold is

$$j_c^i = \frac{eS^2}{Pa^3\hbar}K_\perp\lambda. \tag{58}$$

For device applications, this threshold needs to be lowered by reducing the hard-axis anisotropy and wall width.[33] At the same time, the intrinsic pinning is promising for stable device operations. In fact, in the intrinsic pinning regime, the threshold current and dynamics is insensitive to extrinsic pinning and external magnetic field,[26] as was confirmed experimentally.[34] This is due to the fact that the wall dynamics in the intrinsic pinning regime is governed by a torque (right hand side of the second equation of Eq. (50)), which governs the wall velocity $\dot{X}$, while pinning and magnetic field induce force, which governs $\dot{\phi}$; The forces due to sample irregularity therefore does not modify the motion induced by a torque in the intrinsic pinning regime.

Experimentally, intrinsic pinning is observed in perpendicularly magnetized materials,[34] perhaps due to relatively low intrinsic pinning threshold, while materials with in-plane magnetization mostly are in the extrinsic pinning regime governed by the nonadiabatic parameter $\beta$ and extrinsic pinning. In this regime, the threshold current of the wall motion is given by[35]

$$j_{\mathrm{c}}^{\mathrm{e}} \propto \frac{V_{\mathrm{e}}}{\beta}, \tag{59}$$

where $V_{\mathrm{e}}$ represents strength of extrinsic pinning potential like those generated by geometrical notches and defects. Control of nonadiabaticity parameter is therefore expected to be useful for driving domain walls at low current density.

Of recent interest from the viewpoint of low current operation is to use multilayer structures. For instance, heavy metal layers turned out to lower the threshold current by exerting a torque as a result of spin Hall effect,[36] and synthetic antiferromagnets turned out to be suitable for fast domain wall motion at low current.[37,38]

## 10. Interface spin-orbit effects

Physics tends to focus on infinite systems or bulk system approximated as infinite, as one of the most important objective of physics is to search for beautiful general law supported by symmetries. In the condensed matter physics today, studying such 'beautiful' systems seems to be insufficient anymore. This is because demands to understand physics of interfaces and surfaces has been increasing rapidly as present devices are in nanoscales to meet the needs for fast processing of huge data. Systems with lower symmetry are therefore important subjects of material science today.

Surfaces and interfaces have no inversion symmetry, and this leads to emergence of an antisymmetric exchange interaction (Dzyaloshinskii-

Moriya interaction)[39,40] in magnetism. As for electrons, broken inversion symmetry leads to a peculiar spin-orbit interaction, called the Rashba interaction,[41] whose Hamiltonian is

$$H_{\mathrm{R}} = i\boldsymbol{\alpha}_{\mathrm{R}} \cdot (\boldsymbol{\nabla} \times \boldsymbol{\sigma}), \tag{60}$$

where $\boldsymbol{\sigma}$ is the vector of Pauli matrices and $\boldsymbol{\alpha}_{\mathrm{R}}$ is a vector representing the strength and direction of the interaction. The form of the interaction is the one derived directly from the Dirac equation as a relativistic interaction, but the magnitude can be strongly enhanced in solids having heavy elements compared to the vacuum case.

As is obvious from the form of the Hamiltonian, the Rashba interaction induces electromagnetic cross correlation effects where a magnetization and an electric current are induced by external electric and magnetic field, $\boldsymbol{E}$ and $\boldsymbol{B}$, respectively, like represented as

$$\boldsymbol{M} = \gamma_{ME}(\boldsymbol{\alpha}_{\mathrm{R}} \times \boldsymbol{E}), \qquad \boldsymbol{j} = \gamma_{jB}(\boldsymbol{\alpha}_{\mathrm{R}} \times \boldsymbol{B}), \tag{61}$$

where $\gamma_{ME}$ and $\gamma_{jB}$ are coefficients, which generally depend on frequency. The emergence of spin accumulation from the applied electric field, mentioned in Ref. 41, was studied by Edelstein[42] in detail, and the effect is sometimes called Edelstein effect. The generation of electric current by magnetic field or magnetization, called the inverse Edelstein effect,[43] was recently observed in multilayer of Ag, Bi and a ferromagnet.[44]

### 10.1. *Effective magnetic field*

Equation (60) indicates that when a current density $\boldsymbol{j}$ is applied, the conduction electron has average momentum of $\boldsymbol{p} = \frac{m}{en}\boldsymbol{j}$ ($n$ is electron density), and thus an effective magnetic field of $\boldsymbol{B}_{\mathrm{e}} = \frac{ma^3}{-e\hbar^2\gamma}\boldsymbol{\alpha}_{\mathrm{R}} \times \boldsymbol{j}$, acts on the conduction electron spin ($\gamma(= \frac{|e|}{m})$ is the gyromagnetic ratio). When the *sd* exchange interaction between the conduction electron and localized spin is strong, this field multiplied by the the spin polarization, $P$, is the field acting on the localized spin. Namely, the localized spin feels a current-induced effective magnetic field of

$$\boldsymbol{B}_{\mathrm{R}} = \frac{Pma^3}{-e\hbar^2\gamma}\boldsymbol{\alpha}_{\mathrm{R}} \times \boldsymbol{j}. \tag{62}$$

One may argue more rigorously using field theoretic description. Considering the case of *sd* exchange interaction stronger than the Rashba interaction, we use a unitary transformation to diagonalize the *sd* exchange interaction (Eq. (28)). The Rashba interaction in the field representation then becomes

$$H_{\rm R} = -\int d^3r \frac{m}{\hbar e} \epsilon_{ijk} \alpha_{{\rm R},i} R_{kl} \tilde{j}^l_{{\rm s},j}, \tag{63}$$

where $\tilde{j}^l_{{\rm s},j} \equiv -i\frac{\hbar e}{2m} a^\dagger \overleftrightarrow{\nabla}_j \sigma_l a$ is the spin current in the rotated frame, $R_{ij}$ is given in Eq. (43). Terms containing spatial derivatives of magnetization structure is neglected, considering the slowly-varying structure. In this adiabatic limit, spin current is polarized along the $z$ direction, i.e., $\tilde{j}^l_{{\rm s},j} = \delta_{l,z} j_{\rm s}$. We therefore obtain using $R_{kz} = n_k$,

$$H_{\rm R} = \int d^3r \frac{m}{\hbar e} \boldsymbol{j}_{\rm s} \cdot (\boldsymbol{\alpha}_{\rm R} \times \boldsymbol{n}), \tag{64}$$

which results in the same expression as Eq. (62).

The strength of the Rashba-induced magnetic field is estimated (choosing $a = 2$Å) as $B_{\rm R} = 2 \times 10^{16} \times \alpha_{\rm R}({\rm Jm}) j_{\rm s}({\rm A/m^2})$; For a strong Rashba interaction $\alpha_{\rm R} = 1$ eVÅ like at surfaces,[45] $B_{\rm R} = 4 \times 10^{-2}$ T at $j_{\rm s} = 10^{11}$ A/m$^2$. This field appears not very strong, but is sufficient at modify the magnetization dynamics. In fact, for the domain wall motion, when the Rashba-induced magnetic field is along the magnetic easy axis, the field is equivalent to that of an effective $\beta$ parameter of

$$\beta_{\rm R} = \frac{2m\lambda}{\hbar^2} \alpha_{\rm R}, \tag{65}$$

where $\lambda$ is the wall thickness. If $\alpha_{\rm R} = 1$ eVÅ, $\beta_{\rm R}$ becomes extremely large like $\beta_{\rm R} \simeq 250$ for $\lambda = 50$ nm. Note that $\beta$ arising from spin relaxation is the same order as Gilbert damping constant, namely of the order of $10^{-2}$. Such a large effective $\beta$ is expected to leads to an extremely fast domain wall motion under current.[46,47]

Experimentally, it was argued that fast domain wall motion observed in Pt/Co/AlO was due to the Rashba interaction,[48] but the result is later associated with the torque generated by spin Hall effect in Pt layer.[36] It was recently shown theoretically that strong Rashba-induced magnetic field works as a strong pinning center when introduced locally, and that this Rashba pinning effect is useful for highly reliable control of domain walls in racetrack memories.[49]

### 10.2. *Rashba-induced spin gauge field*

Since the interaction (63) is the one coupling to the spin current, the Rashba interaction is regarded as a gauge field acting on electron spin as far as the

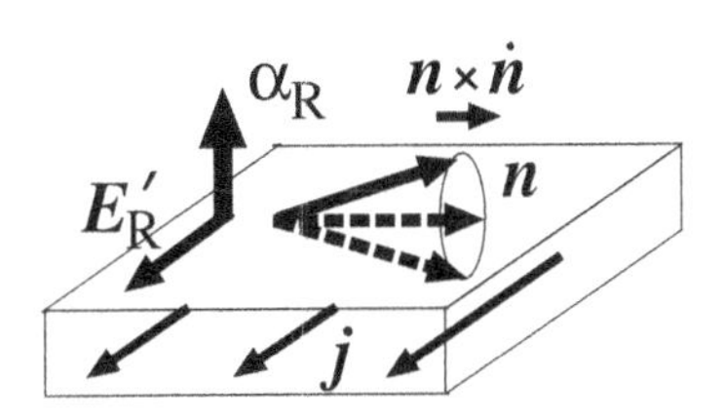


Fig. 11. Schematic figure depicting spin relaxation contribution of Rashba-induced spin electric field $\boldsymbol{E}'_{\mathrm{R}}$ generated by magnetization precession. Electric current $\boldsymbol{j}$ is induced as a result of motive force $\boldsymbol{E}'_{\mathrm{R}}$ in the direction perpendicular to both $\boldsymbol{n} \times \dot{\boldsymbol{n}}$ and Rashba field $\boldsymbol{\alpha}_{\mathrm{R}}$.

linear order concerns. The gauge field defined by Eq. (63) is

$$\boldsymbol{A}_{\mathrm{R}} \equiv -\frac{m}{e\hbar}(\boldsymbol{\alpha}_{\mathrm{R}} \times \boldsymbol{n}). \tag{66}$$

Existence of a gauge field naturally leads to an effective electric and magnetic field[5,7]

$$\begin{aligned} \boldsymbol{E}_{\mathrm{R}} &= -\dot{\boldsymbol{A}}_{\mathrm{R}} = \frac{m}{e\hbar}(\boldsymbol{\alpha}_{\mathrm{R}} \times \dot{\boldsymbol{n}}) \\ \boldsymbol{B}_{\mathrm{R}} &= \boldsymbol{\nabla} \times \boldsymbol{A}_{\mathrm{R}} = -\frac{m}{e\hbar}\boldsymbol{\nabla} \times (\boldsymbol{\alpha}_{\mathrm{R}} \times \boldsymbol{n}). \end{aligned} \tag{67}$$

In the presence of electron spin relaxation, the electric field has a perpendicular component[6]

$$\boldsymbol{E}'_{\mathrm{R}} = \frac{m}{e\hbar}\beta_{\mathrm{R}}[\boldsymbol{\alpha}_{\mathrm{R}} \times (\boldsymbol{n} \times \dot{\boldsymbol{n}})], \tag{68}$$

where $\beta_{\mathrm{R}}$ is a coefficient representing the strength of spin relaxation. For the case of strong Rashba interaction of $\alpha_{\mathrm{R}} = 3$ eVÅ, as realized in Bi/Ag, the magnitude of the electric field is $|E_{\mathrm{R}}| = \frac{m}{e\hbar}\alpha_{\mathrm{R}}\omega = 26$kV/m if the angular frequency $\omega$ of magnetization dynamics is 10 GHz. The magnitude of relaxation contribution is $|E'_{\mathrm{R}}| \sim 260$V/m if $\beta_{\mathrm{R}} = 0.01$. The effective magnetic field in the case of spatial length scale of 10 nm is high as well; $B_{\mathrm{R}} \sim 260$T.

The Rashba-induced electric fields, $\boldsymbol{E}_{\mathrm{R}}$ and $\boldsymbol{E}'_{\mathrm{R}}$, are important from the viewpoint of spin-charge conversion. In fact, results (67), (68) indicates that a voltage is generated by a dynamics magnetization if the Rashba interaction is present, even in the case of spatially uniform magnetization, in sharp contrast to the conventional adiabatic effective electric field from the spin Berry's phase of Eq. (17). In the case of a think film with Rashba interaction perpendicular to the plane and with a precessing magnetization, the component $\boldsymbol{E}_{\rm R} \propto \dot{\boldsymbol{n}}$ has no DC component, while the relaxation contribution $\boldsymbol{E}'_{\rm R}$ has a DC component perpendicular to $\overline{\boldsymbol{n} \times \dot{\boldsymbol{n}}} \parallel \overline{\boldsymbol{n}}$. The geometry of this (spin-polarized) current pumping effect, $\boldsymbol{j} \propto \boldsymbol{E}'_{\rm R} \propto \boldsymbol{\alpha}_{\rm R} \times \overline{\boldsymbol{n}}$, is therefore the same as the one expected in the case of inverse Edelstein effect (Fig. 11). In the present form, there is a difference between the Rashba-induced electric field effect and the system in Ref. 44, that is, the former assumes a direct contact between the Rashba interaction and magnetization while they are separated by a Ag spacer in Ref. 44. It is expected, however, that the Rashba-induced electric field becomes long-ranged and survives in the presence of a spacer if we include the electron diffusion processes. The spin-charge conversion observed in junctions will then be interpreted as due to the Rashba-induced electromagnetic field. For this scenario to be justified, it is crucial to confirm the existence of magnetic component, $\boldsymbol{B}_{\rm R}$, which can be of the order of 100T. In the setup of Fig. 11, $\boldsymbol{B}_{\rm R}$ is along $\overline{\boldsymbol{n}}$. The field can therefore be detected by measuring "giant" in-plane spin Hall effect when a current is injected perpendicular to the plane.

## 11. Application of effective vector potential theory

### 11.1. *Anomalous optical properties of Rashba conductor*

The idea of effective gauge field is useful for extending the discussion to include other degrees of freedom, like optical properties. In fact, the fact that the Rashba interaction coupled with magnetization leads to an effective vector potential $\boldsymbol{A}_{\rm R}$ (Eq. (66)) for electron spin indicates that the existence of intrinsic spin flow. Such intrinsic flow affects the optical properties, as incident electromagnetic waves get Doppler shift when interacting with flowing electrons, resulting in a transmission depending on the direction (directional dichroism), as was theoretically demonstrated in Refs. 50,51. The magnitude of the directional dichroism for the case of wave vector $\boldsymbol{q}$ is given by $\boldsymbol{q} \cdot (\boldsymbol{\alpha}_{\rm R} \times \boldsymbol{n})$. The vector $(\boldsymbol{\alpha}_{\rm R} \times \boldsymbol{n})$ is called in the context of multiferroics the toroidal moment, and it was argued to acts as an effective vector potential for light.[52]

It was shown also that Rashba conductor, even without magnetization, shows peculiar optical properties such as negative refraction as a result of spin-charge mixing effects.[50] In fact, spin-charge mixing effects of Eq. (61) leads to a current generated by applied electric field, $\boldsymbol{E}$, given by

$$j_{\rm IE\cdot E} = -\hbar\gamma\kappa_{\rm EIE}[\boldsymbol{\alpha}_{\rm R} \times (\boldsymbol{\alpha}_{\rm R} \times \boldsymbol{E})], \tag{69}$$

where $\kappa_{\rm EIE}$ is a coefficient (Fig. 12). As it is opposite to the applied field,

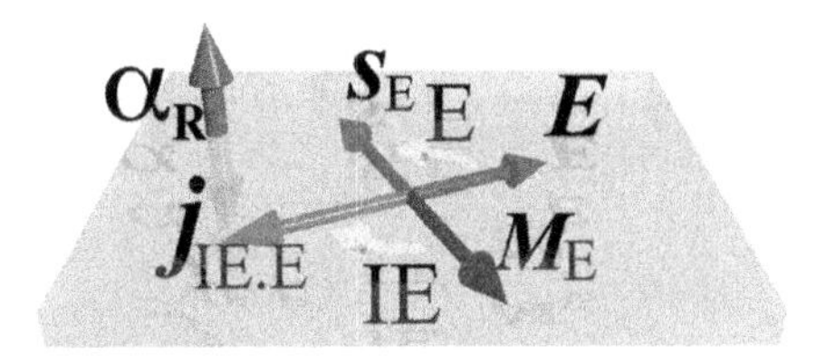


Fig. 12. Schematic figure showing the cross-correlation effects in the plane perpendicular to the Rashba field $\boldsymbol{\alpha}_{\mathrm{R}}$. Edelstein effect (E) generates spin density, $\boldsymbol{s}_{\mathrm{E}}$, from the applied electric field, and inverse Edelstein effect (IE) generates current $\boldsymbol{j}_{\mathrm{IE\cdot E}}$ from magnetization $\boldsymbol{M}_{\mathrm{E}}$.

the mixing effect results in a softening of the plasma frequency as for the $\boldsymbol{E}$ having components perpendicular to $\boldsymbol{\alpha}_{\mathrm{R}}$. The electric permittivity of the system is therefore anisotropic; Choosing $\boldsymbol{\alpha}_{\mathrm{R}}$ along the $z$ axis, we have

$$\varepsilon_z = 1 - \frac{\omega_{\mathrm{p}}^2}{\omega(\omega + i\eta)}, \qquad \varepsilon_x = \varepsilon_y = 1 - \frac{\omega_{\mathrm{R}}^2}{\omega(\omega + i\eta)}, \tag{70}$$

where $\omega_{\mathrm{p}} = \sqrt{e^2 n_{\mathrm{e}}/\varepsilon_0 m}$ is the bare plasma frequency ($n_{\mathrm{e}}$ is the electron density), and $\omega_{\mathrm{R}} \equiv \omega_{\mathrm{p}}\sqrt{1 + \mathrm{Re}C(\omega_{\mathrm{R}})} < \omega_{\mathrm{p}}$ is the plasma frequency reduced by the spin mixing effect.[50] ($C(\omega)$ represents the correlation function representing the Rashba-Edelstein effect, and its real part is negative.) The frequency region $\omega_{\mathrm{R}} < \omega < \omega_{\mathrm{p}}$ is of interest, as the system is insulating ($\varepsilon_z > 0$) in the direction of the Rashba field but metallic in the perpendicular direction ($\varepsilon_x < 0$). The dispersion in this case becomes hyperbolic, and the group velocity and phase velocity along $\boldsymbol{q}$ can have opposite direction, resulting in negative refraction. Rashba system is, therefore a natural hyperbolic metamaterial.[53] A great advantage of Rashba conductors are that the metamaterial behavior arises in the infrared or visible light region, which is not easily accessible in fabricated systems. For instance, in the case of BiTeI with Rashba splitting of $\alpha = 3.85$ eVÅ,[54] the plasma frequency is $\omega_{\mathrm{p}} = 2.5 \times 10^{14}$ Hz (corresponding to a wavelength of 7.5$\mu$m) for $n_{\mathrm{e}} = 8 \times 10^{25}$ m$^{-3}$ and $\epsilon_F = 0.2$ eV.[55] We then have $\omega_{\mathrm{R}}/\omega_{\mathrm{p}} = 0.77$ ($\omega_{\mathrm{R}} = 1.9 \times 10^{14}$ Hz, corresponding to the wavelength of 9.8$\mu$m), and hyperbolic behavior arises in the infrared regime. The directional dichroism arises in the infrared-red light regime.[50]

### 11.2. *Dzyaloshinskii-Moriya interaction*

Another interesting effect of spin gauge field pointed out recently is to induce the Dzyaloshinskii-Moriya interaction. Dzyaloshinskii-Moriya (DM) interaction is an antisymmetric exchange interaction between magnetic atoms that can arise when inversion symmetry is broken. In the continuum limit, it is represented as

$$H_{\rm DM} \equiv \int d^3r D_i^\alpha (\nabla_i \boldsymbol{n} \times \boldsymbol{n})^\alpha, \tag{71}$$

where $D_i^a$ is the strength, $\alpha$ and $i$ denotes the spin and spatial direction, respectively. It was recently discussed theoretically that the interaction is a result of Doppler shift due to an intrinsic spin current generated by broken inversion symmetry.[56] In fact, spin current density, $j_{\rm s}$, which is odd and even under spatial inversion and time-reversal, respectively, is induced by spin-orbit interaction in systems with broken inversion symmetry. Spatial variation of localized spins observed by the flowing electron spin is then described by a covariant derivative,

$$\mathfrak{D}_i \boldsymbol{n} = \nabla_i \boldsymbol{n} + \eta(\boldsymbol{j}_{{\rm s},i} \times \boldsymbol{n}), \tag{72}$$

where $\eta$ is a coefficient. This covariant derivative leads to the magnetic energy generated by the electron of $(\mathfrak{D}_i \boldsymbol{n})^2 = (\nabla \boldsymbol{n})^2 + 2\eta \sum_i \boldsymbol{j}_{{\rm s},i} \cdot (\boldsymbol{n} \times \nabla_i \boldsymbol{n}) + O(\eta^2)$. We see that the second term proportional to $j_{\rm s}$ is the DM interaction, and thus the coefficient is $D_i^a \propto j_{{\rm s},i}^\alpha$.

More rigorous derivation is performed by deriving an effective Hamiltonian. The electrons interacting strongly with localized spin is described by a Lagrangian (38), where $\mathcal{A}_{{\rm s},i}^\alpha$ is an SU(2) gauge field describing the spatial and temporal variation of localized spin. To discuss DM interaction, we include a spin-orbit interaction with broken inversion symmetry,

$$H_{\rm so} = \int d^3r \frac{i}{2} c^\dagger \left[\lambda_i \cdot \boldsymbol{\sigma} \overleftrightarrow{\nabla}_i\right] c, \tag{73}$$

where $\boldsymbol{\lambda}$ is a vector representing the broken inversion symmetry. (Multiorbital cases are treated similarly.[56]) As is obvious from this form linear in spatial derivative and Pauli matrix, the spin-orbit interaction generates a spin current proportional to $\boldsymbol{\lambda}$. From Eq. (38), the effective Lagrangian for localized spin to the linear order in derivative is

$$H_{\rm eff} = \int d^3r \sum_{ia} \tilde{j}_{{\rm s},i}^a \mathcal{A}_{{\rm s},i}^a, \tag{74}$$

where $\tilde{j}_{{\rm s},i}^a \equiv \left\langle \hat{\tilde{j}}_{{\rm s},i}^a \right\rangle$ is the expectation value of the spin current density in the rotated frame. In terms of the spin current in the laboratory frame,

$j^a_{s,i}$, the effective Hamiltonian reads

$$H_{\text{eff}} = \int d^3r D^a_i (\nabla_i \boldsymbol{n} \times \boldsymbol{n})^a, \tag{75}$$

where

$$D^a_i \equiv j^{\perp,a}_{s,i}, \tag{76}$$

and $j^{\perp,a}_{s,i}$ is a component of $j^a_{s,i}$ perpendicular to the local magnetization direction, $\boldsymbol{n}$. We therefore see that the DM coefficient is indeed given by the expectation value of the spin current density of the conduction electrons. The first principles calculation based on this spin current expression turns out to have advantage of shorter calculation time than previous methods[57,58] by evaluating twist energy of magnetization.[56]

It has been noted that spin wave dispersion is modified in the presence of DM interaction, resulting in Doppler shift of spin waves.[59,60] The spin wave Doppler shift is natural from our physical interpretation of DM interaction, as DM interaction itself is a consequence of flowing electron spin current.

## 12. Summary

We have discussed various magnetic and electron transport properties in metallic ferromagnets from the view points of effective gauge field. The concept of gauge field turned out to be highly useful to describe novel electromagnetic cross correlation effects and optical properties.

## References

1. L. Berger, Possible existence of a josephson effect in ferromagnets, *Phys. Rev. B* **33**, 1572 (1986).
2. G. E. Volovik, Linear momentum in ferromagnets, *J. Phys. C: Solid State Phys.* **20**, L83 (1987).
3. R. A. Duine, Spin pumping by a field-driven domain wall, *Phys. Rev. B* **77**, p. 014409 (2008).
4. A. Takeuchi and G. Tatara, Spin damping monopole, *Journal of the Physical Society of Japan* **81**, p. 033705 (2012).
5. K.-W. Kim, J.-H. Moon, K.-J. Lee and H.-W. Lee, Prediction of giant spin motive force due to rashba spin-orbit coupling, *Phys. Rev. Lett.* **108**, p. 217202 (May 2012).
6. G. Tatara, N. Nakabayashi and K.-J. Lee, Spin motive force induced by rashba interaction in the strong *sd* coupling regime, *Phys. Rev. B* **87**, p. 054403 (Feb 2013).

7. N. Nakabayashi and G. Tatara, Rashba-induced spin electromagnetic fields in the strong sd coupling regime, *New Journal of Physics* **16**, p. 015016 (2014).
8. R. Raimondi, Spin-charge coupling effects in a two-dimensional electron gas, *in this lecture series* (2017).
9. X. Waintal and M. Viret, Current-induced distortion of a magnetic domain wall, *Europhys. Lett.* **65**, p. 427 (2004).
10. A. Stern, Berry's phase, motive forces, and mesoscopic conductivity, *Phys. Rev. Lett.* **68**, 1022 (Feb 1992).
11. G. Tatara, H. Kohno and J. Shibata, Microscopic approach to current-driven domain wall dynamics, *Physics Reports* **468**, 213 (2008).
12. G. G. Cabrera and L. M. Falicov, Theory of the residual resistivity of bloch walls. pt. 1. paramagnetic effects, *Phys. Stat. Sol. (b)* **61**, p. 539 (1974).
13. L. Berger, Low-field magnetoresistance and domain drag in ferromagnets, *J. Appl. Phys.* **49**, p. 2156 (1978).
14. G. Tatara and H. Fukuyama, Resistivity due to a domain wall in ferromagnetic metal, *Phys. Rev. Lett.* **78**, 3773 (May 1997).
15. G. Tatara, Domain wall resistance based on landauer's formula, *Journal of the Physical Society of Japan* **69**, 2969 (2000).
16. G. Tatara, Effect of domain-wall on electronic transport properties of metallic ferromagnet, *Int. J. Mod. Phys. B.* **15**, p. 321 (2001).
17. J. C. Slonczewski, Current-driven excitation of magnetic multilayers, *J. Magn Magn Mater.* **159**, p. L1 (1996).
18. J. J. Sakurai, *Modern Quantum Mechanics* (Addison Wesley, 1994).
19. G. Tatara, A. Takeuchi, N. Nakabayashi and K. Taguchi, Monopoles in ferromagnetic metals, *Journal of the Korean Physical Society* **61**, 1331 (2012).
20. S. A. Yang, G. S. D. Beach, C. Knutson, D. Xiao, Q. Niu, M. Tsoi and J. L. Erskine, Universal electromotive force induced by domain wall motion, *Phys. Rev. Lett.* **102**, p. 067201 (2009).
21. K. Tanabe, D. Chiba, J. Ohe, S. Kasai, H. Kohno, S. E. Barnes, S. Maekawa, K. Kobayashi and T. Ono, Spin-motive force due to a gyrating magnetic vortex, *Nat Commun* **3**, p. 845 (May 2012).
22. A. Neubauer, C. Pfleiderer, B. Binz, A. Rosch, R. Ritz, P. G. Niklowitz and P. Böni, Topological hall effect in the *a* phase of mnsi, *Phys. Rev. Lett.* **102**, p. 186602 (May 2009).
23. T. Schulz, R. Ritz, A. Bauer, M. Halder, M. Wagner, C. Franz, C. Pfleiderer, K. Everschor, M. Garst and A. Rosch, Emergent electrodynamics of skyrmions in a chiral magnet, *Nat Phys* **8**, 301 (Apr 2012).

24. K. Taguchi, J.-i. Ohe and G. Tatara, Ultrafast magnetic vortex core switching driven by the topological inverse faraday effect, *Phys. Rev. Lett.* **109**, p. 127204 (Sep 2012).
25. J. Shibata, G. Tatara and H. Kohno, Effect of spin current on uniform ferromagnetism: Domain nucleation, *Phys. Rev. Lett.* **94**, p. 076601 (2005).
26. G. Tatara and H. Kohno, Theory of current-driven domain wall motion: Spin transfer versus momentum transfer, *Phys. Rev. Lett.* **92**, p. 086601 (2004).
27. G. Tatara, H. Kohno, J. Shibata, Y. Lemaho and K.-J. Lee, Spin torque and force due to current for general spin textures, *J. Phys. Soc. Jpn.* **76**, p. 054707 (2007).
28. M. Oogane, T. Wakitani, S. Yakata, R. Yilgin, Y. Ando, A. Sakuma and T. Miyazaki, Magnetic damping in ferromagnetic thin films, *Japanese Journal of Applied Physics* **45**, p. 3889 (2006).
29. S. Zhang and Z. Li, Roles of nonequilibrium conduction electrons on the magnetization dynamics of ferromagnets, *Phys. Rev. Lett.* **93**, p. 127204 (2004).
30. A. Thiaville, Y. Nakatani, J. Miltat and Y. Suzuki, Micromagnetic understanding of current-driven domain wall motion in patterned nanowires, *Europhys. Lett.* **69**, p. 990 (2005).
31. H. Kohno, G. Tatara and J. Shibata, Microscopic calculation of spin torques in disordered ferromagnets, *Journal of the Physical Society of Japan* **75**, p. 113706 (2006).
32. J. C. Slonczewski, Dynamics of magnetic domain walls, *Int. J. Magn.* **2**, p. 85 (1972).
33. S. Fukami, T. Suzuki, N. Ohshima, K. Nagahara and N. Ishiwata, Micromagnetic analysis of current driven domain wall motion in nanostrips with perpendicular magnetic anisotropy, *J. Appl. Phys.* **103**, p. 07E718 (2008).
34. T. Koyama, D. Chiba, K. Ueda, K. Kondou, H. Tanigawa, S. Fukami, T. Suzuki, N. Ohshima, N. Ishiwata, Y. Nakatani, K. Kobayashi and T. Ono, Observation of the intrinsic pinning of a magnetic domain wall in a ferromagnetic nanowire, *Nat Mater* **10**, 194 (Mar 2011).
35. G. Tatara, T. Takayama, H. Kohno, J. Shibata, Y. Nakatani and H. Fukuyama, Threshold current of domain wall motion under extrinsic pinning, $\beta$-term and non-adiabaticity, *Journal of the Physical Society of Japan* **75**, p. 064708 (2006).

36. S. Emori, U. Bauer, S.-M. Ahn, E. Martinez and G. S. D. Beach, Current-driven dynamics of chiral ferromagnetic domain walls, *Nat Mater* **12**, 611 (Jul 2013), Letter.
37. H. Saarikoski, H. Kohno, C. H. Marrows and G. Tatara, Current-driven dynamics of coupled domain walls in a synthetic antiferromagnet, *Phys. Rev. B* **90**, p. 094411 (Sep 2014).
38. S.-H. Yang, K.-S. Ryu and S. Parkin, Domain-wall velocities of up to 750ms$^{-1}$ driven by exchange-coupling torque in synthetic antiferromagnets, *Nat Nano* **10**, p. 221–226 (Feb 2015), Letter.
39. I. Dzyaloshinsky, A thermodynamic theory of ferromagnetism of antiferromagnetics, *Journal of Physics and Chemistry of Solids* **4**, 241 (1958).
40. T. Moriya, Anisotropic superexchange interaction and weak ferromagnetism, *Phys. Rev.* **120**, 91 (Oct 1960).
41. E. Rashba, *Sov. Phys. Solid State* **2**, 1109 (1960).
42. V. Edelstein, Spin polarization of conduction electrons induced by electric current in two-dimensional asymmetric electron systems, *Solid State Communications* **73**, 233 (1990).
43. K. Shen, G. Vignale and R. Raimondi, Microscopic theory of the inverse edelstein effect, *Phys. Rev. Lett.* **112**, p. 096601 (Mar 2014).
44. J. C. R. Sanchez, L. Vila, G. Desfonds, S. Gambarelli, J. P. Attane, J. M. De Teresa, C. Magen and A. Fert, Spin-to-charge conversion using rashba coupling at the interface between non-magnetic materials, *Nat Commun* **4**, p. 2944 (Dec 2013), Article.
45. C. R. Ast, J. Henk, A. Ernst, L. Moreschini, M. C. Falub, D. Pacilé, P. Bruno, K. Kern and M. Grioni, Giant spin splitting through surface alloying, *Phys. Rev. Lett.* **98**, p. 186807 (2007).
46. K. Obata and G. Tatara, Current-induced domain wall motion in rashba spin-orbit system, *Phys. Rev. B* **77**, p. 214429 (2008).
47. A. Manchon and S. Zhang, Theory of spin torque due to spin-orbit coupling, *Phys. Rev. B* **79**, p. 094422 (Mar 2009).
48. I. M. Miron, G. Gaudin, S. Auffret, B. Rodmacq, A. Schuhl, S. Pizzini, J. Vogel and P. Gambardella, Current-driven spin torque induced by the rashba effect in a ferromagnetic metal layer, *Nature Materials* **9**, 230 (2010).
49. G. Tatara, H. Saarikoski and C. Mitsumata, Efficient stopping of current-driven domain wall using a local rashba field, *Applied Physics Express* **9**, p. 103002 (2016).

50. J. Shibata, A. Takeuchi, H. Kohno and G. Tatara, Theory of anomalous optical properties of bulk rashba conductor, *Journal of the Physical Society of Japan* **85**, p. 033701(5pages) (2016).
51. Kawaguchi, *arXiv:1610.01743* (2016).
52. K. Sawada and N. Nagaosa, Optical magnetoelectric effect in multiferroic materials: Evidence for a lorentz force acting on a ray of light, *Phys. Rev. Lett.* **95**, p. 237402 (Dec 2005).
53. E. E. Narimanov and A. V. Kildishev, Metamaterials: Naturally hyperbolic, *Nat Photon* **9**, 214 (Apr 2015), News and Views.
54. K. Ishizaka, M. S. Bahramy, H. Murakawa, M. Sakano, T. Shimojima, T. Sonobe, K. Koizumi, S. Shin, H. Miyahara, A. Kimura, K. Miyamoto, T. Okuda, H. Namatame, M. Taniguchi, R. Arita, N. Nagaosa, K. Kobayashi, Y. Murakami, R. Kumai, Y. Kaneko, Y. Onose and Y. Tokura, Giant rashba-type spin splitting in bulk bitei, *Nat Mater* **10**, 521 (Jul 2011).
55. L. Demkó, G. A. H. Schober, V. Kocsis, M. S. Bahramy, H. Murakawa, J. S. Lee, I. Kézsmárki, R. Arita, N. Nagaosa and Y. Tokura, Enhanced infrared magneto-optical response of the nonmagnetic semiconductor bitei driven by bulk rashba splitting, *Phys. Rev. Lett.* **109**, p. 167401 (Oct 2012).
56. T. Kikuchi, T. Koretsune, R. Arita and G. Tatara, Dzyaloshinskii-moriya interaction as a consequence of a doppler shift due to spin-orbit-induced intrinsic spin current, *Phys. Rev. Lett.* **116**, p. 247201 (Jun 2016).
57. M. I. Katsnelson, Y. O. Kvashnin, V. V. Mazurenko and A. I. Lichtenstein, Correlated band theory of spin and orbital contributions to dzyaloshinskii-moriya interactions, *Phys. Rev. B* **82**, p. 100403 (Sep 2010).
58. F. Freimuth, S. Blügel and Y. Mokrousov, Berry phase theory of dzyaloshinskii-moriya interaction and spin- orbit torques, *Journal of Physics: Condensed Matter* **26**, p. 104202 (2014).
59. Y. Iguchi, S. Uemura, K. Ueno and Y. Onose, Nonreciprocal magnon propagation in a noncentrosymmetric ferromagnet $life_5o_8$, *Phys. Rev. B* **92**, p. 184419 (Nov 2015).
60. S. Seki, Y. Okamura, K. Kondou, K. Shibata, M. Kubota, R. Takagi, F. Kagawa, M. Kawasaki, G. Tatara, Y. Otani and Y. Tokura, Magnetochiral nonreciprocity of volume spin wave propagation in chiral-lattice ferromagnets, *Phys. Rev. B* **93**, p. 235131 (Jun 2016).

# Introduction to the Topic of Jack Polynomials in the Context of Fractional Quantum Hall Effect

B. Kuśmierz* and A. Wójs

*Department of Theoretical Physics, Faculty of Fundamental Problems of Technology*
*Wrocław University of Science and Technology, Wybrzeże Wyspiańskiego 27*
*50-370 Wrocław, Poland.*
**E-mail: bartosz.kusmierz@pwr.edu.pl*

The Jack polynomials $J_\lambda^\alpha$ are a remarkable family of symmetric polynomials. They generalize many families of symmetric polynomials including: Schur polynomials, monomial symmetric polynomials, and elementary symmetric polynomials. The original development of Jack polynomials focused on the case of the real parameter $\alpha > 0$. This assures that the corresponding $J_\lambda^\alpha$ has no pole and is well defied. However for a specific values of $\alpha < 0$, Jack polynomials have no pole as well and reveal interesting properties, useful in physics of fractional quantum Hall effect. This solid state phenomenon involves essentially two-dimensional electrons in extremely low temperatures and strong magnetic field. Aim of this paper is to introduce the Jack polynomials in the context of fractional quantum Hall effect. We give a brief description of classical version of Hall effect and discuss both the integer and fractional quantum Hall effect. Then we introduce the basics of the theory of symmetric functions; in particular we define, describe, and construct the Jack polynomials. This knowledge is later applied and illustrated with numerical generation of the coefficients of "Jack states" (quantum Hall states related to the Jack polynomials). As an illustration we examine overlaps of the Jack state wave functions and the Coulomb ground states for different system sizes.

*Keywords*: Quantum Hall effect, Jack polynomials, Jack states, partitions.

## 1. Physics of quantum Hall effect

We start these notes with a brief introduction to the topic of quantum Hall effect. Firstly we discuss a classical version of the phenomenon, then follow with the integer quantum Hall effect. This includes analysis of one particle model and solutions to the stationary Schrödinger equation. Then, we focus on the fractional quantum Hall effect and introduce the standard tools used to study it, like: Haldane sphere, composite fermions, and trial wave function approach.

### 1.1. *Classical Hall effect*

The Hall effect was discovered by Edwin Herbert Hall in 1879. The phenomenon refers to an electric current passing through a cuboid conductor in a uniform magnetic field. For such systems the electric charge accumulates on the face of cuboid transverse to the applied magnetic field and the electric current.[1–3]

The Hall effect can be explained within frame of classical physics. Consider a flat cuboid conductor with sides $a$, $b$, $c$ placed in a uniform magnetic field $\mathbf{B}$ parallel to the side $c$. Denote the magnitude of magnetic field by $B$. Assume there is a current parallel to $a$, and an average velocity of carriers equals $\mathbf{v_c}$. Then, on average, charge carrier of charge $q$ experiences the magnetic force $\mathbf{F_m}$

$$\mathbf{F_m} = q\mathbf{v_c} \times \mathbf{B}. \tag{1}$$

This force pushes the charge carrier to one face of the cuboid. Hence one can observe charge density difference, and the $y$-component (parallel to the side $b$) of the electric field $E_y$ builds up. Then, the electric carrier is driven by the electric force $\mathbf{F_e}$. Its $y$-component can be derived from a formula $(\mathbf{F_e})_y = qE_y$. For a density of carriers denoted $\rho$, the current $\mathbf{I}$ equals

$$\mathbf{I} = \mathbf{v_c}\rho(b \cdot c)q. \tag{2}$$

Since only the $x$-component of $\mathbf{I}$ is nonzero, one writes $I_x = v_c n(b \cdot c)q$, where $v_c$ is a magnitude of $\mathbf{v_c}$. Then $y$-component of $\mathbf{F_m}$ equals

$$(\mathbf{F_m})_y = B\frac{I_x}{\rho(b \cdot c)}. \tag{3}$$

In an equilibrium net force equals zero therefore

$$0 = \mathbf{F_m} - \mathbf{F_e}. \tag{4}$$

What follows

$$B = \frac{I_x}{\rho(b \cdot c)} = qE_y. \tag{5}$$

For the Hall voltage defined as

$$V_H = E_y b, \tag{6}$$

one obtains

$$V_H = B\frac{I_x}{\rho q c}. \tag{7}$$

The formula 7 shows that Hall voltage is proportional to the magnitude of the magnetic field. The Hall resistance is defined as $R_H = V_H/I_x$. Then

$$R_H = \frac{B}{\rho e c}. \tag{8}$$

## 1.2. *Integer quantum Hall effect*

The formula 7 suggests that Hall resistance is a linear function of magnetic field. However this relation is not fulfilled in two-dimensional, cold electron systems in high magnetic fields. In such systems, $R_H$ is restricted to the specific values $R_H = \nu^{-1}\frac{h}{e^2}$, where $h$ is Planck constant and $e$ is electric charge.

In the case of integer quantum Hall effect (IQHE), $\nu \in \mathbb{N}_+$. Quantity $\nu$ is also known as a filling factor and is defined as

$$\nu = \frac{\rho}{\frac{B}{\phi_0}}, \tag{9}$$

where $\phi_0 = \frac{hc}{e}$ is a flux quantum. IQHE was first observed in 1980 at the High Magnetic Field Laboratory in Grenoble. In 1985 Klaus von Klitzing was awarded Nobel prize for its discovery. Due to the discrete values of the conductance, quantum Hall effect provides an extremely precise determination of the fine structure constant $\alpha = \frac{e^2}{4\pi\varepsilon_0 \hbar c}$. IQHE is closely related to Landau quantization of cyclotron orbits of charged particles in magnetic field. Discrete energy levels of electrons are called the Landau levels (LL) and are highly degenerate.

### 1.2.1. *One particle model: Landau gauge*

In order to explain IQHE one can consider a model of one, non interacting particle of mass $m$ and charge $q$, confined to an area $[0, L_x] \times [0, L_y]$ in an $XY$-plane.[2,3] Let the particle be exposed to a uniform magnetic field $\mathbf{B}$. Then the magnetic field enters the Hamiltonian $H$ of such system through a vector potential $\mathbf{A}$ ($\mathbf{B} = \nabla \times \mathbf{A}$).

$$H = \frac{1}{2m}\left(\mathbf{p} - q\frac{\mathbf{A}}{c}\right)^2, \tag{10}$$

where $\mathbf{p} = (\mathbf{p_x}, \mathbf{p_y}, \mathbf{p_z})$ is a vector of the momentum operators and $c$ is a speed of light. The square of vector is understand as a dot product of this vector with itself. In the Landau gauge

$$\mathbf{A} = (0, B\mathbf{x}, 0), \tag{11}$$

where $\mathbf{x}$ is an operator of $x$-component of a position. Then

$$H = \frac{1}{2m}\left(\mathbf{p_x}^2 + \left(\mathbf{p_y} - \frac{qB\mathbf{x}}{c}\right)^2\right). \tag{12}$$

One looks for solutions of an eigenvector problem ($H\psi = e \cdot \psi$, $e \in \mathbb{R}$) in functions of a form:

$$\psi(x, y) = \exp(iky)\varphi(x). \tag{13}$$

When substituted to the equation 12, the problem reduces to one dimensional equation

$$e \cdot \varphi(x) = \left(-\frac{\hbar^2}{2m}\partial_x^2 + \frac{m\omega_c^2}{2}\left(x + k\ell_B^2\right)^2\right)\varphi(x), \tag{14}$$

for a cyclotron frequency $\omega_c^2 = \frac{qB}{mc}$ and a magnetic length $\ell_B = \sqrt{\frac{\hbar c}{qB}}$. Notice an equation 14 is equivalent to a shifted quantum harmonic oscillator.

An eigenvalue and an eigenvector are labelled by two numbers: $k$ and $n$

$$H\psi_{k,n} = e_{k,n}\psi_{k,n}. \tag{15}$$

Then the eigenvalues are given by[4,5]

$$e_{k,n} = \left(n + \frac{1}{2}\right)\hbar\omega_c. \tag{16}$$

Notice there is no dependence in $k$ in the equation 16. The eigenvectors $\psi_{k,n}$ equal

$$\psi_{k,n} = \exp(iky)H_n\left(\frac{x + k\ell_B^2}{l_B}\right)\exp\left(-\frac{1}{2l_B^2}(x + k\ell_B^2)^2\right), \tag{17}$$

where $H_n$ are the Hermite polynomials

$$H_n(x) = (-1)^n \mathrm{e}^{x^2}\frac{d^n}{dx^n}\mathrm{e}^{-x^2}. \tag{18}$$

The Hermite functions $H_n$, solutions to the quantum harmonic oscillator in natural units, are presented in a form of a plot in Fig. 1.2.1. We also present solutions of the Hamiltonian 12, given in the equation 17 (Figs. 2 and 3).

Particle confined to the plane in the uniform perpendicular magnetic field has been discussed in the absence of an electric field. However when electric field is present, one particle IQH wave function is simply shifted.[4]

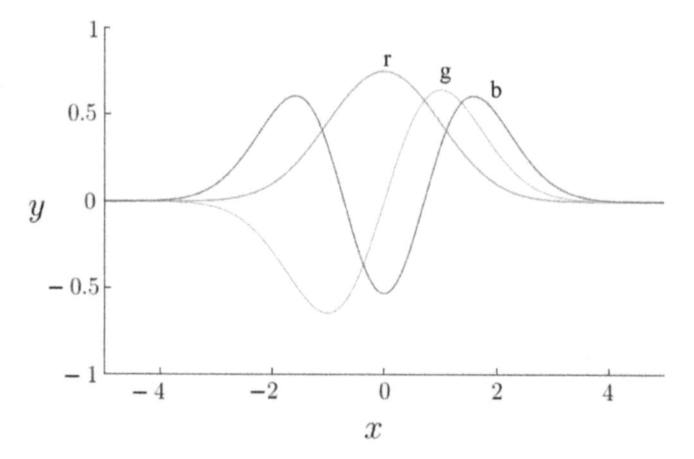


Fig. 1. Three leading Hermite functions normalized to one in $L^2$ norm. $H_0(x) \propto \exp(-x^2/2)$ – red curve, $H_1(x) \propto 2x\exp(-x^2/2)$ – blue curve and $H_2(x) \propto (4x^2 - 2)\exp(-x^2/2)$ – green curve.

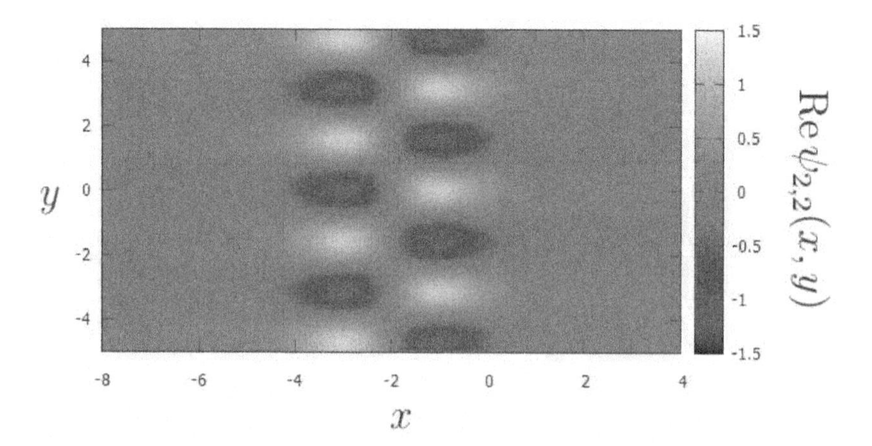


Fig. 2. Illustration of unnormalized solution $\psi_{2,2}(x, y)$ to the eigenvalue problem 12, given in 17 – real part of $\psi_{2,2}(x, y)$ (for magnetic length $\ell_B = 1$).

### 1.2.2. *One particle model: Symmetric gauge*

In a symmetric gauge magnetic vector potential equals

$$\mathbf{A} = \frac{B}{2}(-\mathbf{y}, \mathbf{x}, 0). \tag{19}$$

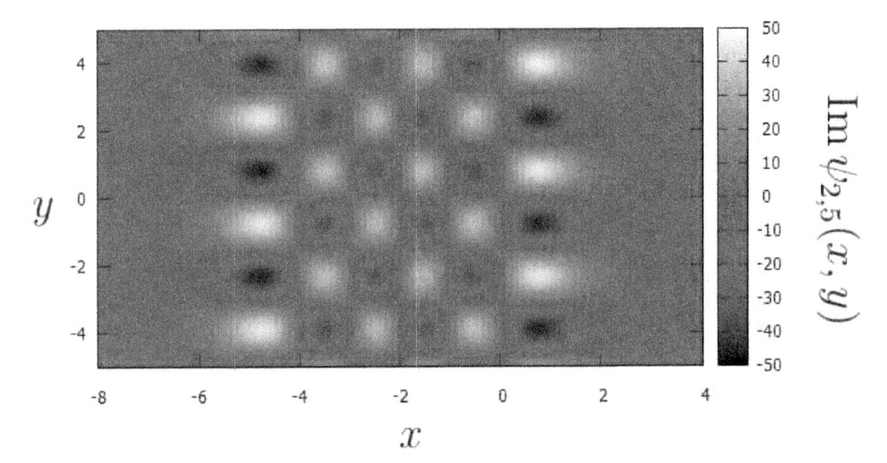


Fig. 3. Illustration of unnormalized solution $\psi_{2,5}(x,y)$ to the eigenvalue problem 12, given in 17 – imaginary part of $\psi_{2,5}$ (for magnetic length $\ell_B = 1$).

Then Hamiltonian may be written with usage of a raising and lowering operators[3,5,6]

$$H = \hbar\omega_c \left( a^\dagger a + \frac{1}{2} \right), \tag{20}$$

for

$$a = \frac{1}{\sqrt{2}} \left( \frac{z}{2\ell_B} + 2\ell_B \partial_{\bar{z}} \right), \quad a^\dagger = \frac{1}{\sqrt{2}} \left( \frac{\bar{z}}{2\ell_B} - 2\ell_B \partial_z \right). \tag{21}$$

In an equation 21 we introduced variable $z \in \mathbb{C}$. It is defined as $z = x - iy$, $\bar{z} = x + iy$. Unconventional definition of $z$ is motivated by a convenient form of wave functions in the lowest Landau level (LLL)[3] Similarly one introduces operators $b, b^\dagger$

$$b = \frac{1}{\sqrt{2}} \left( \frac{\bar{z}}{2\ell_B} + 2\ell_B \partial_z \right), \quad b^\dagger = \frac{1}{\sqrt{2}} \left( \frac{z}{2\ell_B} - 2\ell_B \partial_{\bar{z}} \right). \tag{22}$$

Then

$$[a, a^\dagger] = [b, b^\dagger] = 1. \tag{23}$$

One defines $z$ component of the angular momentum denoted $L_Z$ (in $XY$-plane)

$$L_Z = -\hbar(b^\dagger b - a^\dagger a). \tag{24}$$

One notices

$$[H, L_Z] = 0, \tag{25}$$

thus it is natural to look for the eigenvectors of $H$ in the set of the eigenvectors of $L_Z$. The eigenvalues of $L_Z$ are given by $m\hbar$ for $m \in \mathbb{Z}, -n \leq m$ where $n$ is a number of Landau level.

One notices solutions of the eigenvector problem are indexed by two quantum numbers $n, m$

$$|n, m\rangle = \frac{\left(b^\dagger\right)^{m+n}}{\sqrt{(m+n)!}} \frac{\left(a^\dagger\right)^n}{\sqrt{n!}} |0, 0\rangle, \tag{26}$$

where state $|0, 0\rangle$ corresponds to the Gaussian wave function

$$\Psi_{0,0} = \frac{1}{\sqrt{2\pi}} e^{-|z|^2/4}. \tag{27}$$

One notices

$$a|n, m\rangle = \sqrt{n}|n-1, m\rangle, \quad a^\dagger|n, m\rangle = \sqrt{n+1}|n+1, m\rangle, \tag{28}$$

and

$$b|n, m\rangle = \sqrt{m}|n, m-1\rangle, \quad b^\dagger|n, m\rangle = \sqrt{m+1}|n, m+1\rangle. \tag{29}$$

Moreover

$$a|0, 0\rangle = b|0, 0\rangle = 0. \tag{30}$$

Action of operators $H$ and $L_Z$ on eigenvectors is given by

$$H|n, m\rangle = \left(n + \frac{1}{2}\right) |n, m\rangle. \tag{31}$$

$$L_Z|n, m\rangle = \hbar\,(n - m)\,|n, m\rangle. \tag{32}$$

Direct calculations shows that in the LLL ($n = 0$), states $|0, m\rangle$ correspond to the products of radial Gaussian and certain power of $z$.

### 1.3. *Fractional quantum Hall effect*

The fractional quantum Hall effect (FQHE) was observed for the first time in 1982. Phenomenon occurs in temperatures close to the absolute zero in a presence of a strong magnetic field. Then collective states of quasi two-dimensional electrons reveal plateaus in the Hall resistivity at the filling factors given by rational numbers $\nu = \frac{p}{q} \in \mathbb{Q}$. The source of the FQHE is distinct from its integer counterpart. It does not occur in the one particle model and electron-electron interaction plays significant role in it.

For filling factors $\nu$ less than one, it can be assumed all of the electrons lie within the LLL. Then kinetic energies of all electrons equals and the total kinetic energy can be subtracted from Hamiltonian. Then Hamiltonian for such system of $N$ particles contains only part responsible for Coulomb interaction between electrons and can be written as

$$H = \sum_{1 \le i < j \le N} |\mathbf{r}_i - \mathbf{r}_j|^{-1}. \tag{33}$$

Where summation indexes runs over all particles and $\mathbf{r}_i$ is a position of the $i$-th particle. Since interaction energy cannot be treated as a small perturbation with respect to the kinetic energy standard, perturbation approach cannot be applied.

### 1.3.1. *Trial wave functions*

There is no general method of solving the FQH Hamiltonian (equation 33). Since neither exact nor based on perturbation method solutions can be constructed, phenomenological method, based on trial wave functions were developed. In 1983 Robert Laughlin proposed wave function describing ground state in LLL at filling factor $\nu = 1/q$ ($q$- odd).

Laughlin postulated ground state wave function at filling factor $\nu = 1/q$ should satisfy following conditions.[7]

1. The wave function for the whole system is an antisymmetrised product of single electron wave functions in the LLL.
2. Since the ground state is an eigenstate of the angular momentum operator, the wave function should be a homogeneous polynomial.
3. In order to keep electrons apart the wave function should be multiplied by a Jastrow type factor

$$\prod_{i<j} f(z_i - z_j) \tag{34}$$

where $f$ is a homogeneous polynomial.

Then Laughlin gave an example of such function, now known as the Laughlin wave function.

$$\Phi_L^{1/m} = \prod_{i<j} (z_i - z_j)^m \exp\left(\frac{\sum_i |z_i|^2}{4{\ell_B}^2}\right). \tag{35}$$

The Laughlin wave function is not an exact ground sate of coulomb repulsion Hamiltonian, however theirs overlaps are extremely close to one and it is believed to capture all of physical properties of the phenomenon.

Since the Gaussian factor $\exp\left(\frac{\sum_i |z_i|^2}{4\ell_B{}^2}\right)$ appears in all LLL wave functions it is usually skipped. This allow us to focus on the polynomial part of wave functions, in particular we refer to $\prod_{i<j}(z_i - z_j)^m$ as the Laughlin wave function.

It is postulated that system created from Laughlin ground state by adiabatically inserting a magnetic flux (quasihole state) can be described by a function

$$\Phi_L^{1/m,\mathrm{QH}(z_0)} = \prod_i (z_i - z_0) \prod_{i<j} (z_i - z_j)^m. \tag{36}$$

Where quasihole is located at position $z_0$. Such quasihole is a quantized vortex, since the wave function changes by a phase of $2\pi$ when each electron moves around $z_0$. The creation operator of quasihole (in a center of the system) is given by

$$\prod_i z_i. \tag{37}$$

By conjugation one obtains analogical operator for quasiparticle

$$\prod_i -\partial_{z_i}. \tag{38}$$

When shifted, operators create quasihole and quasiparticle anywhere.

The Laughlin wave function is not the only successful example of trial wave function approach in FQHE. One can discuss the so called Moore-Read (MR) state wave function, sometimes refereed to as the Pfaffian state. The MR state is believed to give approximate description of FQH states in the second LL(for fermions $\nu = 1/2$ in the second LL and for bosons $\nu = 1$ in second LL). MR wave function is well defined for an even number of particles and is given by

$$\Psi_{MR}^m = \mathrm{pf}\left(\frac{1}{z_i - z_j}\right) \prod_{i<j}^N (z_i - z_j)^{m+1}, \tag{39}$$

where $m = 0$ corresponds to the bosonic MR state and $m = 1$ to the fermionic one. Pfaffian $\mathrm{pf}(\cdot)$ is a function of antisymmetric matrix $M$ of $2N \times 2N$ dimension

$$\mathrm{pf}\,(M_{ij}) = \sqrt{\det(M)}, \tag{40}$$

or alternatively

$$\mathrm{pf}(M) = \frac{1}{2^N N!} \sum_{\sigma \in S_{2N}} \mathrm{sgn}(\sigma) \prod_{i=1}^{N} M_{\sigma(2i-1),\sigma(2i)}. \tag{41}$$

The bosonic Moore-Read state is a densest zero energy ground state of the non physical 3-body interaction projection Hamiltonian $H_3^2$

$$H_3^2 = \sum_{i<j<k} P_3^2(i,j,k). \tag{42}$$

Where $P_3^2(i,j,k)$ is a projection operator onto subspace of the relative angular momentum two for each triplet of particles[8]

Other trial wave functions used in FQHE includes functions like the Read-Rezai (RR) associated with $\nu = 3/5$ state; the Gaffnian state (for bosons $\nu = 2/3$ and $\nu = 2/5$ for fermions) or the Haffnian state ($\nu = 1/3$ for fermions).

### 1.4. *Composite fermions*

The composite fermion theory, proposed by Jainendra Jain[1,3,6] explains appearance of many FQH states. Theory postulates existence of the composite fermions – quasiparticles, states of electrons and even number of flux quanta. For a system of $N$ electrons in the perpendicular, uniform magnetic field $B$ when each of them bounds $2p$ fluxes, in the limit of infinite $N$, each of the electrons experience an effective magnetic field $B^*$

$$B^* = B - 2p\rho\phi_0. \tag{43}$$

For a coefficient $\phi$ – number of flux quanta through the sample. Define $N_\phi = \frac{\phi}{\phi_0}$. Then the filling factor may be expressed as $\nu = N/N_\phi$. One gives filing factor of the composite fermions $\nu^*$ in terms of filling factor for the electrons $\nu$

$$(\nu^*)^{-1} = \nu^{-1} - 2p. \tag{44}$$

Moreover

$$\nu = \frac{\nu^*}{2p\nu^* + 1}. \tag{45}$$

Construction of composite fermions gives form of the trial wave function of electrons $\Psi^{FQHE}$ at filling factor $\nu$ in terms of wave function of composite fermions $\Psi^{IQHE}$ at filling factor $\nu^*$

$$\Psi^{FQHE} = P_{LLL}\Psi^{IQHE} \prod_{i<j} (z_i - z_j)^{2p}. \tag{46}$$

Where $P_{LLL}$ is a projection into the LLL. The Jastrow factor $\prod_{i<j}(z_i - z_j)^{2p}$ assures each electron captures $2p$ fluxes.

Example of such construction is Laughlin state $\nu = \frac{1}{3}$ which corresponds to a completely filled Landau level of composite fermions.

### 1.4.1. *Haldane sphere*

Sometimes it is fruitful to examine FQHE when a system of electrons is projected from the plane on the so called Haldane sphere.[1,3,6,9,10] The Haldane sphere is the two dimensional sphere containing electrons. Magnetic field, perpendicular to its surface is provided by a Dirac monopole in a center of the sphere. Notice a magnetic flux through the surface of considered sphere, is quantized to $2Q\phi_0$, where $2Q$ is integer. Spherical geometry resolves problem of the boundary and as such can be used in the study of bulk properties. Moreover Landau levels on the sphere are only finitely degenerate. Such model of FQHE is typically used in numerical study of FQHE (direct diagonalisation of Hamiltonian). Value of the radial magnetic field $B$ on the surface of the sphere of radius $r$ is given by

$$B = \frac{2Q\phi_0}{4\pi r^2}. \tag{47}$$

One introduces coordinates $u, v$ on the Haldane sphere. $u, v$ are more useful in the context of FQHE than standard $\theta$ and $\varphi$. New coordinates can be expressed in terms of standard ones

$$u = \cos(\theta/2)e^{i\varphi/2}, \quad v = \sin(\theta/2)e^{-i\varphi/2}. \tag{48}$$

Consider a Hilbert space of square integrable functions over $\mathbb{C}$ defined on a sphere, with respect to the standard, rotation invariant measure $d\Omega = \sin\theta d\theta d\varphi$. Let $H_S$ denote subspace of Hilbert space, spanned by homogeneous polynomials in $u, v$.[10] Set of functions $\{e_{Q,m}\}_{Q,m}$ provide an orthonormal basis in $H_S$

$$e_{Q,m} = \sqrt{\frac{2Q+1}{4\pi}\binom{2Q}{Q+m}}\, u^{Q+m}v^{Q-m}, \tag{49}$$

where $m \in \{-Q, -Q+1, \ldots, Q\}$.

Transition from the space $H_S$ to the space of wave functions in the LLL is established by a function $\Gamma$

$$\Gamma\left(\sum_{i=0}^{2Q} c_i u^{2Q-i}v^i\right) = \sum_{i=1}^{2Q} c_i z^k, \tag{50}$$

where $c_i \in \mathbb{C}$ are coefficients.

$\Gamma$ is multiplicative and one can extend this mapping onto many particle wave functions.

We give an example how function $\Gamma$ acts on polynomials by mapping the Jastrow factor from the plane

$$\Gamma^{-1}\left(\prod_{i<j}(z_i - z_j)^m\right) = \prod_{i<j}(u_i v_j - v_j u_i)^m. \tag{51}$$

One writes angular momentum operators on the Haldane sphere

$$L^X = \frac{1}{2}\left(v\partial_u + u\partial_v\right), \tag{52a}$$

$$L^Y = \frac{i}{2}\left(v\partial_u - u\partial_v\right), \tag{52b}$$

$$L^Z = \frac{1}{2}\left(u\partial_u - v\partial_v\right). \tag{52c}$$

There are standard raising and lowering operators

$$L^+ = L^X + iL^Y = -u\partial_v, \tag{53a}$$

$$L^- = L^X - iL^Y = -v\partial_u. \tag{53b}$$

## 2. Theory of symmetric functions

Theory of symmetric functions is a field of mathematics with many applications. Objects developed in this theory appear in physics and various branches of mathematics. Symmetric polynomials occur in the Galois theory, specific symmetric functions naturally arise in the asymptotic group representation theory, combinatorics and algebraic combinatorics. Schur polynomials are characters of irreducible representations of the general linear groups. The main object of our interest in this paper – Jack polynomials are one parameter deformations of Schur polynomials. Moreover Jack polynomials occur in the study of free probability and in physics of many body systems.

We give a brief introduction to the symmetric functions theory. We define and discus fundamental objects and concepts of this theory, starting with partitions. Then we give construction of the ring of symmetric polynomials and define important families of symmetric and antisymmetric polynomials.

### 2.1. *Partitions*

The partition $\lambda$[11,12] is a sequence

$$\lambda = (\lambda_1, \lambda_2, .., \lambda_j, \dots), \tag{54}$$

of non negative integers in non increasing order

$$\lambda_1 \geq \lambda_2 \geq \cdots \geq \lambda_j \geq \dots. \tag{55}$$

Partition can be an infinite sequence, however only finitely many elements can be nonzero. Usually when partition indexes polynomial, sequence length corresponds to the number of variables of polynomial. The non zero $\lambda_i$ are called the *parts* of $\lambda$, the number of parts is the *length* of $\lambda$ and it is denoted by $\ell(\lambda)$. The sum of the parts of $\lambda$ is called the *weight* and is denoted $|\lambda|$

$$|\lambda| = \lambda_1 + \lambda_2 + \cdots + \lambda_j + \dots. \tag{56}$$

$m(\lambda, i)$ is a number of parts of $\lambda$ equal $i$. Sometimes one does not distinguish between two sequences varying only by a string of zeros at the end.[11] In such situation one regard $(3, 3, 1)$ and $(3, 3, 1, 0)$ as the same partition.

To keep notation short one can represent partition in its *frequency representation.* Frequency representation indicates how many times given number occurs in partition

$$\lambda = \left(1^{m(\lambda,1)}\ 2^{m(\lambda,2)}\ \dots\ i^{m(\lambda,i)}\ \dots\right). \tag{57}$$

Information about number of occurrences of parts of partition is very useful in physics of fractional quantum Hall effect. However due to the small indexes of frequency representation, it is more practical to use the *occupation number configuration*[20]

$$\lambda = [m(\lambda, 0)\ m(\lambda, 1)\ m(\lambda, 2)\ \dots]. \tag{58}$$

For example partition $(3, 3, 1)$ in the standard notation, is equivalent to $(1^1 2^0 3^2)$ in the frequency representation and to $[0\ 1\ 0\ 2]$ in the occupational representation.

For two partitions there is a natural operation of addition

$$\begin{aligned} \lambda + \mu &= (\lambda_1, \lambda_2, .., \lambda_j, \dots) + (\mu_1, \mu_2, .., \mu_j, \dots) \\ &= (\lambda_1 + \mu_1, \lambda_2 + \mu_2, .., \lambda_j + \mu_2, \dots). \end{aligned} \tag{59}$$

The *natural order* of partitions is defined as follows

$$\lambda \geq \mu \Leftrightarrow \forall i : \lambda_1 + \lambda_2 + \cdots + \lambda_i \geq \mu_1 + \mu_2 + \cdots + \mu_i. \tag{60}$$

Then one says $\lambda$ *dominates* $\mu$. For example $(3,3,1) \geq (3,2,2)$. The natural order is not a total order and incomparable partitions exists. For example $(3,1,1,1) \not\leq (2,2,2,0)$ because for the first parts $3 > 2$ and $(3,1,1,1) \not\geq (2,2,2,0)$ since the sums of first three parts give $5 < 6$. Total order consistent with the natural order is the *reverse lexicographic order* denoted with the symbol "$\overset{R}{\geq}$".[14] One writes $\lambda \overset{R}{\geq} \mu$ when either $\lambda = \mu$ or the first non vanishing difference $\lambda_i - \mu_i$ is positive. Considering the previous example $(3,1,1,1) \overset{R}{\geq} (2,2,2,0)$.

### 2.1.1. *Diagrams*

Partition may be represented in graphical form as a *Young diagram* (Young tableau). A Young diagram of a partition $\lambda$ is a set of the $|\lambda|$ boxes. Boxes are arranged in rows and columns in a way that indicating consecutive parts of the partition. In the so called English notation, there are $\lambda_i$ left aligned boxes in $i$-th row starting from the top. In this notation number of rows equals $\ell(\lambda)$ (see Fig. 4).

Formally one can define a Young diagram as a subset of a plane composed out of $1 \times 1$ boxes. Center of each box is placed in an integer point $\langle i,j \rangle \in \mathbb{Z}^2$ and first box (right-top) in a point $\langle 0,0 \rangle$. The empty set is also Young diagram, it represents the partition of weight 0.

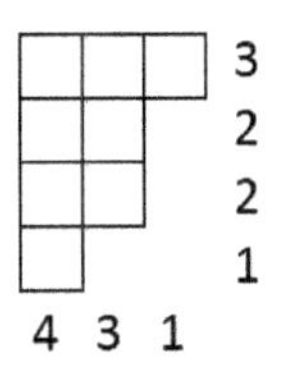


Fig. 4. Young diagram in English notation of a partition $(3,2,2,1)$ and numbers of the boxes in each row and column.

For a given partition $\lambda$ one defines a conjugated partition $\lambda'$ by giving its parts. $i$-th part of the $\lambda'$ equals to the quantity of boxes in $i$-th column of the Young diagram of $\lambda$. As such the Young diagram of the $\lambda'$ can be obtained from the diagram of $\lambda$ by taking mirror reflection with respect to the diagonal line passing through points $\{\langle i,i \rangle\}_{i\in\mathbb{Z}}$. For example $(3,2,2,1)' = (4,3,1)$ (see Fig. 4).

Conjugated partition can be defined without referring to Young diagrams. Parts of conjugated partition are given by a formula

$$\lambda_i' = |\{j : \lambda_j \geq i\}|. \tag{61}$$

Simple properties of the conjugation may be derived: $(\lambda')' = \lambda$, $\lambda_1' = \ell(\lambda)$, $\lambda_1 = \ell(\lambda')$ and $m(\lambda, i) = \lambda_i' - \lambda_{i+1}'$. Moreover conjugation reverses the natural order

$$\lambda \geq \mu \Leftrightarrow \lambda' \leq \mu'. \tag{62}$$

Denote

$$\delta_N = (N-1, N-2, \ldots, 1, 0). \tag{63}$$

When the length of $\delta$ is clear we skip the index.

Conciser a vector of integers $a = (a_1, a_2, \ldots, a_N) \in \mathbb{Z}^N$. There is at least one permutation of vector components that rearranges those components into non increasing order. Such rearranged vector is unique and it is denoted $a^*$.

For any natural numbers $1 \leq i < j \leq N$ and $\ell$ the *raising operator* $R_{i,j}^\ell : \mathbb{Z}^N \to \mathbb{Z}^N$ acts on a vector $a \in \mathbb{Z}^N$ as follows

$$R_{i,j}^\ell(a_1, a_2, \ldots, a_i, \ldots a_j, \ldots a_N) = (a_1, a_2, \ldots, a_i - \ell, \ldots a_j + \ell, \ldots a_N). \tag{64}$$

## 2.2. *The ring of symmetric polynomials*

The polynomial $P(x_1, x_2, \ldots, x_N)$ in $N$ variables is symmetric if it is invariant under all of the permutations of variables i.e. for any permutation $\sigma \in S_N$

$$P(x_1, x_2, \ldots, x_N) = P(x_{\sigma(1)}, x_{\sigma(2)}, \ldots, x_{\sigma(N)}). \tag{65}$$

Similarly the polynomial $A(x_1, x_2, \ldots, x_N)$ in $N$ variables is antisymmetric if for any permutation $\sigma \in S_N$

$$A(x_1, x_2, \ldots, x_N) = \operatorname{sgn}(\sigma) A(x_{\sigma(1)}, x_{\sigma(2)}, \ldots, x_{\sigma(N)}). \tag{66}$$

Let $\Lambda_N^k$ denote the additive group of symmetric, homogeneous polynomials in $N$ variables and degree $k$, with zero polynomial (by the convention, 0 is homogeneous polynomial of every degree). Then $\Lambda_N$ equipped with standard addition and multiplication of polynomials is a graded ring of symmetric polynomials in $N$ variables[11]

$$\Lambda_N = \bigoplus_{k \geq 0} \Lambda_N^k. \tag{67}$$

Presented construction can be performed for polynomials over various fields e.g. $\mathbb{Z}, \mathbb{Q}, \mathbb{R}$. Usually the choice of field is not significant.

Analogical construction can be performed for for infinitely many variables with an usage of an inverse limit. Ring of the symmetric functions in infinitely many variables is denoted $\Lambda$. Formally elements of $\Lambda$ (unlike those of $\Lambda_N$) are no longer polynomials but series of polynomials and the term "function" is being used instead of polynomial.

## 2.3. *Monomial symmetric functions* $m_\lambda$

One defines the *monomial symmetric functions* - $m_\lambda$ as

$$m_\lambda(x_1, x_2, \ldots, x_N) = \sum_{\text{distinct permutations } \alpha \text{ of } \lambda} x_1^{\alpha_1} \cdot x_2^{\alpha_2} \cdot \ldots \cdot x_N^{\alpha_N}, \tag{68}$$

or equivalently

$$m_\lambda(x_1, x_2, \ldots, x_N) = F(\lambda) \cdot \sum_{\sigma \in S_n} x_1^{\lambda_{\sigma(1)}} \cdot x_2^{\lambda_{\sigma(2)}} \cdot \ldots \cdot x_N^{\lambda_{\sigma(N)}}. \tag{69}$$

For $F(\lambda)$ – the normalizing factor, given by

$$F(\lambda) = \frac{1}{m(\lambda, 0)! \cdot m(\lambda, 1)! \cdot \ldots}. \tag{70}$$

For example

$$m_{(3,1,1)}(x, y, z) = x^3yz + xy^3z + xyz^3 = (x^2 + y^2 + z^2)xyz,$$

$$m_{(2,1)}(x_1, \ldots, x_N) = \sum_{i \neq j} {x_i}^2 x_j.$$

The monomial symmetric functions are the $\mathbb{Z}$-basis of $\Lambda_N$.[11]

## 2.4. *Schur polynomials*

### 2.4.1. *Vandermonde determinant*

A square Vandermonde matrix in variables $x_1, x_2, \ldots, x_N$, is a matrix with the terms of a geometric progression in each row

$$V = \begin{bmatrix} 1 & x_1 & x_1^2 & \ldots & x_N^{N-1} \\ 1 & x_2 & x_2^2 & \ldots & x_N^{N-1} \\ 1 & x_3 & x_3^2 & \ldots & x_N^{N-1} \\ \vdots & \vdots & \vdots & \ddots & \vdots \\ 1 & x_N & x_N^2 & \ldots & x_N^{N-1} \end{bmatrix}. \tag{71}$$

Or shorter $V_{ij} = x_i^{j-1}$. The determinant of this matrix is denoted $D(x_1,\ldots,x_N)$ or simply $D$ when variables are known. Vandermonde determinant is an object of great importance in linear algebra and the theory of symmetric functions. It can be represented as

$$D(x_1,\ldots,x_N) = \sum_{\sigma\in S_n} \mathrm{sgn}(\sigma)\cdot x_{\sigma(1)}^{N-1}\cdot x_{\sigma(2)}^{N-2}\cdot\cdots\cdot x_{\sigma(N-1)}^{1}\cdot x_{\sigma(N)}^{0} = \prod_{1\le i<j\le N}(x_i-x_j). \tag{72}$$

Every antisymmetric polynomial is $N$ variables is divisible by the Vandermonde determinant. $D(x_1,\ldots,x_N)$ has a degree $\binom{N}{2}$.

### 2.4.2. *Slater determinants*

Slater determinants $\mathrm{sl}_\nu$ plays similar role in the ring of antisymmetric functions as monomials in the ring of symmetric functions. Slater determinants $\mathrm{sl}_\nu$ (sometimes called antisymmetric polynomials $a_\nu$) are indexed by a partition $\nu$ and given by a following formula

$$\mathrm{sl}_\nu(x_1,x_2,\ldots,x_N) = \sum_{\sigma\in S_N} \mathrm{sgn}(\sigma)\cdot x_{\sigma(1)}^{\nu_1}\cdot x_{\sigma(2)}^{\nu_2}\cdot\cdots\cdot x_{\sigma(N)}^{\nu_N}. \tag{73}$$

Assume the partition $\nu$ has two parts equal, then the transposition interchanging those two parts would not affect it. Since transpositions are odd permutations, polynomial $\mathrm{sl}_\nu$ would equal $-\mathrm{sl}_\nu$ therefore $\mathrm{sl}_\nu = 0$. Thus in order for $\mathrm{sl}_\nu$ to be nonzero, no part of $\nu$ can occur more than once. Let $\delta$ denote the partition from formula 63: $\delta = (N-1,\ldots,0)$. Then a partition $\nu$ with no repeated parts can be represented as $\nu = \delta + \lambda$ for some partition $\lambda$.

Slater determinants are $\mathbb{Z}$-basis of $A_N$, where $A_N$ is a space of antisymmetric polynomials of $N$ variables.[11] One can notice $\mathrm{sl}_\delta$ is a Vandermonde determinant. For example

$$\mathrm{sl}(x,y,z)_\delta = \mathrm{sl}(x,y,z)_{(2,1,0)} = D(x,y,z) = x^2y-x^2z+y^2z-y^2x+z^2x-z^2y.$$

### 2.4.3. *Definition of Schur polynomials*

The Schur polynomials $s_\lambda$ are important family of polynomials in the ring of the symmetric polynomials, they are defined as a quotient of antisymmetric polynomials

$$s_\lambda = \frac{\mathrm{sl}_{\lambda+\delta}}{\mathrm{sl}_\delta}. \tag{74}$$

The multiplication by $D = \mathrm{sl}_\delta$ is an isomorphism of $\Lambda_N$ onto $A_N$. Since $\{s_\lambda\}_\lambda$ are transformed in basis $\{\mathrm{sl}_{\lambda+\delta}\}_\lambda$ one concludes Schur polynomials form $\mathbb{Z}$-basis of $\Lambda_N$.[11] Schur polynomial $s_\lambda$ is homogeneous symmetric polynomial of a degree $|\lambda|$. For example

$$s_{(2,1,1)}(x,y,z) = \frac{a_{(4,2,1)}}{a_{(2,1,0)}} = xyz(x+y+z).$$

Representation of Schur polynomials in monomial basis is triangular i.e. satisfy

$$s_\lambda = m_\lambda + \sum_{\mu<\lambda} K_{\lambda\mu} m_\mu, \tag{75}$$

for the coefficients $K_{\lambda\mu} \geq 0$ – Kostka numbers.[13]

## 3. Jack polynomials

For a real number $\alpha \in \mathbb{R}$, consider a field $\mathbb{Q}(\alpha)$ of rational functions in $\alpha$

$$\mathbb{Q}(\alpha) = \left\{ \frac{p(\alpha)}{q(\alpha)} \in \mathbb{R} :\ q(\alpha) \neq 0;\ p(\alpha), q(\alpha) \text{ - polynomials over rationals} \right\}. \tag{76}$$

$\mathbb{Q}(\alpha)$ is the smallest (in a sense of inclusion) field containing rational numbers and $\alpha$. For $\alpha \in \mathbb{Q}$ field reduces to the rational numbers $\mathbb{Q}(\alpha) = \mathbb{Q}$. Let $\Lambda_N \otimes \mathbb{Q}(\alpha)$ denote the ring of all symmetric polynomials in $N$ variables over the field $\mathbb{Q}(\alpha)$.[14]

The Hamiltonian in the Calogero-Sutherland-Moser model[15] (also called Hamiltonian of the Calogero-Sutherland model[16] or Laplace-Beltrami operator[14]) is defined as follows $H^0(\alpha) : \Lambda_N \otimes \mathbb{Q}(\alpha) \to \Lambda_N \otimes \mathbb{Q}(\alpha)$:

$$H^{CSM}(\alpha) = \alpha \sum_{i=1}^{N} (x_i \partial_i)(x_i \partial_i) + \sum_{1 \leq i < j \leq N} \left( \frac{x_i + x_j}{x_i - x_j} \right) (x_i \partial_i - x_j \partial_j). \tag{77}$$

For a real number $\alpha$ and a partition $\lambda$ a Jack polynomial $J^\alpha_\lambda$ is a polynomial over $\mathbb{Q}(\alpha)$.[17] It is defined (up to normalization) as an eigenvector of $H^{CSM}(\alpha)$

$$H^{CSM} J^\alpha_\lambda = d_\lambda(\alpha) \cdot J^\alpha_\lambda, \tag{78}$$

where eigenvalue $d_\lambda(\alpha)$ equals

$$d_\lambda(\alpha) = \sum_{i=1}^{N} \left( \alpha \lambda_i^2 + (N+1-2i)\lambda_i \right). \tag{79}$$

When parameter $\alpha$ is known one skips it and writes simply $H^{CSM}$ and $d_\lambda$.

Analogically to the Schur polynomials The Jack polynomials (Jacks) are triangular in monomial basis i.e.

$$J_\lambda^\alpha = m_\lambda + \sum_{\mu<\lambda} v_{\lambda\mu}(\alpha) m_\mu. \tag{80}$$

For $\alpha > 0$ coefficients $v_{\lambda\mu}(\alpha)$ are positive. Moreover coefficients can be expressed as an inverse of polynomials in $\alpha$ with no nonnegative root.

**Remark 3.1.** Form of expansion in equation 80 fixes problem of normalization of Jack polynomials.

**Remark 3.2.** Jack polynomials form a complete set of the eigenvectors of $H_0$.

Few examples of Jack polynomials

$$J_{(1,1)}^\alpha = m_{(1,1)},$$

$$J_{(2,1)}^\alpha = m_{(2,1)} + \frac{6}{2+\alpha} m_{(1,1,1)},$$

$$J_{(3)}^\alpha = m_{(3)} + \frac{3}{2+\alpha} m_{(2,1)} + \frac{6}{(1+\alpha)(2+\alpha)} m_{(1,1,1)},$$

$$J_\lambda^1 = s_\lambda, \quad J_\lambda^\infty = m_\lambda.$$

Jack polynomials are generalization of many types of the symmetric polynomials including: Schur polynomials $\alpha = 1$, zonal polynomials $\alpha = 2$, quaternion zonal polynomial $\alpha = \frac{1}{2}$, monomial symmetric functions $\alpha = \infty$ and elementary symmetric functions $\alpha = 0$. Jack polynomials are special, limit case of Macdonald polynomials $P_\lambda^{q,t}$ for $q = t^\alpha$, $t \to 1$.[11–13]

In order to construct Jack polynomials in monomial basis we use two following theorems.

**Theorem 3.1.** ***Action of $H^{CSM}$ on the monomial symmetric functions (K. Sogo)***
*Action of $H^{CSM}$ on a monomial function is given by*

$$H^{CSM}(m_\lambda) = d_\lambda m_\lambda + \sum_{\mu<\lambda} C_{\lambda\mu}\, m_\mu. \tag{81}$$

*For eigenvalues $d_\lambda$ defined in (79), coefficients $C_{\lambda\mu} \in \mathbb{N}$ are nonzero only when there exist some $i, j, \ell$, such $i < j$, $1 \le \ell \le \left\lfloor \frac{\lambda_i - \lambda_j}{2} \right\rfloor$ and $\left(R_{i,j}^\ell(\lambda)\right)^* =$*

$\mu$, *then*

$$C_{\lambda\mu} = \begin{cases} (\lambda_i - \lambda_j)m(\mu, \lambda_i - \ell)(m(\mu, \lambda_i - \ell) - 1) & \text{if } \lambda_i - \ell = \lambda_j + \ell \\ (\lambda_i - \lambda_j)2m(\mu, \lambda_i - \ell)m(\mu, \lambda_j + \ell) & \text{if } \lambda_i - \ell \neq \lambda_j + \ell \end{cases}. \tag{82}$$

Proof of the theorem 3.1 can be found in Ref. 15.

**Remark 3.3.** Theorem 3.1 reveals that action of an operator $H^{CSM}$ on the monomials is triangular i.e. a result of $H^{CSM}m_\lambda$ is a sum of the monomials indexed by the partitions dominated by $\lambda$. This property is analogical to an action of the triangular matrix. Then coefficients $d_\lambda$ are diagonal elements of the operator $H^{CSM}$.

**Theorem 3.2.** ***Determinantal expression for Jack Polynomials (K. Sogo, L. Lapointe, A. Lascoux, J. Morse)***
*Let* $\mu^{(1)} \overset{R}{<} \mu^{(2)} \overset{R}{<} \ldots \overset{R}{<} \mu^{(n)} = \lambda$ *be a reverse lexicographic ordering of all partitions dominated by* $\lambda$. *Then an eigenvector of* $H^{CSM}$ *corresponding to* $\lambda$ *- Jack polynomial* $J_\lambda^\alpha$ *is proportional to the determinant of quasi-triangular matrix (one line below diagonal is nonzero).*

$$J_\lambda^\alpha \propto \det \begin{bmatrix} m_{\mu^{(1)}} & m_{\mu^{(2)}} & \cdots\cdots & m_{\mu^{(n-1)}} & m_{\mu^{(n)}} \\ d_{\mu^{(1)}} - d_{\mu^{(n)}} & C_{\mu^{(2)}\mu^{(1)}} & \cdots\cdots & C_{\mu^{(n-1)}\mu^{(1)}} & C_{\mu^{(n)}\mu^{(1)}} \\ 0 & d_{\mu^{(2)}} - d_{\mu^{(n)}} & \cdots\cdots & C_{\mu^{(n-1)}\mu^{(2)}} & C_{\mu^{(n)}\mu^{(2)}} \\ 0 & 0 & \ddots & \vdots & \vdots \\ \vdots & \vdots & \ddots\;\ddots & \vdots & \vdots \\ 0 & 0 & \ldots\; 0 & d_{\mu^{(n-1)}} - d_{\mu^{(n)}} & C_{\mu^{(n)}\mu^{(n-1)}} \end{bmatrix}. \tag{83}$$

*The eigenvalues* $d_\lambda$ *are defined in equation (79). The matrix is filled with monomials in the first row and numbers elsewhere. The elements below diagonal are differences of eigenvalues of* $H^{CSM}$ *defined in formula 79. Rest of the elements are coefficients* $C_{\mu^{(i)}\mu^{(j)}}$ *defined in the formula 82.*

Proof of the theorem 3.2 can be found in Ref. 16.

**Remark 3.4.** When combined with known properties of quasi diagonal matrix[16] theorem 3.2 (Determinantal expression for Jack Polynomials) gives recursion formula for expansion of Jack polynomials in monomial basis.

Rewrite equation 80:

$$J^\alpha_\lambda = m_\lambda + \sum_{\mu<\lambda} v_{\lambda\mu} m_\mu,$$

for $\mu^{(1)} \overset{R}{<} \mu^{(2)} \overset{R}{<} \ldots \overset{R}{<} \mu^{(n)} = \lambda$ a reverse lexicographic ordering of all partitions dominated by $\lambda$, one has to put $v_{\lambda\lambda}(\alpha) = v_{\mu^{(n)}\mu^{(n)}}(\alpha) = 1$ and obtain other coefficients from the recursion formula

$$v_{\lambda\mu^{(j)}} = \frac{1}{d_\lambda - d_{\mu^{(j)}}} \sum_{j=i+1}^{n} C_{\mu^{(j)}\mu^{(i)}} v_{\lambda\mu^{(i)}}. \tag{84}$$

### 3.1. *Fermionic Jack polynomials*

The Jack fermionic polynomials $S^\alpha_\mu$ [18,19] are antisymmetric analogue of Jack symmetric polynomials. They are defined as a product of Jack and Vandermonde determinant

$$S^\alpha_{\lambda+\delta}(x_1,\ldots,x_N) = J^\alpha_\lambda(x_1,\ldots,x_N) D(x_1,\ldots,x_N), \tag{85}$$

Slater determinants $\mathrm{sl}_{\lambda+\delta}$ are special case of fermionic Jack polynomials ($\alpha = 1$). Therefore $S^\alpha_{\lambda+\delta}$ may be treated as a continuous deformation of antisymmetric polynomials. Obviously fermionic Jack polynomial $S^\alpha_{\lambda+\delta}$ is well defined as long as Jack polynomial $J^\alpha_\lambda$ is well defined. In particularity fermionic Jack polynomial has only finitely many negative poles. One can expand $S^\alpha_{\lambda+\delta}$ in Slater determinant basis

$$S^\alpha_{\lambda+\delta} = \mathrm{sl}_{\lambda+\delta} + \sum_{\mu<\lambda} b_{\lambda\mu}(\alpha)\mathrm{sl}_{\mu+\delta}, \tag{86}$$

for $b_{\lambda\mu}(\alpha) \in \mathbb{Q}(\alpha)$. When $\alpha$ is clear we write simply $b_{\lambda\mu}$.

Fermionic Jack polynomials are eigenvectors of fermionic Laplace-Beltrami operator [18,19] $H^F_{LB}(\alpha)$

$$H^F_{LB}(\alpha) = \sum_{i=1}^{N}(x_i\partial_i)(x_i\partial_i) + \left(\frac{1}{\alpha} - 1\right) \sum_{1\le i<j\le N} \frac{x_i + x_j}{x_i - x_j}(x_i\partial_i - x_i\partial_j) - 2\frac{{x_i}^2 + {x_j}^2}{(x_i - x_j)^2}. \tag{87}$$

$H^F_{LB}(\alpha) S_{\lambda+\delta} = E_\lambda(\alpha) S_{\lambda+\delta}$, where the eigenvalues are given by

$$E_\lambda(\alpha) = \sum_{i=1}^{N} \lambda_i \left(\lambda_i - 2\left(\frac{1}{\alpha} - 1\right) i\right) + \left(\frac{1}{\alpha} - 1\right)\left((N+1)\sum_{i=1}^{N} \lambda_i - N(N-1)\right). \tag{88}$$

When parameter $\alpha$ is clear we skip it and write simply $H^F_{LB}, E_\lambda$.

Recursion formula for fermionic Jacks in terms of antisymmetric polynomials basis had been derived by Bernevig and Regnault.[18,19] Bernevig and Regnault used the same method of constructing determinetal and later recursion form of eigenvectors of certain operators as presented in subsections above. Recursion formula holds. For $\mu^{(1)} \stackrel{R}{<} \mu^{(2)} \stackrel{R}{<} \ldots \stackrel{R}{<} \mu^{(n)} = \lambda$ a reverse lexicographic ordering of all partitions dominated by $\lambda$, one has to put $b_{\lambda\lambda} = b_{\mu^{(n)}\mu^{(n)}} = 1$ and obtain other coefficients from the recursion formula

$$b_{\lambda\mu^{(j)}} = \frac{2\left(\frac{1}{\alpha} - 1\right)}{E_\lambda - E_{\mu^{(j)}}} \sum_{j=i+1}^{n} C^F_{\mu^{(j)}\mu^{(i)}} b_{\lambda\mu^{(i)}}. \tag{89}$$

For coefficients $C^F_{\lambda\mu} \in \mathbb{Z}$ being nonzero only when there exist some $i, j, \ell$, such $i < j$, $1 \leq \ell \leq \left\lfloor \frac{\lambda_i - \lambda_j}{2} \right\rfloor$, $\left(R^\ell_{i,j}(\lambda)\right)^* = \mu$ and $\lambda_i - \ell \neq \lambda_j + \ell$, then

$$C^F_{\lambda\mu} = (\lambda_i - \lambda_j - 2\ell)(-1)^{N_{SW}}. \tag{90}$$

Where $N_{SW}$ is the number of the rising operators of a form $R^1_{i,i+1}$ or $R^1_{i,i-1}$ (swaps) needed, to obtain partition $\mu$ from $\lambda$.

### 3.2. *Jack polynomials indexed by negative α parameter*

Originally Jack polynomials were defied only for positive values of $\alpha$[17] (definition involved a scalar product $\langle \cdot, \cdot \rangle_\alpha$ which had sense only for $\alpha > 0$.[14]) Nonetheless further development allowed for alternative definitions of Jack polynomials, including one given in this section. As one notices recursion construction of Jack polynomials show coefficients of Jacks in monomial basis as inverse polynomials in $\alpha$ that could be easily extended to negative $\alpha$. Let us rewrite already given example of Jack polynomial

$$J^\alpha_{(3)} = v_{(3),(3)}(\alpha) m_{(3)} + v_{(3),(2,1)}(\alpha) m_{(2,1)} + v_{(3),(1,1,1)}(\alpha) m_{(1,1,1)}$$

$$= m_{(3)} + \frac{3}{2+\alpha} m_{(2,1)} + \frac{6}{(1+\alpha)(2+\alpha)} m_{(1,1,1)}.$$

Fig. 5 shows values of $v_{(3),(3)}(\alpha), v_{(3),(2,1)}(\alpha), v_{(3),(1,1,1)}(\alpha)$ coefficients as functions of $\alpha$. Fig. 6 shows absolute values of the same coefficients normalized in a way that sum of theirs squares sum up to one.

One can notice (see Fig. 5) Jack polynomial $J^\alpha_{(3)}$ has two poles at $\alpha = -2, -1$. For $\alpha = -1$ coefficient $v_{(3),(1,1,1)}$ diverges to the infinity while

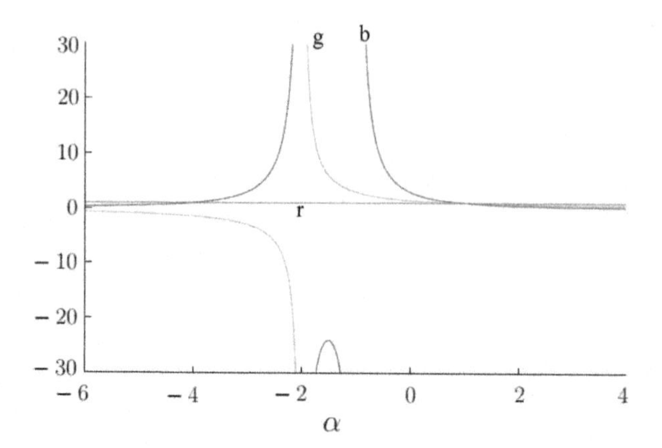


Fig. 5. Values of $v_{(3),(3)}(\alpha)$ (red), $v_{(3),(2,1)}(\alpha)$ (green), $v_{(3),(1,1,1)}(\alpha)$ (blue) coefficient as functions of $\alpha$.

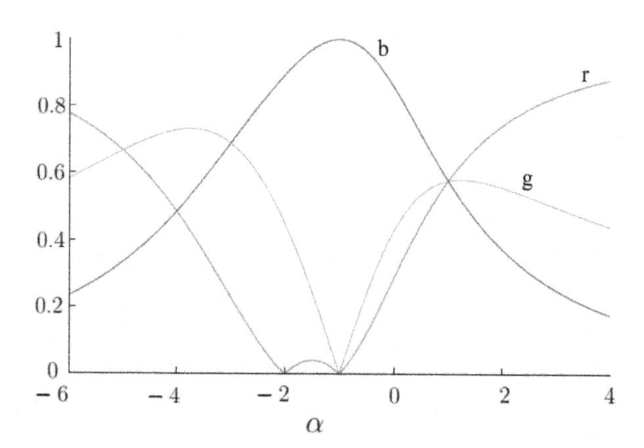


Fig. 6. Absolute values of $v_{(3),(3)}(\alpha)$ (red), $v_{(3),(2,1)}(\alpha)$ (green), $v_{(3),(1,1,1)}(\alpha)$ (blue) coefficient normalized to one in a square, as functions of $\alpha$.

other remain finite. For $\alpha = -2$ coefficients $v_{(3),(1,1,1)}$ and $v_{(3),(2,1)}$ diverges while $v_{(3),(3)} = 1$. Moreover

$$\lim_{\alpha \to -2} \frac{v_{(3),(1,1,1)}(\alpha)}{v_{(3),(2,1)}(\alpha)} = -2. \tag{91}$$

Considerations of those limit justify definition of $J^{\alpha}_{(3)}$ at points $\alpha = -2, -1$ up to normalization as

$$J^{-2}_{(3)} \propto m_{(2,1)} - 2m_{(1,1,1)}, \quad J^{-1}_{(3)} \propto m_{(1,1,1)}.$$

Such approach gives definition of Jack polynomials even when $\alpha$ is its pole

$$J_\lambda^\alpha = \sum_{\mu \leq \lambda} w_{\lambda\mu} m_\mu, \tag{92}$$

where

$$w_{\lambda\mu} = \lim_{x \to \alpha} \frac{v_{\lambda\mu}(x)}{v_{\lambda\nu}(x)} \tag{93}$$

and $v_{\lambda\mu}$ is the coefficients that diverges in the highest power. Notice that for such definition first nonzero coefficient $w_{\lambda\mu}$ always equals one.

### 3.3. *Jack states of quantum Hall effect*

Now we a give brief overview of the Jack states – FQH states related to Jack polynomials. It was pointed out[20–22] that analysis of the angular momentum operators on the sphere can be useful in determination of Jack states. By stereographic projection, operators can be applied to functions on the plane. Then operators take form

$$L^+ = E_0, \tag{94a}$$

$$L^- = N_\Phi Z - E_2, \tag{94b}$$

$$L^Z = \frac{1}{2} N N_\Phi - E_1, \tag{94c}$$

where

$$E_n = \sum_{i=1}^{N} z_i^n \partial_i, \tag{95a}$$

$$Z = \sum_{i=1}^{N} z_i, \tag{95b}$$

and $N_\Phi$ is a strength of a magnetic monopole inside Haldane sphere. Papers 20–25 gave necessary conditions for both partition and real parameter of the Jack polynomial to be a candidate for FQH wave function. Bernevig and Haldane required a Jack wave function $\psi$, to be annihilated by raising and lowering operators on the sphere $L^+\psi = 0, L^-\psi = 0$ (highest weight – HW and lowest weight – LW conditions, respectively).[26] This analysis reveals the condition (necessary but not sufficient)

$$N - \ell(\lambda) + 1 + \alpha(\lambda_{\ell(\lambda)} - 1) = 0. \tag{96}$$

The equation 96 and further considerations, implies the real parameter equals $\alpha \equiv \alpha_{k,r} = -(k+1)/(r-1)$ for $(k+1)$ and $(r-1)$ both positive integers and coprime.[20–22] Partition indexing Jack polynomial is of a form

$$\lambda = [n_0, 0^{s(r-1)}, k, 0^{r-1}, k, 0^{r-1}, k, \ldots, k], \tag{97}$$

where $0^{r-1}$ means a sequence of $r-1$ zeros and $n_0 = (k+1)s - 1$. Such partition is denoted $\lambda^0_{k,r,s}$. The case $s = 1$ was recognized to provide many FQH ground states at filling factor $\nu = k/r$. Cases $s > 1$ are related to the quasiparticle states. Denote partition $\lambda^0_{k,r,s=1} = \lambda^0_{k,r}$.[20–22]

As we already pointed out Jack polynomials indexed by $\lambda^0_{k,r}$ and $\alpha_{k,r}$ are related to bosonic FQH states of filling factors $\nu = k/r$. Now we discuss few explicit Jack states. The bosonic Laughlin wave function for state $\nu = 1/r$ ($r$- even) can be represented as a product of Gaussian and symmetric Jack polynomial

$$\Phi_L^{1/r} = \prod_{i<j}^{N} (z_i - z_j)^r = J^{\alpha_{1,r}}_{\lambda^0(1,r)}. \tag{98}$$

As trivially follows fermionic Laughlin wave functions for a state $1/r$ also are Jack states for partition $\lambda^0(1,r)$ and real parameter $\alpha_{1,(r-1)}$.

The Moore-Read state (for fermions $\nu = 5/2, 1/2$ in the second LL and for bosons $\nu = 1$ in second LL) is also a Jack state

$$\Psi^0_{MR} = J^{\alpha_{2,2}}_{\lambda^0_{(2,2)}}, \quad \Psi^1_{MR} = S^{\alpha_{2,2}}_{\lambda^0_{(2,2)}+\delta}. \tag{99}$$

The Moore-Read state along the Laughlin $\nu = 1/3$ is a part of Read-Rezai parafermion sequence of states $\nu = k/(2+kM)$. Whole series can be given as Jack states (bosonic case $M = 0$ – $J^{\alpha_{k,2}}_{\lambda^0(k,2)}$, fermionic case $M = 1$ – $S^{\alpha_{k,2}}_{\lambda^0(k,2)+\delta}$). Such states are known to be densest ground states of $(k+1)$-body repulsion Hamiltonian. Among noted Jack states one can also find Gaffnian state. THe bosonic Gaffnian wave function may be expressed as $J^{\alpha_{2,3}}_{\lambda^0(2,3)}$ and in the fermionic case as $S^{\alpha_{2,3}}_{\lambda^0(2,3)+\delta}$. The Gaffnian is also characterized as a ground state of non physical four-body interaction Hamiltonian.

## 4. Results and conclusions: numerical generation of Jack polynomials

In the section 3 we discussed recursion formula for coefficients of Jack polynomials in monomial basis and fermionic Jack coefficients in Slater determinant basis. Both formulas can be adapted to the form of computer program and used for generation of large Jack wave functions.

Partitions analyzed in the program may be stored as a vector of integers (we recommend storing partitions in occupational representation). There is a problem of generation of a list of all partitions of fixed weight, dominated by a selected partition $\lambda$, listed according to a reverse lexicographic order. It may be solved as follows: one generates a list of all partitions smaller than $\lambda$ in the reverse lexicographic order and then remove partitions not dominated by the $\lambda$. When the list is obtained, one focuses on rather straightforward calculations of the eigenvalues and coefficients $C_{\lambda\mu}$ (bosonic case), $C^F_{\lambda\mu}$ (fermionic case). Next step is to execute recursion formula given in (84) and (89). Finally if one wishes to get coefficients of Jack wave function on a Haldane sphere one uses inverse of stereographic projection and map coefficients onto the sphere.

As an example of properly executed algorithm, we give results of numerical calculations of Jack polynomials in a form of Tables 1 and 2.

Table 1. Normalized coefficients of the Jack polynomial $J^{-2}_{(4,2,0)}(x, y, z)$ in the monomial basis. The coefficient are also expansion of three particle bosonic Laughlin wave function for $\nu = \frac{1}{2}$ state.

| $m_\lambda$ | coefficient |
|---|---|
| $m_{(4,2,0)}$ | 0.14285 |
| $m_{(4,1,1)}$ | −0.28571 |
| $m_{(3,3,0)}$ | −0.28571 |
| $m_{(3,2,1)}$ | 0.28571 |
| $m_{(2,2,2)}$ | −0.85714 |

Table 2. Normalized coefficients of the Jack polynomial $J^{-\frac{2}{5}}_{(6,0)}(x, y)$ in the monomial basis. The coefficient are also expansion of three particle bosonic Laughlin wave function for $\nu = \frac{1}{6}$ state.

| $m_\lambda$ | coefficient |
|---|---|
| $m_{(6,0)}$ | 0.03886 |
| $m_{(5,1)}$ | −0.23319 |
| $m_{(4,2)}$ | 0.58299 |
| $m_{(3,3)}$ | −0.77732 |

Both Tables 1 and 2 show coefficients of Jack wave function in small Hilbert spaces. However capabilities of computer program are greater and one can generate Jacks in spaces of bigger dimensionality. In the Table 3 we examine overlaps of fermionic Jacks representing Laughlin $\nu = 1/3$ wave functions mapped onto the Haldane sphere and ground states of Coulomb interaction Hamiltonian, for different sizes of a system.

Table 3. Overlaps of the indicated of Jack states wave function $\nu = 1/3$ with LLL Coulomb ground states. Consecutive columns are electron number $N$, magnetic flux on the sphere $2Q$, dimension of the relevant $N$-body subspace, and the overlaps with Coulomb states in the LLL in GaAs.

| $N$ | $2Q$ | dim | overlap |
|---|---|---|---|
| 11 | 30 | $1 \cdot 10^6$ | 0.9922 |
| 12 | 33 | $8 \cdot 10^6$ | 0.9909 |
| 13 | 36 | $4 \cdot 10^7$ | 0.9898 |
| 14 | 39 | $3 \cdot 10^8$ | 0.9887 |

In conclusion we introduced basics of FQHE and standard tools used in its analysis like Haldane sphere or composite fermions. Then we followed with brief introduction to the theory of symmetric functions. We motivated adaptation of Jack polynomials in the context of FQHE and presented method of numerical obtaining of coefficients of Jacks and fermionic Jacks. As an illustration of discussed recursion formulas we gave both values of coefficients of Jack wave functions and overlaps of fermionic Jacks and Coulomb ground states. The paper should be treated as merely an introduction to the topic of Jack polynomials in FQHE. Not all advantages of Jack-based approach have been discussed. As we stated, coefficients of states related to Jack polynomials can be obtained with recursion formula rather than tedious Hamiltonian diagonalization. Moreover, the combinatorial structure of Jack polynomials is of use in analytical analysis. Definition of Jacks as eigenvectors of $H^{CSM}$ can be a great tool, other properties like triangularity of expansion in monomial basis has not been explored yet. Moreover topic of more general symmetric functions in FQHE like Macdonald polynomials is not sufficiently examined.

## References

1. J. J. Quinn, K.-S. Yi, *Solid State, Physics Principles and Modern Applications* (Springer 2009).

2. T. Chakraborty, *The Quantum Hall Effects, Integral and Fractional* (Springer 1995).
3. J. K. Jain, *Composite fermions* (Cambridge University Press 2007).
4. A. Bohm, A. Mostafazadeh, H. Koizumi, Q. Niu, J. Zwanziger, *The geometric phase in quantum systems: foundations, mathematical concepts, and applications in molecular and condensed matter physics* (Springer 2003).
5. B. Hall, *Quantum Theory for Mathematicians* (Springer 2013).
6. J. K. Jain, *Composite Fermions And The Fractional Quantum Hall Effect: A Tutorial*, Lecture notes (2011).
7. R. Laughlin, *Phys. Rev. Lett.* **50**, 1395 (1983).
8. M. V. Milovanovic, Th. Jolicoeur, and I. Vidanovic, *Phys. Rev. B* **80**, 155324 (2009).
9. F. D. M. Haldane, *Phys. Rev. Lett.* **51**, 605 (1983).
10. G. Fano, F. Ortolani, and E. Colombo, *Phys. Rev. B.* **34**, 2670 (1986).
11. I. G. Macdonald, *Symmetric Functions and Hall Polynomials*, 2nd ed. (Oxford University Press, 1997).
12. I. G. Macdonald, Publ. I.R.M.A. Strasbourg, Actes 20e Seminaire Lotharingien, 131-171 (1988).
13. S. Kerov, Asymptotic Representation Theory of the Symmetric Group and its Applications in Analysis, Translation of Mathematical Monographs, v.219. AMS (2003).
14. R. P. Stanley, *Adv. Math.* **77**, 76-115 (1988).
15. F K. Sogo, *J. Math. Phys.* **35**, 22822296 (1994).
16. L. Lapointe, A. Lascoux, and J. Morse, *Elec. Jour. Combin.* **7**, N1 (2000).
17. H. Jack, *Proc. R. Soc. Edin. Sect. A* **69**, 1 (1970/1971)
18. A. Bernevig and N. Regnault, *Phys. Rev. Lett.* **103**, 206801 (2009).
19. R. Thomale, B. Estienne, N. Regnault, and A. Bernevig, *Phys. Rev. B* **84**, 045127 (2011).
20. B. A. Bernevig and F. D. M. Haldane, *Phys. Rev. B* **77**, 184502 (2008).
21. B. A. Bernevig and F. D. M. Haldane, *Phys. Rev. Lett.* **100**, 246802 (2008).
22. B. A. Bernevig and F. D. M. Haldane, *Phys. Rev. Lett.* **102**, 066802 (2009).
23. B. Kuśmierz, Y.-H. Wu, A. Wójs, *Acta Phys. Pol. A* **126**, 1134 (2014).
24. B. Kuśmierz, Y.-H. Wu, A. Wójs, *Acta Phys. Pol. A* **129**, A-73 (2016).
25. B. Kuśmierz, A. Wójs, *Acta Phys. Pol. A* **130**, 607 (2016).
26. W. Baratta, P. J. Forrester, *Nucl. Phys. B* **843**, 362-381 (2011).

# Zitterbewegung (Trembling Motion) of Electrons in Narrow Gap Semiconductors

Wlodek Zawadzki

*Institute of Physics, Polish Academy of Sciences*
*Al. Lotników 32/46, 02-688 Warsaw, Poland*
*E-mail: zawad@ifpan.edu.pl*

Tomasz M. Rusin

*Orange Poland sp. z o. o.*
*Al. Jerozolimskie 160, 02-326 Warsaw, Poland*
*E-mail: tmr@vp.pl*

We describe recent research on Zitterbewegung (ZB, trembling motion) of electrons in semiconductors. A brief history of the subject is presented. Zitterbewegung of charge carriers in monolayer and bilayer graphene is elaborated in some detail. Effects of an external magnetic field on ZB are described using monolayer graphene as an example. Nature of electron ZB in crystalline solids is explained. It is proposed to trigger the coherent trembling motion of many electrons by exciting them from the valence to the conduction band by laser pulses and detecting the resulting ZB by nonlinear optical effects. Simulations of the trembling motion in vacuum and in semiconductors are reviewed. A recent experimental observation of the trembling motion in semiconducting material InGaAs is mentioned.

*Keywords*: Zitterbewegung, semiconductors, two-band model.

## 1. Introduction and history

This article article is concerned with a somewhat mysterious phenomenon known under the German name of "Zitterbewegung" (trembling motion). Both the phenomenon and its name were devised by Schrodinger who, in 1930, observed that in the Dirac equation, describing relativistic electrons in vacuum, the $4 \times 4$ operators corresponding to components of relativistic velocity do not commute with the free-electron Hamiltonian.[1] In consequence, the electron velocity is not a constant of the motion also in absence of external fields. Such an effect must be of a quantum nature as it does not obey Newton's first law of classical motion. Schrodinger calculated the resulting time dependence of the electron velocity and position concluding that, in addition to classical motion, they experience very fast periodic oscillations which he called Zitterbewegung (ZB). Schrodinger's idea stimulated numerous theoretical investigations but no experiments since the predicted frequency $\hbar\omega_Z \simeq 2m_0c^2 \simeq 1$ MeV and the amplitude of about $\lambda_c = \hbar/m_0c \simeq 3.86 \times 10^{-3}$Å are not accessible to current experimental techniques. Huang[2] calculated averages of velocity and position operators. It was understood that the ZB is due to an interference of states corresponding to the positive and negative electron energies resulting from the Dirac equation.[3–5] Lock[7] showed that, if an electron is represented by a wave packet, its ZB has a transient character, i.e. it disappears with time.

It was recognized years later that the trembling electron motion should occur also in crystalline solids if their band structure could be represented by a two-band model reminiscent of the Dirac equation. In 1970 Lurie and Cremer,[8] considered superconductors, in which the energy-wave vector dependence is similar to the relativistic relation. Similar approach was applied to semiconductors twenty years later using a model of two energy bands.[9–12] However, an intense interest in ZB of electrons in semiconductors was launched only in 2005. Zawadzki[13] used a close analogy between the $\mathbf{k} \cdot \mathbf{p}$ theory of energy bands in narrow gap semiconductors (NGS) and the Dirac equation for relativistic electrons in vacuum to show that one should deal with the electron ZB in NGS which would have much more favorable frequency and amplitude characteristics than those in a vacuum. On the other hand, Schliemann *et al.*[14] demonstrated that the spin splitting of energies linear in $k$, caused by the inversion asymmetry in semiconductor systems (the Bychkov-Rashba splitting), also leads to a ZB-type of motion if the electron wave packet has a non-vanishing initial momentum. The above contributions triggered a wave of theoretical considerations for various semiconductor and other systems. It was recognized that the phenomenon of ZB occurs every time one deals with two or more interacting energy bands.[15–17]

It was confirmed that, indeed, when the electron is represented by a wave packet, the ZB has a transient character.[18] Considering graphene in a magnetic field it was demonstrated that, if the electron spectrum is discrete, ZB contains many frequencies and is sustained in time.[19] It was pointed out that the trembling electrons should emit electromagnetic radiation if they are not in their eigenstates.[20] The physical origin of ZB was analyzed and it turned out that, at least in its "classical" solid state version analogous to the ZB in a vacuum, the trembling motion represents oscillations of electron velocity due to the energy conservation as the particle moves in a periodic potential.[22]

As mentioned above, in vacuum the ZB characteristics are not favorable. In solids, the ZB characteristics are much better but it is difficult to observe the motion of a single electron. However, Gerritsma *et al.*[21] simulated experimentally the Dirac equation and the resulting electron Zitterbewegung with the use of trapped ions and laser excitations. The power of the simulation method is that one can adjust experimentally the essential parameters of the Dirac equation: $m_0c^2$ and $c$, and thus achieve more favorable values of the ZB frequency and amplitude. The experimental results obtained by Gerritsma *et al.* agreed well with the predictions of Zawadzki and Rusin.[22] Interestingly, it turned out that analogues of ZB can occur also in classical wave propagation phenomena. Several predictions were made, but in two systems, namely macroscopic sonic crystals,[23] and photonic superlattices,[24] the ZB-like effects were actually observed. Finally, there has been growing recognition that the mechanisms responsible for ZB in solids are related to their other properties, for example to the electric conductivity.

Thus the subject of ZB is quite universal. From an obscure, perplexing and somewhat marginal effect that would probably never be observed in vacuum the Zitterbewegung has grown into a universal, almost ubiquitous phenomenon that was experimentally simulated in its quantum form and directly observed in its classical version. Our article summarizes this intensive development which can be characterized as the "Sturm und Drang" period, to use another pertinent German term. We concentrate mostly on the ZB in semiconductors but mention other systems, in particular different ZB simulations by trapped ions and atoms. We also briefly review important papers describing the ZB of free relativistic electrons in a vacuum since they inspired early considerations concerning solids. The subject of Zitterbewegung has been until now almost exclusively theoretical, but we propose realistic ways of its observation. At the end we mention very recent experimental confirmation of electron ZB in semiconducting InGaAs.

Our article is organized in the following way. In Sec. II we present descriptions of ZB for semi-relativistic, spin, and nearly-free electron Hamiltonians and quote papers on other model systems. Section III treats the trembling motion in bilayer graphene, monolayer graphene and carbon nanotubes. In Sec. IV we consider the ZB in the presence of an external magnetic field and quote related works. Section V is concerned with the origin of ZB in crystalline solids. There follows short section VI in which we mention work relating ZB to calculations of electric conductivity. In Sec. VII we describe papers on the ZB of free relativistic electrons, necessary to understand the trembling motion in solids. Section VIII contains a very brief introduction to simulations of the Dirac equation and the resulting ZB in absence of fields and in a magnetic field. The section is completed by summaries of related papers. In Sec. IX we describe wave ZB-like effects in non-quantum periodic systems. We terminate by discussion and conclusions.

## 2. ZB in Model Systems

We begin our considerations by using the so called relativistic analogy.[13] This way we can follow simultaneously the procedure of Schrodinger and derive corresponding relations for narrow gap semiconductors. It was noted in the past that the $E(k)$ relation between the energy $E$ and the wave vector $k$ for electrons in NGS is analogous to that for relativistic electrons in vacuum.[12,25,26] The semi-relativistic phenomena appear at electron velocities of $v \simeq 10^7 - 10^8$ cm/s, much lower than the light velocity $c$. The reason is that the maximum velocity $u$ in semiconductors, which plays the role of $c$ in a vacuum, is about $10^8$ cm/s. To be more specific, we use the $\mathbf{k} \cdot \mathbf{p}$ approach to InSb-type semiconductors.[27] Taking the limit of of large spin-orbit energy, the resulting dispersion relation for the conduction and the light-hole bands is $E = \pm E_p$, where

$$E_p = \left[ \left( \frac{E_g}{2} \right)^2 + E_g \frac{p^2}{2m_0^*} \right]^{1/2} . \tag{1}$$

Here $E_g$ is the energy gap and $m_0^*$ is the effective mass at the band edge. This expression is identical to the relativistic relation for electrons in a vacuum, with the correspondence $2m_0c^2 \to E_g$ and $m_0 \to m_0^*$. The electron velocity $\boldsymbol{v}$ in the conduction band described by (1) reaches a saturation value as $p$ increases. This can be seen directly by calculating $v_i = \partial E_p / \partial p_i$ and taking the limit of large $p_i$, or by using the analogy $c = (2m_0c^2/2m_0)^{1/2} \to (E_g/2m_0^*)^{1/2} = u$. Taking the experimental parameters $E_g$ and $m_0^*$ one calculates a very similar value of $u \simeq 1.3 \times 10^8$ cm/s for different semiconductor compounds. Now we define an important quantity for electrons in NGS

$$\lambda_z = \frac{\hbar}{m_0^* u} . \tag{2}$$

We note that it corresponds to the Compton wavelength $\lambda_c = \hbar / m_0 c$ for relativistic electrons in vacuum.

Next we consider the band Hamiltonian for NGS. It is derived within the model including $\Gamma_6$ (conduction), $\Gamma_8$ (light and heavy hole), and $\Gamma_7$ (split-off) bands and it represents an $8 \times 8$ operator matrix.[27] We assume, as before, $\Delta \gg E_g$ and omit the free electron terms since they are negligible for NGS. The resulting $6 \times 6$ Hamiltonian has $\pm E_g/2$ terms on the diagonal and linear $\hat{p}_i$ terms off the diagonal, just like in the Dirac equation for free electrons. However, the three $6 \times 6$ matrices $\hat{\alpha}_i$ multiplying the momentum components $\hat{p}_i$ do not have the properties of $4 \times 4$ Dirac matrices, which considerably complicates calculations. For this reason, with only a slight loss of generality, we take $\hat{p}_z \neq 0$ and $\hat{p}_x = \hat{p}_y = 0$. In the $\hat{\alpha}_3$ matrix, two rows and columns corresponding to the heavy holes contain only zeros and they can be omitted. The remaining Hamiltonian for the conduction and the light hole bands reads

$$\hat{H} = u\hat{\alpha}_3\hat{p}_z + \frac{1}{2}E_g\hat{\beta}, \tag{3}$$

where $\hat{\alpha}_3$ and $\hat{\beta}$ are the well-known $4 \times 4$ Dirac matrices.[28] The Hamiltonian (3) has the form appearing in the Dirac equation and in the following we can use the procedures of relativistic quantum mechanics (RQM). The electron velocity is $\dot{\hat{z}} = (1/i\hbar)[\hat{z}, \hat{H}] = u\hat{\alpha}_3$. The eigenvalues of $\hat{\alpha}_3$ are $\pm 1$, so that the eigenvalues of $\dot{\hat{z}}$ are $\pm u$. In order to determine $\hat{\alpha}_3(t)$ we calculate $\dot{\hat{\alpha}}_3(t)$ by commuting $\hat{\alpha}_3(t)$ with $\hat{H}$ and integrating the result with respect to time. This gives $\dot{\hat{z}}(t)$ and we calculate $\hat{z}(t)$ integrating again. The final result is

$$\hat{z}(t) = \hat{z}(0) + \frac{u^2\hat{p}_z}{\hat{H}}t + \frac{i\hbar u}{2\hat{H}}\hat{A}_0\left[\exp\left(\frac{-2i\hat{H}t}{\hbar}\right) - 1\right], \tag{4}$$

where $\hat{A}_0 = \hat{\alpha}_3(0) - u\hat{p}_z/\hat{H}$. There is $1/\hat{H} = \hat{H}/E_p^2$. The first two terms of (4) represent the classical electron motion, while the third term describes time dependent oscillations with the frequency of $\omega_Z \simeq E_g/\hbar$. Since $\hat{A}_0 \simeq 1$, the amplitude of oscillations is $\hbar u/2\hat{H} \simeq \hbar/2m_0^*u = \lambda_z/2$. In RQM the analogous oscillations are called Zitterbewegung. The expression obtained by Schrodinger for ZB of free relativistic electrons in vacuum is identical to that given by (4) with the use of the above relativistic analogy. In RQM it is demonstrated that ZB is a result of interference between states of positive and negative electron energies.[3–5] Clearly, one can say the same of ZB in semiconductors calculated according to the above model. However, as we show below, the origin of ZB in crystalline solids can be interpreted in simple, more physical terms. The magnitude of $\lambda_z$ is essential. There is $\lambda_z = \lambda_c(c/u)(m_0/m_0^*) \simeq 0.89(m_0/m_0^*)$Å since, as mentioned above, $u = \simeq 1.3 \times 10^8$ cm/s for various materials. We estimate: for GaAs ($m_0^* \simeq 0.067m_0$) $\lambda_z \simeq 13$Å, for InAs ($m_0^* \simeq 0.024m_0$) $\lambda_z \simeq 37$Å, for InSb ($m_0^* \simeq 0.014m_0$) $\lambda_z \simeq 64$Å. Thus, in contrast to vacuum, the length of ZB in semiconductors can be quite large. However, one should bear in mind that the above derivations, as well as the original procedure of Schrodinger's, use only operator considerations, whereas physical observables are given by quantum averages. We show below that, if one calculates such averages using electron wave packets, the amplitude of ZB may be considerably smaller than $\lambda_z$.

To demonstrate universality of ZB in the two-band situation we consider the well-known case of nearly free electrons in which the periodic lattice potential $V(\boldsymbol{r})$ is treated as a perturbation (see Ref. 17). Near the Brillouin zone boundary the Hamiltonian has, to a good approximation, a $2 \times 2$ form (spin is omitted)

$$\hat{H} = \begin{pmatrix} E_{\boldsymbol{k}+\boldsymbol{q}} & V_{\boldsymbol{q}} \\ V_{\boldsymbol{q}}^* & E_{\boldsymbol{k}} \end{pmatrix}, \tag{5}$$

where $V_{\boldsymbol{q}}^* = V_{-\boldsymbol{q}}$ are the Fourier coefficients in the expansion of $V(\boldsymbol{r})$, and $E_{\boldsymbol{k}} = \hbar^2 k_z^2/2m_0$ is the free electron energy. The $2 \times 2$ quantum velocity $\hat{v}_z$ can now be calculated and the acceleration $\dot{\hat{v}}_z$ is computed in the standard way. Finally, one calculates the displacement matrix $\hat{z}_{ij}$.

Since the ZB is by its nature not a stationary state but a dynamical phenomenon, it is natural to study it with the use of wave packets. These have become a practical instrument when femtosecond pulse technology emerged. Thus, in a more realistic picture the electrons are described by wave packets

$$\psi(z) = \frac{1}{\sqrt{2\pi}} \frac{d^{1/2}}{\pi^{1/4}} \int_{\infty}^{\infty} \exp\left(-\frac{1}{2} d^2 (k_z - k_{z0})^2\right) \times \\ \exp(ik_z z) dk_z \begin{pmatrix} 1 \\ 0 \end{pmatrix}. \tag{6}$$

The electron displacement is calculated as an average of the position operator $\hat{z}$ over the above wave packet, see Figure 1 and Ref. 29. The essential result is that, in agreement with Lock general predictions,[7] the ZB oscillations of the average electron position have a *transient* character, i.e. they disappear with time on a femtosecond scale. The frequency of oscillations is $\omega_Z = E_g/\hbar$, where $E_g = 2|V_{\boldsymbol{q}}|$.

Cserti and David[15] observed that the Hamiltonians describing the Bychkov-Rashba and the Dresselhaus spin splitting, monolayer and bilayer

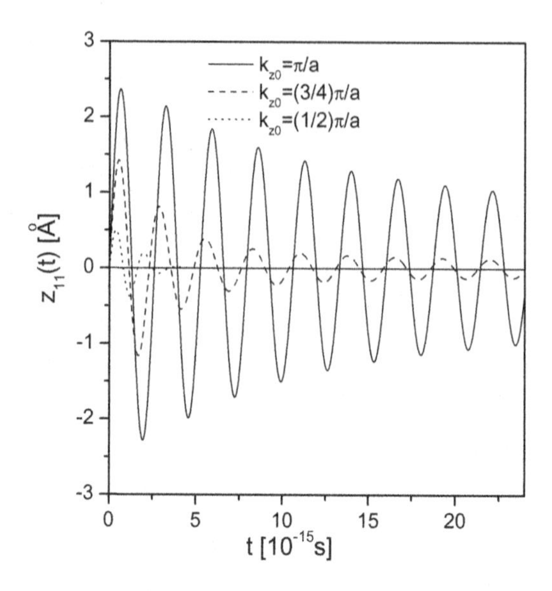


Fig. 1. Transient Zitterbewegung oscillations of nearly-free electrons *versus* time, calculated for a very narrow wave packet centered at various $k_{z0}$ values. The band parameters correspond to GaAs. After Ref. 29.

grapheme, nearly-free electrons, electrons in superconductors, etc., can be represented in a unified general form. One can use this form to calculate ZB in several cases. For the Bychkov-Rashba coupling and for monolayer graphene ZB can be interpreted as a consequence of conservation of the total angular momentum. This is in analogy to the results discussed below where ZB is shown to be a consequence of the energy conservation. Wilamowski *et al.*[31] investigated microwave absorption in asymmetric Si quantum wells in an external magnetic field and detected a spin-dependent component of the Joule heating at the spin resonance. The observation was explained in terms of the Bychkov-Rashba spin-splitting due to the structure inversion asymmetry with the resulting current-induced spin precession and the ZB at the Larmor frequency.

## 3. ZB in graphene

Now we study the Zitterbewegung of mobile charge carriers in modern materials: bilayer graphene and monolayer graphene.[18]

### 3.1. *Bilayer graphene*

We first present the results for bilayer graphene since they can be obtained in the analytical form, which allows one to see directly important features of the trembling motion. Two-dimensional Hamiltonian for bilayer graphene is well approximated by[32]

$$\hat{H}_B = -\frac{1}{2m^*}\begin{pmatrix} 0 & (\hat{p}_x - i\hat{p}_y)^2 \\ (\hat{p}_x + i\hat{p}_y)^2 & 0 \end{pmatrix}, \tag{7}$$

where $m^* = 0.054m_0$. The energy spectrum is $\mathcal{E}_k = \pm E_k$, where $E_k = \hbar^2k^2/2m^*$, i.e. there is no energy gap between the conduction and valence bands. The position operator in the Heisenberg picture is a $2 \times 2$ matrix $\hat{x}(t) = \exp(i\hat{H}_Bt/\hbar)\hat{x}\exp(-i\hat{H}_Bt/\hbar)$. One calculates

$$x_{11}(t) = x(0) + \frac{k_y}{k^2}\left[1 - \cos\left(\frac{\hbar k^2 t}{m^*}\right)\right], \tag{8}$$

where $k^2 = k_x^2 + k_y^2$. The third term represents the Zitterbewegung with the frequency $\hbar\omega_Z = 2\hbar^2k^2/2m^*$, corresponding to the energy difference between the upper and lower energy branches for a given value of $k$. We want to calculate ZB of a charge carrier represented by a two-dimensional wave packet centered at $\boldsymbol{k}_0 = (0, k_{0y})$ and characterized by the width $d$. An average of $\hat{x}_{11}(t)$ is a two-dimensional integral which can be calculated analytically

$$\bar{x}_{11}(t) = \langle\psi(\boldsymbol{r})|\hat{x}(t)|\psi(\boldsymbol{r})\rangle = \bar{x}_c + \bar{x}_Z(t) \tag{9}$$

where $\bar{x}_c = (1/k_{0y})\left[1 - \exp(-d^2k_{0y}^2)\right]$, and

$$\bar{x}_Z(t) = \frac{1}{k_{0y}}\left[\exp\left(-\frac{\delta^4d^2k_{0y}^2}{d^4+\delta^4}\right)\cos\left(\frac{\delta^2d^4k_{0y}^2}{d^4+\delta^4}\right) - \exp(-d^2k_{0y}^2)\right], \tag{10}$$

in which $\delta = \sqrt{\hbar t/m^*}$ contains the time dependence. In Figure 2a we show the ZB of the electron position $\bar{x}_{11}$ as given in (9) and (10).

We enumerate the main features of ZB following from (9) and (10). First, in order to have ZB in the direction $x$ one needs an initial transverse momentum $\hbar k_{0y}$. Second, the ZB frequency depends only weakly on the packet width: $\omega_Z = (\hbar k_{0y}^2/m^*)(d^4/(d^4+\delta^4))$, while its amplitude is strongly dependent on the width $d$. Third, the ZB has a transient character since it is attenuated by the exponential term. For small $t$ the amplitude of $\bar{x}_Z(t)$ diminishes as $\exp(-\Gamma_Z^2t^2)$ with

$$\Gamma_Z = \frac{\hbar k_{0y}}{m^*d}. \tag{11}$$

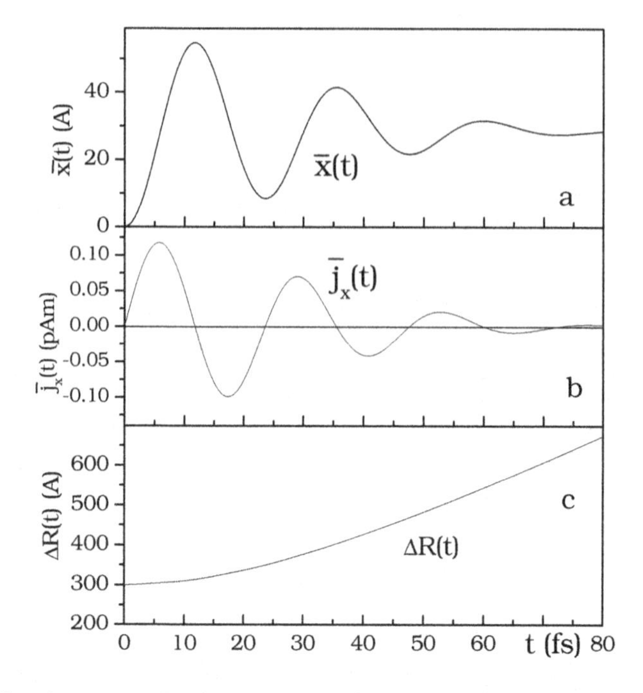


Fig. 2. Zitterbewegung of a charge carrier in bilayer graphene *versus* time, calculated for a gaussian wave packet width $d = 300\text{Å}$ and $k_{0y} = 3.5 \times 10^8\text{m}^{-1}$: a) position, b) electric current, c) dispersion $\Delta R(t)$. After the ZB disappears a constant shift remains. After Ref. 18.

Fourth, as $t$ (or $\delta$) increases, the cosine term tends to unity and the first term in (10) cancels out with the second term, which illustrates the Riemann-Lebesgue theorem (see Ref. 7). After the oscillations disappear, the charge carrier is displaced by the amount $\bar{x}_c$, which is a "remnant" of ZB. Fifth, for very wide packets ($d \to \infty$) the exponent in (10) tends to unity, the oscillatory term is $\cos(\delta^2 k_{0y}^2)$ and the last term vanishes. In this limit one recovers undamped ZB oscillations.

Next, we consider other quantities related to ZB, beginning by the current. The latter is given by the velocity multiplied by charge. The velocity is simply $\bar{v}_x = \partial \bar{x}_Z / \partial t$, where $\bar{x}_Z$ is given by (10). The calculated current is plotted in Figure 2b, its oscillations are a direct manifestation of ZB. The transient character of ZB is accompanied by a temporal spreading of the wave packet. In fact, the question arises whether the attenuation of ZB is not simply *caused* by the spreading of the packet. The calculated packet width $\Delta R$ is plotted versus time in Figure 2c. It is seen that during the initial 80 femtoseconds the packet's width increases only twice compared to its initial value, while the ZB disappears almost completely. We conclude that the spreading of the packet is *not* the main cause of the transient character of ZB.

Looking for physical reasons behind the transient character of ZB it was shown that one can decompose the wave packet representing the electron into two subpackets moving in the opposite directions. As long as the subpackets overlap, their spatial interference results in the ZB. When the subpackets cease to overlap and to interfere, the ZB disappears. This process determines the decay time of ZB.

### 3.2. *Monolayer graphene*

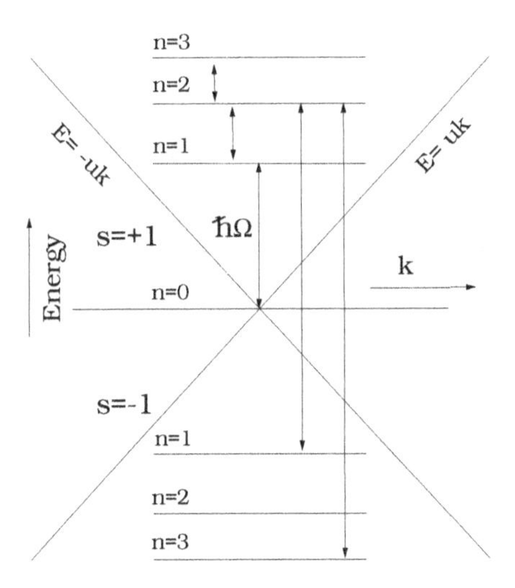


Fig. 3. The energy dispersion $E(k)$ and the Landau levels for monolayer graphene in a magnetic field (schematically). Intraband (cyclotron) and interband (ZB) energies for $n' = n \pm 1$ are indicated. The basic energy is $\hbar\Omega = \sqrt{2}\hbar u/L$. After Ref. 19.

Now we turn to monolayer graphene. The two-dimensional band Hamiltonian describing its band structure is[33–36]

$$\hat{H}_M = u \begin{pmatrix} 0 & \hat{p}_x - i\hat{p}_y \\ \hat{p}_x + i\hat{p}_y & 0 \end{pmatrix}, \tag{12}$$

where $u \approx 1 \times 10^8$cm/s. The resulting energy dispersion is linear in momentum: $\mathcal{E} = \pm u\hbar k$, where $k = \sqrt{k_x^2 + k_y^2}$. The quantum

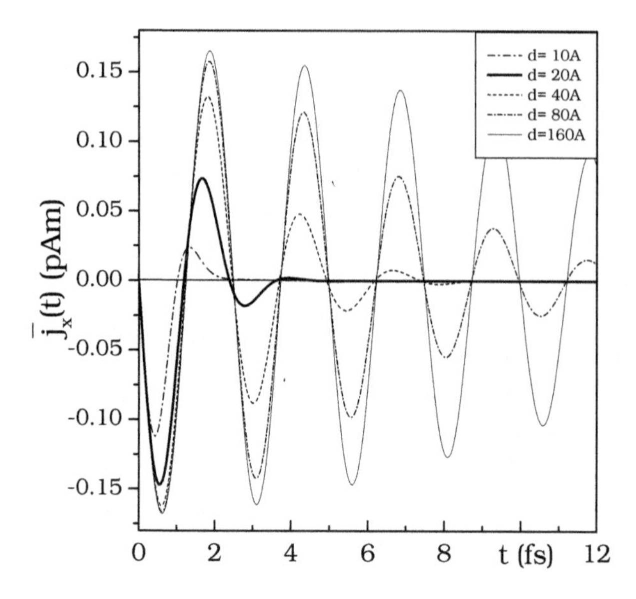


Fig. 4. Oscillatory electric current in the $x$ direction caused by the ZB in monolayer graphene *versus* time, calculated for a gaussian wave packet with $k_{0y} = 1.2 \times 10^9 \mathrm{m}^{-1}$ and various packet widths $d$. Transient character of ZB is clearly seen. After Ref. 18.

velocity in the Schrodinger picture is $\hat{v}_i = \partial H_M / \partial \hat{p}_i$, it does not commute with the Hamiltonian (12). In the Heisenberg picture we have $\hat{\boldsymbol{v}}(t) = \exp(i\hat{H}_M t/\hbar)\hat{v}\exp(-i\hat{H}_M t/\hbar)$. Using (12) one calculates

$$v_x^{(11)} = u\frac{k_y}{k}\sin(2ukt). \tag{13}$$

The above equation describes the trembling motion with the frequency $\omega_Z = 2uk$, determined by the energy difference between the upper and lower energy branches for a given value of $k$. As before, ZB in the direction $x$ occurs only if there is a non-vanishing momentum $\hbar k_y$. One calculates an average velocity (or current) taken over a two-dimensional wave packet with nonzero initial momentum $k_{0x}$. The results for the current $\bar{j}_x = e\bar{v}_x$ are plotted in Figure 4 for different realistic packet widths $d$. It is seen that the ZB frequency does not depend on $d$ and is nearly equal to $\omega_Z$ given above for the plane wave. On the other hand, the amplitude of ZB does depend on $d$ and we deal with decay times of the order of femtoseconds. For small $d$ there are almost no oscillations, for very large $d$ the ZB oscillations are undamped. These conclusions agree with our analytical results for bilayer graphene. The behavior of ZB depends quite critically on the values of $k_{0y}$ and $d$, which is reminiscent of the damped harmonic oscillator. In the limit $d \to \infty$ the above results for the electric current resemble those of Katsnelson[37] for ZB in graphene obtained with the use of plane wave representation.

## 4. ZB in a magnetic field

The trembling motion of charge carriers in solids has been described above for no external potentials. Now we consider the trembling motion of electrons in the presence of an external magnetic field.[19] The magnetic field is known to cause no interband electron transitions, so the essential features of ZB are expected not to be destroyed. On the other hand, introduction of an external field provides an important parameter affecting the ZB behavior. This case is special because the electron spectrum is fully quantized. We consider a graphene monolayer in an external magnetic field parallel to the $z$ axis. The Hamiltonian for electrons and holes at the $K_1$ point is[33,34]

$$\hat{H} = u \begin{pmatrix} 0 & \hat{\pi}_x - i\hat{\pi}_y \\ \hat{\pi}_x + i\hat{\pi}_y & 0 \end{pmatrix}, \tag{14}$$

where $u \approx 1 \times 10^8$ cm/s is the characteristic velocity, $\hat{\boldsymbol{\pi}} = \hat{\boldsymbol{p}} - q\hat{\boldsymbol{A}}$ is the generalized momentum, in which $\hat{\boldsymbol{A}}$ is the vector potential and $q$ is the electron charge. Using the Landau gauge, we take $\hat{\boldsymbol{A}} = (-By, 0, 0)$, and for an electron $q = -e$ with $e > 0$. We take the wave function in the form $\Psi(x, y) = e^{ik_x x}\Phi(y)$. Introducing the magnetic radius $L = \sqrt{\hbar/eB}$, the variable $\xi = y/L - k_x L$, and defining the standard raising and lowering operators for the harmonic oscillator $\hat{a} = (\xi + \partial/\partial\xi)/\sqrt{2}$ and $\hat{a}^\dagger = (\xi - \partial/\partial\xi)/\sqrt{2}$, the Hamiltonian becomes

$$\hat{H} = -\hbar\Omega \begin{pmatrix} 0 & \hat{a} \\ \hat{a}^\dagger & 0 \end{pmatrix}, \tag{15}$$

where the frequency is $\Omega = \sqrt{2}u/L$. Next one determines the eigenstates and eigenenergies of the Hamiltonian $\hat{H}$. The energy is $E_{ns} = s\hbar\Omega\sqrt{n}$. Here $n = 0, 1, \ldots$, and $s = \pm 1$ for the conduction and valence bands, respectively. The above energies were confirmed experimentally. The complete wave function is

$$|\mathrm{n}\rangle \equiv |nk_x s\rangle = \frac{e^{ik_x x}}{\sqrt{4\pi}} \begin{pmatrix} -s|n-1\rangle \\ |n\rangle \end{pmatrix} \tag{16}$$

where $|n\rangle$ are the harmonic oscillator functions.

We want to calculate the velocity of charge carriers described by a wave packet. We first calculate matrix elements $\langle f|\mathrm{n}\rangle$ between an arbitrary two-component function $f = (f^u, f^l)$ and eigenstates (16). A straightforward manipulation gives $\langle f|\mathrm{n}\rangle = -sF^u_{n-1} + F^l_n$, where

$$F^j_n(k_x) = \frac{1}{\sqrt{2LC_n}} \int g^j(k_x, y) e^{-\frac{1}{2}\xi^2} \mathrm{H}_n(\xi) dy, \tag{17}$$

in which

$$g^j(k_x, y) = \frac{1}{\sqrt{2\pi}} \int f^j(x, y) e^{ik_x x} dx. \tag{18}$$

The superscript $j = u, l$ stands for the upper and lower components of the function $f$. The Hamilton equations give the velocity components: $\hat{v}_i(0) = \partial\hat{H}/\partial\hat{p}_i$, with $i = x, y$. We want to calculate averages of the time-dependent velocity operators $\hat{v}_i(t)$ in the Heisenberg picture taken on the function $f$. The averages are

$$\bar{v}_i(t) = \sum_{\mathrm{n,n'}} e^{iE_{\mathrm{n'}}t/\hbar} \langle f|\mathrm{n'}\rangle \langle \mathrm{n'}|v_i(0)|\mathrm{n}\rangle \langle \mathrm{n}|\mathrm{f}\rangle \mathrm{e}^{-\mathrm{i}\mathrm{E_n}\mathrm{t}/\hbar}, \tag{19}$$

where the energies and eigenstates are given in (16). The summation in (19) goes over all the quantum numbers: $n, n', s, s', k_x, k'_x$. The only non-vanishing matrix elements of the velocity components are for the states states $n' = n \pm 1$. One finally obtains after some manipulation

$$\begin{aligned}\bar{v}_y(t) = &\, u\sum_{n=0}^{\infty} V_n^+ \sin(\omega_n^c t) + u\sum_{n=0}^{\infty} V_n^- \sin(\omega_n^Z t) + iu\sum_{n=0}^{\infty} A_n^+ \cos(\omega_n^c t) \\ &+ iu\sum_{n=0}^{\infty} A_n^- \cos(\omega_n^Z t),\end{aligned} \tag{20}$$

$$\begin{aligned}\bar{v}_x(t) = &\, u\sum_{n=0}^{\infty} B_n^+ \cos(\omega_n^c t) + u\sum_{n=0}^{\infty} B_n^- \cos(\omega_n^Z t) + iu\sum_{n=0}^{\infty} T_n^+ \sin(\omega_n^c t) \\ &+ iu\sum_{n=0}^{\infty} T_n^- \sin(\omega_n^Z t),\end{aligned} \tag{21}$$

where $V_n^\pm$, $T_n^\pm$, $A_n^\pm$ and $B_n^\pm$ are given by combinations of $U^{\alpha,\beta}_{m,n}$ integrals

$$U^{\alpha,\beta}_{m,n} = \int F^{\alpha *}_m(k_x) F^\beta_n(k_x) dk_x. \tag{22}$$

The superscripts $\alpha$ and $\beta$ refer to the upper and lower components, see Ref. 19. The time dependent sine and cosine functions come from the exponential terms in (19). The frequencies in (20) and (21) are $\omega_n^c = \Omega(\sqrt{n+1} - \sqrt{n})$, $\omega_n^Z = \Omega(\sqrt{n+1} + \sqrt{n})$, where $\Omega$ is given in (15). The frequencies $\omega_n^c$ correspond to the intraband energies while frequencies $\omega_n^Z$ correspond to the interband energies, see Figure 3. The interband frequencies are characteristic of the Zitterbewegung. The intraband (cyclotron) energies are due to the band quantization by the magnetic field and th do not appear in field-free situations.

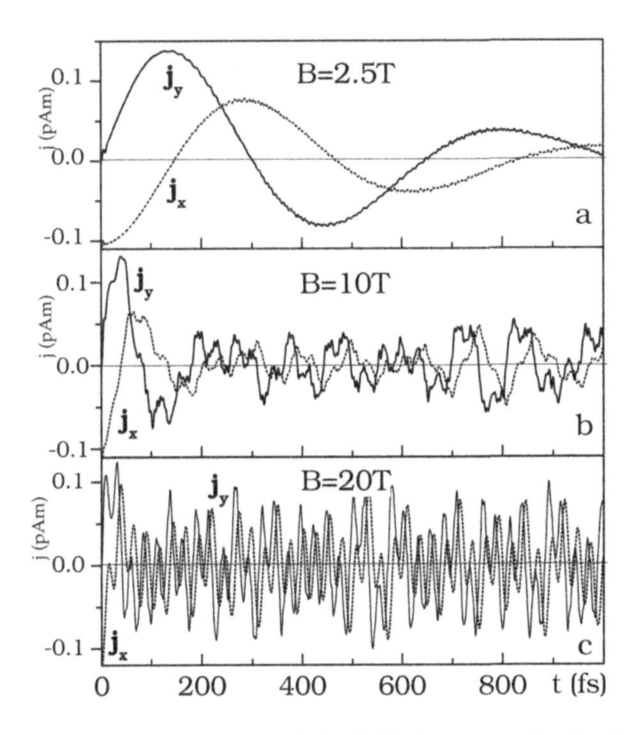


Fig. 5. Contribution of the $K_1$ point of the Brillouin zone to the electric current in graphene versus time, calculated for a Gaussian wave packet of the width $d_x = d_y =$ 81.13Å and $k_{0x} = 0.035$Å$^{-1}$ and at different magnetic fields.

Final calculations were carried out for a two-dimensional Gaussian wave packet centered around the wave vector $\mathbf{k_0} = (k_{0x}, 0)$ and having two non-vanishing components. In this case one can obtain analytical expressions for $U_{m,n}^{\alpha,\beta}$. The main frequency of oscillations is $\omega_0 = \Omega$, which can be interpreted either as $\omega_0^c = \Omega(\sqrt{n+1} - \sqrt{n})$ or $\omega_0^Z = \Omega(\sqrt{n+1} + \sqrt{n})$ for $n = 0$. Frequency $\omega_0^c$ belongs to the intraband (cyclotron) set, while $\omega_0^Z$ belongs to the interband set (see Figure 3). The striking feature is, that ZB is manifested by several frequencies simultaneously. This is a consequence of the fact that in graphene the energy distances between the Landau levels diminish with $n$, which results in different values of frequencies $\omega_n^c$ and $\omega_n^Z$ for different $n$. It follows that it is the presence of an external quantizing magnetic field that introduces various frequencies into ZB. It turns out that, after the ZB oscillations seemingly die out, they actually reappear at higher times. Thus, *for all* $k_{0x}$ *values (including* $k_{0x} = 0$*), the ZB oscillations have a permanent character*, that is they do not disappear in time. This feature is due to the discrete character of the electron spectrum caused by a magnetic field. The above property is in sharp contrast to the no-field cases considered above, in which the spectrum is not quantized and the ZB of a wave packet has a transient character. In mathematical terms, due to the discrete character of the spectrum, averages of operator quantities taken over a wave packet are sums and not integrals. The sums do not obey the Riemann-Lebesgues theorem for integrals which guaranteed the damping of ZB in time for a continuous spectrum (see Ref. 7).

All in all, the presence of a quantizing magnetic field has the following important effects on the trembling motion. (1) For $B \neq 0$ the ZB oscillation are permanent, while for $B = 0$ they are transient. The reason is that for $B \neq 0$ the electron spectrum is discrete. (2) For $B \neq 0$ many ZB frequencies appear, whereas for $B = 0$ only one ZB frequency exists. (3) For $B \neq 0$ both interband and intraband (cyclotron) frequencies appear in ZB; for $B = 0$ there are no intraband frequencies. (4) Magnetic field intensity changes not only the ZB frequencies but the entire character of ZB spectrum.

The Zitterbewegung should be accompanied by electromagnetic dipole radiation emitted by the trembling electrons. The oscillations $\bar{\boldsymbol{r}}(t)$ are related to the dipole moment $-e\bar{\boldsymbol{r}}(t)$, which couples to the electromagnetic radiation. One can calculate the emitted electric field from the electron acceleration $\ddot{\bar{\boldsymbol{r}}}(t)$ and takes its Fourier transform to determine the emitted frequencies. In Figure 6 we plot the calculated intensities of various emitted lines. The strong peak corresponds to oscillations with the basic frequency $\omega = \Omega$. The peaks on the high-frequency side correspond to the interband excitations and are characteristic of ZB. The peaks on the lower frequency side correspond to the intraband (cyclotron) excitations. In absence of ZB the emission spectrum would contain only the intraband (cyclotron) frequencies. Thus the interband frequencies $\omega_n^Z$ shown in Figure 6 are a direct signature of the trembling motion. It can be seen that the $\omega_z^Z$ peaks are not much weaker than the central peak at $\omega = \Omega$, which means

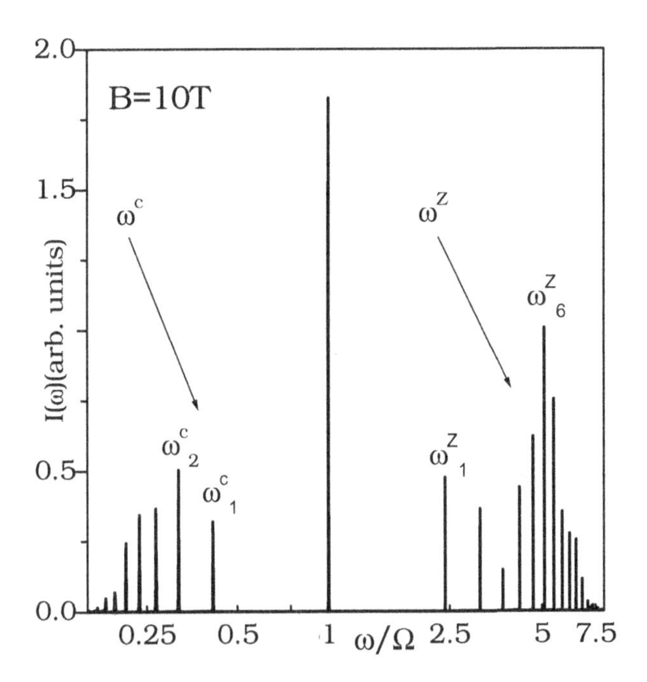


Fig. 6. Intensity spectrum versus frequency during the first 20 ps of motion of an electron described by a Gaussian wave packet having $k_{0x} = 0.035\text{Å}^{-1}$ in monolayer graphene. After Ref. 19.

that there exists a reasonable chance to observe them. Generally speaking, the excitation of the system is due to the nonzero momentum $\hbar k_{0x}$ given to the electron. It can be provided by accelerating the electron in the band or by exciting the electron with a nonzero momentum by light from the valence band to the conduction band. The electron can emit light because the Gaussian wave packet is not an eigenstate of the system described by the Hamiltonian (14). The energy of the emitted light is provided by the initial kinetic energy related to the momentum $\hbar k_{0x}$. Once this energy is completely used, the emission will cease. Radiation emitted by the trembling electrons in monolayer graphene excited by femtosecond laser pulses is described in Ref. 20. This problem is not trivial since it is difficult to prepare an electron in a solid in the form of a Gaussian wave packet. On the other hand, a formation of a light wave packet is mastered by present technics. It was shown that, when the Landau levels are broadened by scattering or defects, the light emission is changed from sustained to decaying in time.

## 5. Nature of ZB in solids

In spite of the great interest in the phenomenon of ZB its physical origin remained mysterious. As mentioned above, it was recognized that the ZB in vacuum is due to an interference of states with positive and negative electron energies. Since the ZB in solids was treated by the two-band Hamiltonian similar to the Dirac equation, its interpretation was also similar. This did not explain its origin, it only provided a way to describe it. For this reason we consider the fundamentals of electron propagation in a periodic potential trying to elucidate the nature of electron Zitterbewegung in solids. The physical origin of ZB is essential because it contributes to the question of its observability. The second purpose is to decide whether the two-band $\mathbf{k} \cdot \mathbf{p}$ model of the band structure, used to describe the ZB in semiconductors, is adequate.

One should keep in mind that we described above *various kinds* of ZB. Every time one deals with two interacting energy bands, an interference of the lower and upper states results in electron oscillations. In particular, one deals with ZB related to the Bychkov-Rashba-type spin subbands[14] or to the Luttinger-type light and heavy hole subbands.[16,30] However, the problem of our interest here is the simplest electron propagation in a periodic potential. The trembling motion of this type was first treated in Refs. 13,18. It is often stated that an electron moving in a periodic potential behaves like a free particle characterized by an effective mass $m^*$. The above picture suggests that, if there are no external forces, the electron moves in a crystal with a constant velocity. This, however, is clearly untrue because the electron velocity operator $\hat{v}_i = \hat{p}_i/m_0$ does not commute with the Hamiltonian $\hat{H} = \hat{\mathbf{p}}^2/2m_0 + V(\mathbf{r})$, so that $\hat{v}_i$ is not a constant of the motion. In reality, as the electron moves in a periodic potential, it accelerates or slows down keeping its total energy constant. This situation is analogous to that of a roller-coaster: as it goes down losing its potential energy, its velocity (i.e. its kinetic energy) increases, and when it goes up its velocity decreases.

We first consider the trembling frequency $\omega_Z$.[22] The latter is easy to determine if we assume, in the first approximation, that the electron moves with a constant average velocity $\bar{v}$ and the period of the potential is $a$, so $\omega_Z = 2\pi\bar{v}/a$. Putting typical values for GaAs: $a = 5.66$Å, $\bar{v} = 2.3 \times 10^7$cm/s, one obtains $\hbar\omega_Z = 1.68$eV, i.e. the interband frequency since the energy gap is $E_g \simeq 1.5$eV. This interband frequency is in fact typical for the ZB in solids.

Next, we describe the velocity oscillations classically assuming for simplicity a one-dimensional periodic potential of the form $V(z) = V_0 \sin(2\pi z/a)$. The first integral of the motion expressing the total energy is: $E = m_0 v_z^2/2 + V(z)$. Thus the velocity is

$$\frac{dz}{dt} = \sqrt{\frac{2E}{m_0}} \left[1 - \frac{V(z)}{E}\right]^{1/2} . \tag{23}$$

One can now separate the variables and integrate each side in the standard way. In the classical approach $V_0$ must be smaller than $E$. In general, the integration of Eq. (23) leads to elliptical integrals. However, trying to obtain an analytical result we assume $V_0(z) \simeq E/2$, expand the square root retaining the first two terms and solve the remaining equation by iteration taking in the first step a constant velocity $v_{z0} = (2E/m_0)^{1/2}$. This gives $z = v_{z0}t$ and

$$v_z(t) \simeq v_{z0} - \frac{v_{z0}V_0}{2E} \sin\left(\frac{2\pi v_{z0}t}{a}\right) . \tag{24}$$

Thus, as a result of the motion in a periodic potential, the electron velocity oscillates with the expected frequency $\omega_Z = 2\pi v_{z0}/a$ around the average value $v_{z0}$. Integrating with respect to time we get an amplitude of ZB: $\Delta z = V_0 a/(4\pi E)$. Taking again $V_0 \simeq E/2$, and estimating the lattice constant to be $a \simeq \hbar p_{cv}/(m_0 E_g)$ (see Luttinger and Kohn[39]), we have finally $\Delta z \simeq \hbar p_{cv}/(8\pi m_0 E_g)$, where $p_{cv}$ is the interband matrix element of momentum. This should be compared with an estimation obtained previously from the two-band $\mathbf{k} \cdot \mathbf{p}$ model:[13] $\Delta z \simeq \lambda_z = \hbar/m^* u = \hbar(2/m^* E_g)^{1/2} \simeq 2\hbar p_{cv}/m_0 E_g$. Thus the classical and quantum results depend in the same way on the fundamental parameters, although the classical approach makes no use of the energy band structure. We conclude that the Zitterbewegung in solids is simply due to the electron velocity oscillations assuring the energy conservation during motion in a periodic potential.

Now we describe ZB using a quantum approach. We employ the Kronig-Penney delta-like potential since it allows one to calculate explicitly the eigenenergies and eigenfunctions.[40,41] In the extended zone scheme the Bloch functions are $\psi_k(z) = e^{ikz} A_k(z)$, where

$$A_k(z) = e^{-ikz} C_k \left\{ e^{ika} \sin[\beta_k z] + \sin[\beta_k (a - z)] \right\} , \tag{25}$$

in which $k$ is the wave vector, $C_k$ is a normalizing constant and $\beta_k = \sqrt{2m_0 E}/\hbar$ is a solution of the equation

$$Z \frac{\sin(\beta_k a)}{\beta_k a} + \cos(\beta_k a) = \cos(ka), \tag{26}$$

with $Z > 0$ being an effective strength of the potential. In the extended zone scheme, the energies $E(k)$ are discontinuous functions for $k = n\pi/a$, where $n = \ldots -1, 0, 1 \ldots$. In the Heisenberg picture the time-dependent velocity averaged over a wave packet $f(z)$ is

$$\langle \hat{v}(t) \rangle = \frac{\hbar}{m_0} \iint dk dk' \langle f|k\rangle \langle k| \frac{\partial}{i\partial z} |k'\rangle \langle k'|f\rangle e^{i(E_k - E_{k'})t/\hbar}, \tag{27}$$

where $|k\rangle$ are the Bloch states. The matrix elements of momentum $\langle k|\hat{p}|k'\rangle = \hbar\delta_{k',k+k_n} K(k,k')$ are calculated explicitly. The wave packet $f(z)$ is taken in a Gaussian form of the width $d$ and centered at $k_0$. Figure 7 shows results for the electron ZB, as computed for a superlattice. The electron velocity and position are indicated. It is seen that for a superlattice with the period $a = 200$Å the ZB displacement is about $\pm 50$Å, i.e. a fraction of the period, in agreement with the rough estimations given above. The period of oscillations is of the order of several picoseconds.

The oscillations of the packet velocity calculated directly from the periodic potential have many similarities to those computed on the basis of the two-band $\mathbf{k} \cdot \mathbf{p}$ model. The question arises: does one deal with *the same* phenomenon in the two cases? To answer this question we calculate ZB using the two methods for the same periodic potential. We calculate the packet velocity near the point $k_0 = \pi/a$ for a one-dimensional Kronig-Penney periodic Hamiltonian using the Luttinger-Kohn (LK) representation.[39] The LK functions $\chi_{nk}(z) = e^{ikz} u_{nk_0}(z)$, where $u_{nk_0}(z) = u_{nk_0}(z + a)$, also form a complete set and we can calculate the velocity using a formula similar to (27). The two-band model is derived by the $\mathbf{k} \cdot \mathbf{p}$ theory with the result

$$\hat{H}_{kp} = \begin{pmatrix} \hbar^2 q^2/2m + E_1 & \hbar q P_{12}/m \\ \hbar q P_{21}/m & \hbar^2 q^2/2m + E_2 \end{pmatrix}, \tag{28}$$

where $E_1$ and $E_2$ are the energies at band extremes, $P_{12} = \hbar/m \langle u_{1k_0}|\partial/i\partial x|u_{2k_0}\rangle$, and $q = k - \pi/a$. The band gap $E_g = E_2 - E_1$ and the matrix elements $P_{12}$ are calculated from the same Kronig-Penney potential, see inset of Figure 8. Apart from the small free-electron terms on the diagonal, equation (28) simulates the 1+1 Dirac equation for free relativistic electrons in a vacuum.

In Figure 8 we compare the ZB oscillations of velocity calculated using: (a) real $E(k)$ dispersions resulting from the Kronig-Penney model and the corresponding Bloch functions of (25); (b) two-band $E(k)$ dispersions and the corresponding LK functions. It is seen that the two-band $\mathbf{k} \cdot \mathbf{p}$ model

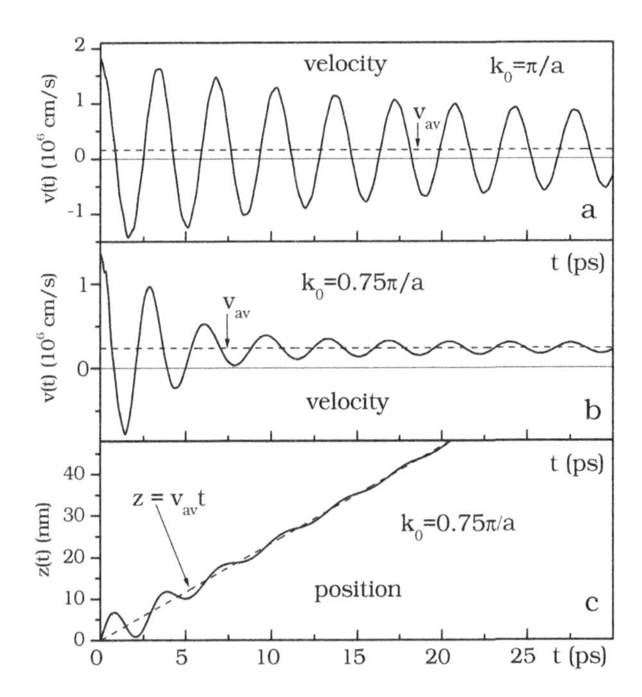


Fig. 7. Calculated electron ZB velocities and displacement in a superlattice versus time. The packet width is $d = 400$Å, Kronig-Penney parameter is $Z = 1.5\pi$, superlattice period is $a = 200$Å. (a) Packet centered at $k_0 = \pi/a$; (b) and (c) packet centered at $k_0 = 0.75\pi/a$. The dashed lines indicate motions with average velocities. After Ref. 22.

gives an excellent description of ZB for instantaneous velocities. This agreement demonstrates that the theories based on: (a) the periodic potential and (b) the band structure, describe *the same* trembling motion of the electron. The procedure based on the energy band structure is more universal since it also includes cases like the Bychkov-Rashba-type spin subbands or the Luttinger-type light and heavy hole subbands which do not exhibit an energy gap and do not seem to have a direct classical interpretation. The distinctive character of the situation we considered is that it has a direct spatial interpretation and it is in a direct analogy to the situation first considered by Schrodinger for a vacuum.

The main conclusion of the above considerations is that the electron Zitterbewegung in crystalline solids is not an obscure and marginal phenomenon but the basic way of electron propagation in a periodic potential. The ZB oscillations of electron velocity are simply due to the total energy

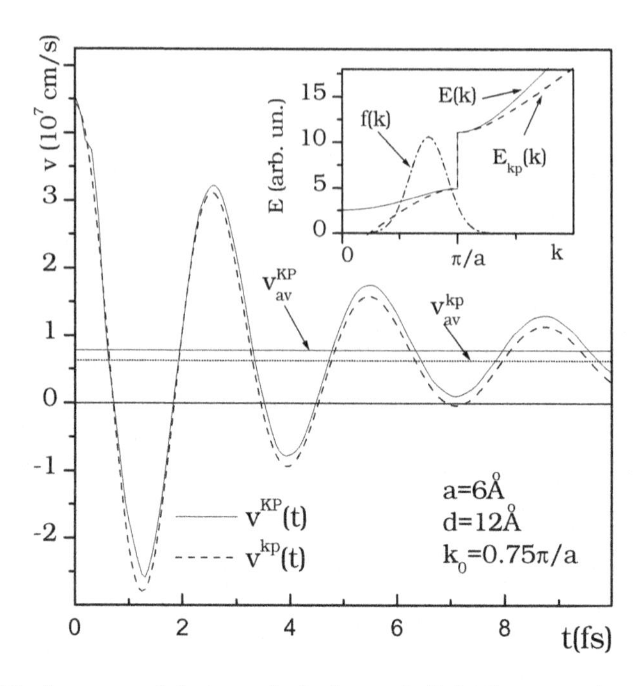


Fig. 8. Zitterbewegung of electron velocity in a periodic lattice versus time. Solid line: the Kronig-Penney model, dashed line: the two-band $\mathbf{k}\cdot\mathbf{p}$ model. Inset: Calculated bands for the Kronig-Penney (solid line) and the two-level $\mathbf{k}\cdot\mathbf{p}$ model (dashed line) in the vicinity of $k = \pi/a$. The wave packet $f(k)$ centered at $k_0 = 0.75\pi/a$ is also indicated (not normalized). After Ref. 22.

conservation. The trembling motion can be described either as a mode of propagation in a periodic potential or, equivalently, by the two-band $\mathbf{k}\cdot\mathbf{p}$ model of band structure. The latter gives very good results because, using the effective mass and the energy gap, it reproduces the main features of the periodic potential. According to the two-band model, the ZB is related to the interference of positive and negative energy components, while the direct periodic potential approach reflects real character of this motion. The established nature of ZB in solids indicates that the latter should be observable. We mention that in their early paper Ferrari and Russo[10] wrote: "The motion of Zitterbewegung and the resulting formalism ... is applied to describe the acceleration of a non-relativistic electron moving in a crystal, due to the periodic force experienced (...). The resulting Zitterbewegung is a real effect just because it follows from a real force."

## 6. Relativistic electrons in vacuum

The subject of ZB for free relativistic electrons is vast. We mention below a few papers in this domain which contributed to the understanding of ZB in solids. The main idea of Schrodinger's pioneering work is given in (4) because the initial equation considered above (3) is the same as the Dirac equation (DE) with changed parameters. Details of the original Schrodinger derivations were given by Barut and Bracken.[1] Considerations showing that the ZB is caused by the interference of electron states related to positive and negative electron energies are quoted in most books on relativistic quantum mechanics, see e.g. Refs. 3–5. Feschbach and Villars[42] argued that, in addition to the so-called Darwin term, the spin-orbit term in the standard $v^2/c^2$ expansion of DE can also be related to ZB.

Huang[2] went beyond the operator considerations of ZB calculating averages of the electron position and angular momentum with the use of wave packets. According to this treatment the electron magnetic moment may be viewed as a result of ZB, see also Ref. 43. Sasabe showed explicitly in a recent paper that the spin magnetic moment of a free relativistic electron is directly related to its Zitterbewegung.[54] Foldy and Wouthuysen[45] (see also Refs. 43,44) found a unitary transformation that separates the states of positive and negative electron energies in the free-electron DE. They showed that such states do not exhibit the ZB. Lock[7] remarked that, in order to talk seriously about observing ZB, one should consider a localized electron since "it seems to be of limited practicality to speak of rapid fluctuations in the average position of a wave of infinite extent". He then showed that, if an electron is represented by a localized wave packet, its Zitterbewegung is transient, i.e. it decays in time. This prediction was subsequently confirmed by many descriptions (beginning with Refs. 18,46 and in experimental simulation[21]). Lock further showed that, if the electron spectrum is discrete, the resulting ZB is sustained in time. This property was confirmed for graphene in the presence of an external magnetic field,[19] as well as for relativistic electrons in a vacuum when the spectrum is quantized into Landau levels.[47]

It was pointed out, see e.g. Refs. 4,48, that according to the DE not only the velocity and position operators experience ZB, but also the angular momentum $\hat{\boldsymbol{L}}$, the spin $\hat{\boldsymbol{S}}$, and the operator $\hat{\beta}$ exhibit the trembling time dependence. On the other hand, the total angular momentum $\hat{\boldsymbol{J}} = \hat{\boldsymbol{L}} + \hat{\boldsymbol{S}}$ is a constant of the motion, which can be shown directly by its vanishing commutator with the free electron Hamiltonian $\hat{\mathcal{H}}_D$. Thaller[46] computed and simulated the time behavior of relativistic Gaussian wave packets according to the one-dimensional DE. For a packet with vanishing average momentum, the packet position shows ZB that decays with time very slowly. For a non-vanishing average momentum the decaying of ZB is much faster. This is caused by the fact that the ZB arises due to an interference of positive and negative energy sub-packets which in this case move in opposite directions and cease to overlap relatively quickly. This process is mentioned above in connection with ZB in bilayer graphene.

## 7. Simulations

As we said above, the electron Zitterbewegung in a vacuum or in a solid is difficult to observe. The characteristics of electron ZB in semiconductors are much more favorable than in a vacuum but it is difficult to follow the motion of a single electron; one would need to follow motion of many electrons moving in phase. Recently, however, there appeared many propositions to *simulate* the Dirac equation and the resulting phenomena with the use of other systems. We want to enumerate below these propositions but we are not in a position to explain all the underlying ideas. It will suffice to say that many (not all) ideas make use of trapped atoms or ions interacting with laser light. There are two essential advantages of such simulations. First, it is possible to follow the interaction of laser light with few or even single atoms or ions. Second, when simulating the DE it is possible to modify its two basic parameters: $m_0c^2$ and $c$, in order to make the ZB frequency much lower and its amplitude much larger than in vacuum. In consequence, they become measurable with current experimental techniques.

As a matter of example we will briefly consider a simulation of DE with the use of Jaynes-Cummings model[49] known from the quantum and atomic optics, see Refs. 50–52. The Dirac equation contains electron momenta, so the essential task is to simulate $\hat{p}_l$. The common types of light interactions with ions and vibronic levels are used to that purpose: a carrier interaction $\hat{H}^c = \hbar\Omega(\sigma^+ e^{i\phi_c} + \sigma^- e^{-i\phi_c})$, the Jaynes-Cummings interaction $\hat{H}^{\phi_r}_{JC} = \hbar\eta\tilde{\Omega}(\sigma^+\hat{a}e^{i\phi_r} + \sigma^-\hat{a}^+e^{-i\phi_r})$, and the anti-Jaynes-Cummings interaction $\hat{H}^{\phi_b}_{AJC} = \hbar\eta\tilde{\Omega}(\sigma^+\hat{a}^+e^{i\phi_b} + \sigma^-\hat{a}e^{-i\phi_b})$. Here $\sigma^\pm = \sigma_x \pm i\sigma_y$ are the raising and lowering ionic spin-1/2 operators, $\hat{a}$ and $\hat{a}^+$ are the creation and annihilation operators associated with the motional states of the ion, $\eta$ is the so called Lamb-Dicke parameter, $\Omega$ and $\tilde{\Omega}$ are the Rabi frequencies. The basic idea is to use proper light phases in $\hat{H}^{\phi_r}_{JC}$ and $\hat{H}^{\phi_r}_{AJC}$ in order to obtain the momentum from the relation $\hat{p}_l = i\hbar(\hat{a}^+_l - \hat{a}_l)/\Delta$, which results in $\hat{H}^{p_l}_{\sigma_j} = \pm i\hbar\eta\tilde{\Omega}\sigma_j(\hat{a}^+_l - \hat{a}_l)$. One can show that the simulated parameters of DE are

$$c \to 2\eta\Delta\tilde{\Omega}, \qquad m_0c^2 \to \hbar\Omega, \tag{29}$$

where $\Delta$ is the spread in the position of the ground ion wave function. If the dynamics created by the $3+1$ Dirac equation is to be reproduced in an experiment with a four-level ion system using Raman beams, it requires 14 pairs of Raman lasers. One needs to control their phases independently.

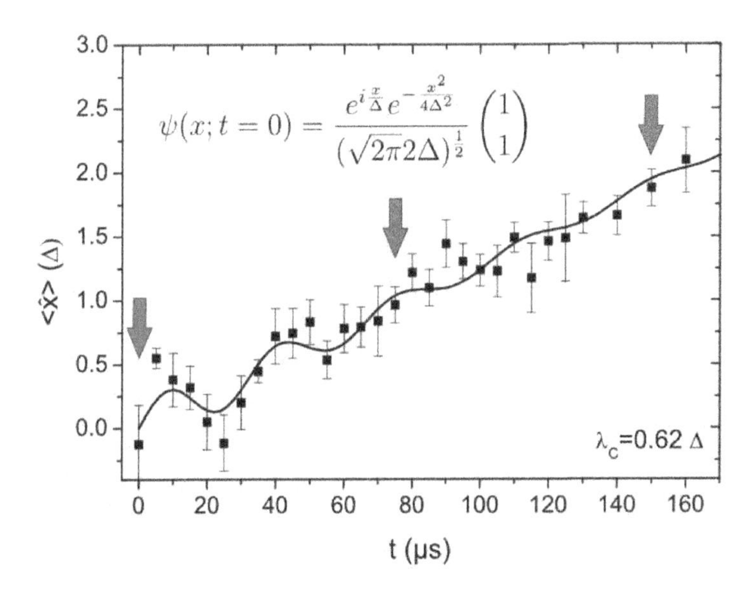


Fig. 9. Zitterbewegung for a state with nonzero average momentum. The solid curve represents a numerical simulation. (b) Measured (filled areas) and numerically calculated (solid lines) probability distributions $|\psi(x)|^2$ at the times $t = 0$, 75 and 150 $\mu$s (as indicated by the arrows). The probability distribution corresponding to the state $|1\rangle$ is inverted for clarity. The vertical solid line represents $\langle\hat{x}\rangle$. The two dashed lines are the expectation values for the positive and negative energy parts of the spinor. Error bars $1\sigma$. After Ref. 21.

In Figure 9 we show experimental results of Gerritsma *et al.*[21], who simulated for the first time the 1+1 Dirac equation with the resulting one-dimensional Zitterbewegung using $^{40}Ca^+$ trapped ions. It can be seen that the results agree very well with the predictions of Ref. 22, see our Figure 7c. The reason of this agreement is that the theory of Ref. 22, while concerned with solids, also uses an effective Dirac equation, see (28) and the inset of

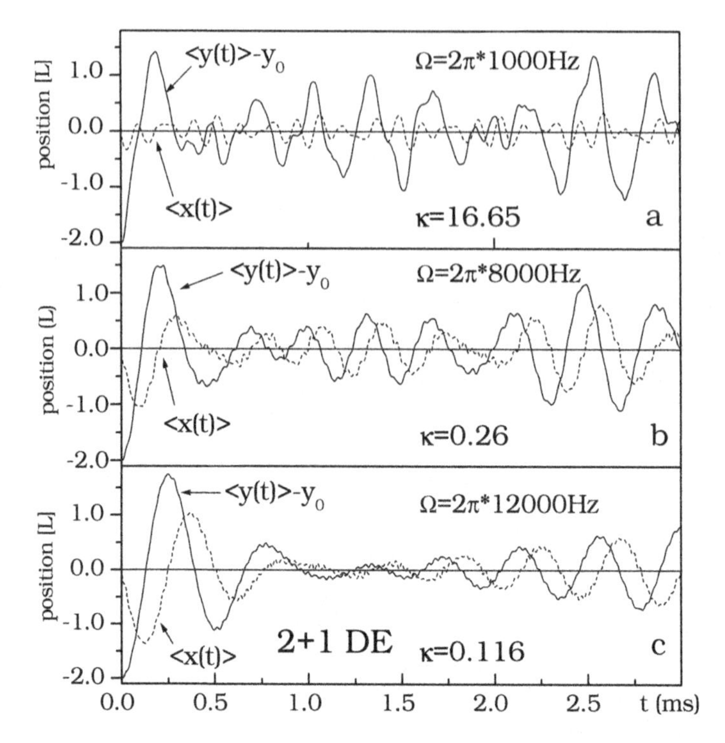


Fig. 10. Calculated motion of two-component wave packet simulated by trapped $^{40}Ca^{+}$ ions for three values of effective rest energies $\hbar\Omega$. Simulations correspond to $\kappa = \hbar\omega_c/2m_0c^2 = 16.65$ (a), 0.26 (b), 0.116 (c), respectively. Positions are given in $L = \sqrt{2}\Delta$ units. Oscillations do not decay in time. After Ref. 47.

Figure 8. Gerritsma *et al.* showed that, if the wave packet does not have the initial momentum, the decay time of ZB is much slower than that seen in Figure 9. This agrees with theoretical results for carbon nanotubes, see also Ref. 46.

The problem of ZB for free relativistic electrons in a magnetic field was described by Rusin and Zawadzki.[47] The main experimental problem in investigating the ZB phenomenon in an external magnetic field is the fact that for free relativistic electrons the basic ZB (interband) frequency corresponds to the energy $\hbar\omega_Z \simeq 1$ MeV, whereas the cyclotron frequency for a magnetic field of 100 T is $\hbar\omega_c \simeq 0.1$ eV, so that the magnetic effects in ZB are very small. However, it is possible to simulate the Dirac equation including an external magnetic field with the use of trapped ions interacting with laser radiation. This gives a possibility to modify the ratio $\hbar\omega_c/(2m_0c^2)$ making its value much more advantageous. If the magnetic field $\boldsymbol{B}$ is directed along the $z$ direction, it can be described by the vector potential $\boldsymbol{A} = (-By, 0, 0)$. Then the main modification introduced to the DE is that, instead of the momentum $\hat{p}_x$, one has $\hat{\pi}_x = \hat{p}_x - eBy$ which leads to the appearance of raising and lowering operators $\hat{a}_y = (\xi + \partial/\partial\xi)/\sqrt{2}$ and $\hat{a}_y^+ = (\xi - \partial/\partial\xi)/\sqrt{2}$ with $\xi = y/L - k_xL$. The simulation of DE proceeds as for a free particle with the difference that $\hat{a}_y$ and $\hat{a}_y^+$ can be simulated by a *single* JC or AJC interaction. It can be shown that the crucial ratio is

$$\kappa = \frac{\hbar eB}{m_0(2m_0c^2)} \Rightarrow \left(\frac{\eta\tilde{\Omega}}{\Omega}\right)^2, \tag{30}$$

where $\eta$, $\Omega$ and $\tilde{\Omega}$ were defined above. Therefore, by adjusting frequencies $\Omega$ and $\tilde{\Omega}$ one can simulate different regimes of $\kappa = \hbar\omega_c/2m_0c^2$. In Figure 10 we show the calculated ZB for three values of $\kappa$: 16.65, 1.05, 0.116. It is seen that, as $\kappa$ gets larger (i.e. the field intensity increases or the effective gap decreases), the frequency spectrum of ZB becomes richer. This means that more interband and intraband frequency components contribute to the spectrum. Both intraband and interband frequencies correspond to the selection rules $n' = n \pm 1$ so that, for example, one deals with ZB (interband) energies between the Landau levels $n = 0$ to $n' = 1$, and $n = 1$ to $n' = 0$, as the strongest contributions. For high magnetic fields the interband and intraband components are comparable. Qualitatively, the results shown in Figure 10 are similar to those obtained for graphene in a magnetic field, see Ref. 19.

## 8. Zitterbewegung created by laser pulses

It follows from our considerations that the choice of initial wave packet, i.e. its shape, initial momentum and components, is decisive for the resulting properties of ZB. As to the common choice of packets, which requires the initial value of $\hbar\boldsymbol{k}_0$, it is not clear how to create this momentum. If one uses light to create the electron packet, the initial wave vector $\boldsymbol{k}_0$ will be small since photons do not carry much momentum. If one uses acoustic phonons to trigger the ZB motion, the initial energy will be small. In view of the above difficulties and also to make the project more realistic experimentally, it was proposed not to assume anything a priori about the electron wave packet, but to determine it as a result of a realistic laser pulse and then use it for a calculation of ZB oscillations, see Ref. 65 and

20. Carbon nanotubes (CNT) were chosen as a suitable electronic system, see Refs. 18,38, their band structure was calculated with the use of tight binding scheme by Saito *et al.*.[63,64]

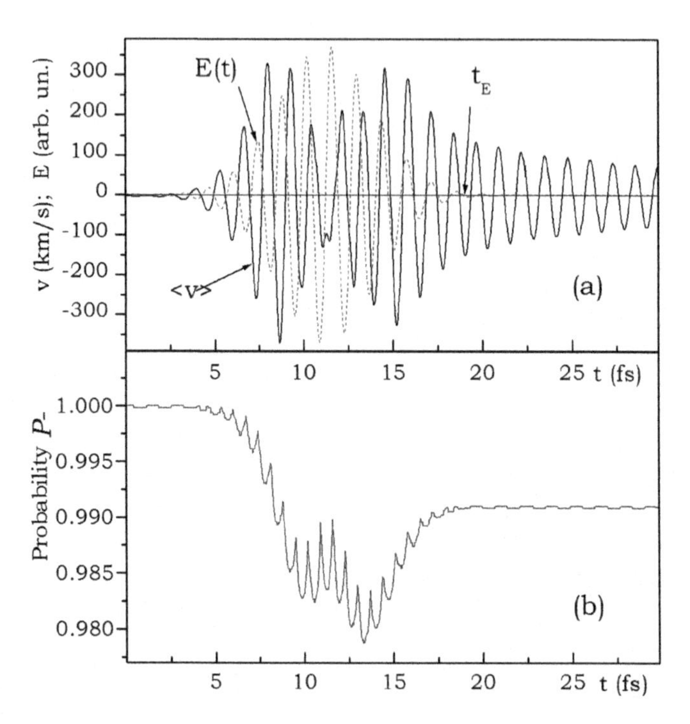


Fig. 11. (a) Calculated average packet velocity versus time calculated for CNT and laser pulse parameters listed in table 1. Solid line: packet velocity, dashed line: electric field of the laser pulse (in arbitrary units). Arrow indicates time $t_E$ at which the electric field of the pulse vanishes. For $t > t_E$ the amplitude of oscillations decays as $t^{-1}$. (b) Normalized probability $\mathcal{P}^-$ of finding the electron in states with negative energy versus time.

The light is introduced by the vector potential $\boldsymbol{A} = \boldsymbol{A}(t)$ and the scalar potential $\phi = 0$. Then $\boldsymbol{E} = -\partial \boldsymbol{A}(t)/\partial t$ and $\boldsymbol{B} = \boldsymbol{\nabla} \times \boldsymbol{A}(t) = 0$. By choosing $\boldsymbol{A}(t) = [A(t), 0, 0]$ we have approximately

$$A(t) \simeq \frac{E_0}{\omega_L} \exp\left(-b\frac{t'^2}{\tau^2}\right) \cos(\omega_L t'). \tag{31}$$

As a result of the illumination by light one deals with the generalized quasi momentum $\boldsymbol{q}(t) = \boldsymbol{k}(e/\hbar)\boldsymbol{A}(t)$ and the time-dependent electron

Hamiltonian. One calculates the time dependent velocity and its average velocity with the use of corresponding wave functions. The results are quoted in Figure 11 In Figure 11(a) we plot the calculated average electron velocity versus time for realistic material and pulse parameters. The solid line shows velocity oscillations of the electron packet excited by the pulse, while the dashed line shows the electric field of the pulse (in arbitrary units). At the initial time $t = 0$ both the electron velocity and electric field are zero. Then, within the first 19 fs, the amplitude if electron velocity grows and decreases similarly but not identically to the field amplitude. After $t_E \simeq 19$ fs the electric field of the pulse disappears, but there persist oscillations slowly decaying in time. These oscillations resemble the ZB oscillations in CNT described in Ref. 18 and they have a similar character, slowly decaying as $t^{-\alpha}$ with $\alpha \simeq 1$. In Figure 11(b) we show the calculated time-dependent probability of negative-energy component of the packet. Initially, for $t = 0$, the electron occupies only the valence states and there is $\mathcal{P}^{-}(0) = 1$. For $t \geq t_E$, after the pulse disappears, the final electron state has a nonzero admixture of states with positive energies which is a necessary condition for the appearance of ZB oscillations.[5,6]

Calculated Fourier transform of the calculated velocity has a sharp maximum at the interband frequency $\Omega_Z = 4.9$ fs$^-1$, corresponding to a specific pair of bands in the CNT. These results strongly indicate that one deals with the ZB phenomenon for times $t \geq t_E \simeq 19$ fs, i.e. when the external light signal is already absent. It should be emphasized that the laser light must be roughly tuned to the energy gap between proper pairs of energy bands in CNT. All in all, the above results show that that one can produce the ZB phenomenon using light and avoiding thereby the problems of producing an electron wave packet with nonrealistic parameters.

We mention that a similar calculation was carried out for monolayer graphene where one deals with the pair of linear bands without the energy gap. The results did not exhibit the ZB-like oscillations after the light pulse is terminated.[55]

The main difficulty in observing the ZB in solids is the fact that the "trembling electrons" move in a crystal with different directions and phases, so that the oscillations may average to zero. Thus, in order to observe the trembling directly, one would need to follow the motion of a single electron. In order to overcome the above difficulties, example of the Bloch oscillator (BO) was followed. The Bloch oscillator is another phenomenon which had been proposed a long time ago and it took many years to observe it. The phenomena of ZB and BO are basically similar, their nature is different, but they are both characterized by electron oscillations with an inherent frequency: for BO it is determined by an external electric field and for ZB by an energy gap between the conduction and valence bands. In both situations various electrons oscillate with different phases. The Bloch oscillator was finally investigated and observed by means of nonlinear optics. Von Plessen and Thomas[57] used the third-order perturbation expansion in electric field to calculate a two-photon echo (2PE) signal resulting from Bloch oscillations in superlattices. In this scheme the sample is illuminated by two subsequent laser pulses with the wave vectors $\boldsymbol{k}_1$ and $\boldsymbol{k}_2$ at times $t = 0$ and $t = \tau_D$, respectively. Light is then emitted into the background-free direction $\boldsymbol{k}_3 = 2\boldsymbol{k}_2 - \boldsymbol{k}_1$ due to the nonlinear optical interaction in the sample.[58] The above method, with some variations, was successfully used to observe the Bloch oscillations in GaAs/GaAlAs superlattices, see e.g. Refs. 59–61. The essential point that makes the techniques of nonlinear optics so powerful is the condition of phase matching $\boldsymbol{k}_1 + \boldsymbol{k}_2 + \boldsymbol{k}_3 = 0$. When this condition is fulfilled, the individual dipoles created by oscillating electrons are properly phased so that the field emitted by each dipole adds coherently in the $\boldsymbol{k}_3$ direction, which strongly enhances the emitted signal.[62]

Following the path taken in above works we turned to the nonlinear optics and, more specifically, to the two-photon echo technique.[58] We showed that such an experiment can be used to detect the ZB phenomenon in solids and more specifically in CNT.[56] In this approach, we did not make any assumption about the electron wave packet, but determined it from excitation by a laser pulse. In calculating the medium polarization we did not use perturbation expansions, but computed exact solutions from the time-dependent Hamiltonian. The density matrix (DM) formalism was used in order to include effects of relaxation and decoherence.

The band structure of CNT was computed by the tight binding method, as indicated above. The laser field was introduced as before, by the vector potential which is added to the wave vector of the charge carrier. This makes the Hamiltonian of CNT time-dependent. The time-dependent velocity is then calculated and averaged over the eigenfunctions of the Hamiltonian.

The above results suggest that, by measuring the time-dependent medium polarization, one can directly observe electron ZB oscillations in CNT. However, in real systems one deals with *many* electrons, which are excited with random phases depending on their initial phases as well as on collisions with other electrons or lattice defects. Thus, the polarization created by the laser shot will be destroyed and no net polarization will be

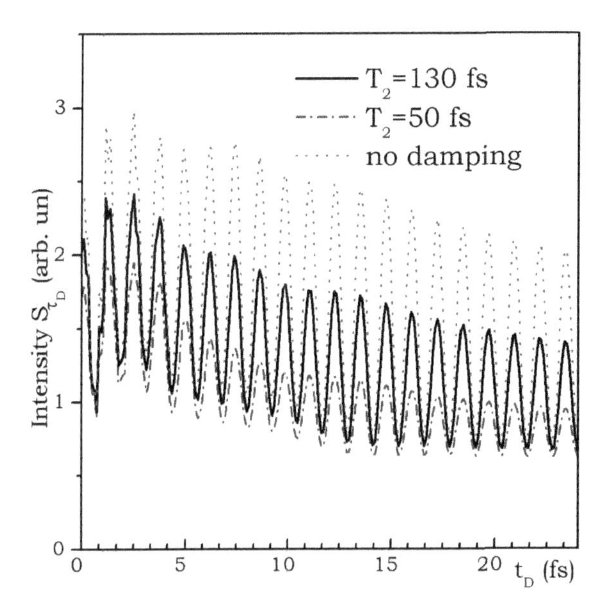


Fig. 12. Time-integrated two-photon echo signal $S_{\tau_D}$ as given in (32) calculated versus delay between the two pulses using non-perburbative method. Upper dotted line: damping excluded; solid line: damping times listed in Table 1, $T_2 = 130$ fs obtained experimentally in Ref. 66, lowest dashed line: short artificial damping times $T_1 = T_2 = 50$ fs. The ZB oscillations correspond to $\tau_D > 13.5$ fs, see text.

measured. As mentioned in the Introduction, the way to overcome this difficulty is to use nonlinear laser spectroscopy,[58] in which two or more laser pulses allow one to observe nonlinear polarization components and to select signals with proper phase-matching conditions.[62] Common methods to measure electron coherence are the two-photon echo (2PE) and the degenerate four-wave mixing spectroscopies (DFWM). In the following we calculate the 2PE signal corresponding to the ZB oscillations induced by two laser pulses.

The configuration of the two-photon echo experiment is described in Ref. 67. We consider two incident laser beams characterized by two non-parallel wave vectors $\boldsymbol{k}_1$ and $\boldsymbol{k}_2$. The first laser pulse creates the medium polarization [as shown in Figure 11(a)], which propagates in time after pulse disappearance. The second pulse probes the electron state and leads to a coherent emission of the signal. In the homodyne detection scheme the two-photon echo signal is measured in the background-free direction $2\boldsymbol{k}_2 - \boldsymbol{k}_1$.

The signal intensity $S_{\tau_D}$ depends on the delay between two pulses, and on the polarization component $\widetilde{P}_{2\boldsymbol{k}_2-\boldsymbol{k}_1}(t)$ in the $2\boldsymbol{k}_2 - \boldsymbol{k}_1$ direction. Then there is[58]

$$S_{\tau_D} \propto \int_{-\infty}^{\infty} \left| \widetilde{P}_{2\boldsymbol{k}_2-\boldsymbol{k}_1}(t) \right|^2 dt. \tag{32}$$

To extract the polarization $\widetilde{P}_{2\boldsymbol{k}_2-\boldsymbol{k}_1}(t)$ from the total polarization $P(t)$, we apply a non-perturbative method proposed by Seidner and coworkers[68,69].

The results of our calculations are presented in Figure 12 where we plot the signal $S_{\tau_D}$ as a function of $\tau_D$ for three sets of dephasing times. The dotted line corresponds to the no-damping case, the solid line corresponds to the relaxation time $T_1$ and experimental decoherence time $T_2$ listed in Table 1, while the dashed line corresponds to artificially short times $T_1 = T_2 = 50$ fs. The common feature of the three results shown in Figure 12 is their oscillating character with the same period $T_Z \simeq 1.25$ fs, corresponding to the frequency $\omega_Z = 2\pi/T_Z = 5.03$ fs$^{-1}$. This value is close to the interband frequency $\omega_4$e 1. The amplitudes of the signals depend on $T_1$ and $T_2$, but in all cases the oscillations are clearly visible. This indicates that the Zitterbewegung is a robust phenomenon, not sensitive to the details of model parameters.

The results shown in Figure 12 indicate a possibility of the experimental observation of ZB motion with the use of the two-photon echo experiment. Each element of such an experiment is available within the current techniques. All in all, it seems possible to observe the ZB oscillations using the method proposed above.

## 9. Discussion and conclusions

An important recognition won after the considerable effort of the last years is, that the Zitterbewegung is not a marginal, obscure and probably unobservable effect of interest to a few esoteric theorists, but a real and universal phenomenon that often occurs in both quantum and non-quantum systems. Clearly, the ZB in a vacuum proposed by Schrodinger[1] stands out as an exception since it is supposed to occur without any external force. However, in its original form it is probably not directly observable for years to come and one has to recourse to its simulations. On the other hand, manifestations of ZB in crystalline solids and other periodic systems turned out to be quite common and they are certainly observable. A universal background underlying the phenomenon of ZB in any system (including a vacuum) is an interference of states belonging to positive and negative energies.

The positive and negative energies belong usually to bands but they can also be discrete levels, as shown for electrons in graphene in a magnetic field.

As we said above, the ZB amplitude is around $\lambda_z = \hbar/(m_0^* u)$. Let us suppose that we confine an electron to the dimension $\Delta z \simeq \lambda_z/2$. Then the uncertainty of momentum is $\Delta p_z \geq \hbar/\Delta z$ and the resulting uncertainty of energy $\Delta E \simeq (\Delta p_z)^2/(2m_0^*)$ becomes $\Delta E \geq 2m_0^* u^2 = \mathcal{E}_g$. Thus the electron confined to $\Delta z \simeq \lambda_z/2$ has the uncertainty of energy larger than the gap. For electrons in a vacuum the restriction $\Delta z \simeq \lambda_c/2$ is not significant, but for electrons in narrow-gap semiconductors the restriction $\Delta z \simeq \lambda_z/2$ should be taken seriously since $\lambda_z$ is of the order of tens of Angstroms, so that this confinement is not difficult to realize experimentally by quantum wells or magnetic fields. The question arises, what happens if the electron is confined to $\Delta z < \lambda_z/2$, so that the trembling motion is strongly perturbed by the confinement. We showed above that an electron in a magnetic field radiates interband ZB frequencies and their contribution to the motion increases with the increasing field, see also Ref. 47. It is possible that this effect is just a manifestation of the perturbation of the trembling motion by magnetic confinement. Also, it was shown that an effective one-band semirelativistic Hamiltonian in a narrow-gap semiconductor contains the so-called Darwin term which can be traced back to the ZB. The Darwin term can lead to measurable effects for ground impurity states.[13]

An important question arises: what should be called "Zitterbewegung"? It seems that the signature of ZB phenomenon is its *interband frequency*, in which the term interband has the meaning "between interacting bands". Thus, for example, the ZB resulting from the Bychkov-Rashba spin splitting (or the so-called linear Dresselhaus spin splitting) is not characterized by a truly interband frequency, since in this situation there is no gap, but the frequency corresponding to the energy difference between the two spin branches of the same band. Another illustration is the ZB of holes in the valence bands of $\Gamma_8$ symmetry, where the ZB frequency is given by the energy difference of light and heavy hole bands. Finally, an instructive example is provided by graphene in a magnetic field, where the electron motion contains both intraband and interband frequencies. We believe that only the interband contributions should be called ZB, while the intraband ones are simply the cyclotron components. It appears that the second signature of ZB is the actual *motion* which, for instance, distinguishes it from the Rabi oscillations. The above considerations indicate that an unambiguous definition of ZB is not obvious.

If an electron is prepared in the form of a wave packet, and if the electron spectrum is not completely quantized, the ZB has a *transient* character, i.e. it decays in time. This was predicted by Lock[7] and confirmed by many specific calculations, as well as by experimental simulation.[21,23] One can show that the decay time is inversely proportional to the momentum spread $\Delta k$ of the wave packet. Physically, the transient character of ZB comes about as a result of waning interference of the two sub-packets belonging to positive and negative energies as they go apart because of their different speeds. On the other hand, if the electron spectrum is discrete, ZB persists in time, sometimes in the form of collapse-revival patterns.

Clearly, one should ask the question about possible observation of Zitterbewegung in solids. It seems that an effective way to observe ZB is to excite electrons across the gap by laser pulses and detect the electromagnetic radiation emitted by the trembling electrons, as proposed above. The emission is possible because, if the electrons are prepared in form of wave packets or they respond to light wave packets, they are not in their eigenstates. The effective way is to use the two-pulse spin echo and observe ZB at an appropriate angle. The phenomenon of ZB should not be confused with the Bloch oscillations of charge carriers in superlattices. The Bloch oscillations are basically a one-band phenomenon and they require an external electric field driving electrons all the way to the Brillouin zone boundary. On the other hand, the ZB needs at least two bands and it is a no-field phenomenon. Narrow gap superlattices can provide a suitable system for its observation. In the near future one can expect theoretical predictions of ZB in new systems as well its quantum and classical simulations.

## 10. Experimental confirmation of Zitterbewegung

As we mentioned previously, Wilamowski *et al.*[31] investigated experimentally and theoretically the spin resonance in asymmetric silicon structures in a magnetic field. The findings were analyzed in terms of the Rashba spin splitting that causes non-commutativity of the velocity and Hamiltonian operators. The precession of electron spin with the Larmor frequency results via the spin-orbit interaction in an ac current. The latter is a source of the Joule heat which is manifested in additional effects in the spin resonance.

Very recently, a coherent Zitterbewegung of electrons was observed experimentally in n-type InGaAs in the presence of a magnetic field by Stepanov *et al.*[70] The Zitterbewegung of electron velocity originates in the interference between two spin states split by the magnetic field, while non-commutativity of the Hamiltonian and velocity operators is related to the spin-orbit interaction manifested in the Bychkov-Rashba and Dresselhaus interactions. Many electrons tremble with the same phase being all excited across the InGaAs gap with laser pulses to the same spin state. The ZB motion is measured as an ac current. The amplitude of ZB oscillations is estimated to be about 20 nm and their frequency at the field $B = 3$ T is 26 GHz. Some features of the observation agree with the general theoretical predictions presented above: the frequency is determined by the energy gap of the two level system and electrons are excited by laser pulses. However, the investigated system contains in addition the spin-orbit interaction which "mediates" between the precession of electron spin in a magnetic field and the ac current observed experimentally. It can be seen that the physical considerations in the two above works are quite similar. The work of Stepanov *et al.* uses time-resolved measurements of the current representing the latest progress in the field.

## References


1. E. Schrodinger, *Sitzungsber. Preuss. Akad. Wiss. Phys. Math. Kl.* **24**, 418 (1930). Schrodinger's derivation is reproduced in A. O. Barut and A. J. Bracken, *Phys. Rev. D* **23**, 2454 (1981).
2. K. Huang, *Am. J. Phys.* **20**, 479 (1952).
3. J. D. Bjorken and S. D. Drell, *Relativistic Quantum Mechanics* (New York: McGraw-Hill, 1964).
4. J. J. Sakurai, *Modern Quantum Mechanics* (New York: Addison-Wesley, 1987).
5. W. Greiner, *Relativistic Quantum Mechanics* (Berlin: Springer, 1994).
6. A. Wachter, *Relativistic Quantum Mechanics* (Berlin: Springer, 2010).
7. J. A. Lock, *Am. J. Phys.* **47**, 797 (1979).
8. D. Lurie and S. Cremer, *Physica* **50**, 224 (1970).
9. F. Cannata, L. Ferrari and G. Russo, *Sol. St. Comun.* **74**, 309 (1990).
10. L. Ferrari and G. Russo, *Phys. Rev. B* **42**, 7454 (1990).
11. S. V. Vonsovskii, M. S. Svirskii and L. M. Svirskaya, *Teor. Matem. Fizika* **85**, 211 (1990), (in Russian).
12. W. Zawadzki, *High Magnetic Fields in the Physics of Semiconductors II*, eds. G. Landwehr and W. Ossau, (Singapore: World Scientific, 1997), p 755.

13. W. Zawadzki, *Phys. Rev. B* **72**, 085217 (2005), (arXiv:0411488).
14. J. Schliemann, D. Loss and R. M. Westervelt, *Phys. Rev. Lett.* **94**, 206801 (2005), (arXiv:0410321).
15. J. Cserti and G. David, *Phys. Rev. B* **74**, 172305 (2006), (arXiv:0604526).
16. R. Winkler, U. Zulicke and J. Bolte, *Phys. Rev. B* **75**, 205314 (2007), (arXiv:0609005).
17. T. M. Rusin and W. Zawadzki, *J. Phys. Cond. Matter* **19**, 136219 (2007), (arXiv:0605384).
18. T. M. Rusin and W. Zawadzki, *Phys. Rev. B* **76**, 195439 (2007), (arXiv:0702425).
19. T. M. Rusin and W. Zawadzki, *Phys. Rev. B* **78**, 125419 (2008), (arXiv:0712.3590).
20. T. M. Rusin and W. Zawadzki, *Phys. Rev. B* **80**, 045416 (2009), (arXiv:0812.4773).
21. R. Gerritsma, G. Kirchmair, F. Zahringer, E. Solano, R. Blatt and C. F. Roos, *Nature* **463**, 68 (2010), (arXiv:0909.0674).
22. W. Zawadzki and T. M. Rusin, *Phys. Lett. A* **374**, 3533 (2010), (arXiv:0909.0463).
23. X. Zhang and Z. Liu, *Phys. Rev. Lett.* **101**, 264303 (2008), (arXiv:0804.1978).
24. F. Dreisow, M. Heinrich, R. Keil, A. Tunnermann, S. Nolte, S. Longhi and A. Szameit, *Phys. Rev. Lett.* **105**, 143902 (2010).
25. W. Zawadzki and B. Lax, *Phys. Rev. Lett.* **16**, 1001 (1966).
26. W. Zawadzki *Optical Properties of Solids*, ed. E. D. Haidemenakis (New York: Gordon and Breach, 1970) p. 179.
27. E. O. Kane, *J. Phys. Chem. Solids* **1**, 249 (1957).
28. P. A. M. Dirac, *The Principles of Quantum Mechanics* (Oxford: Clarendon Press, 1958).
29. W. Zawadzki and T. M. Rusin, *J. Phys. Cond. Matter* **20**, 454208 (2008), (arXiv:0805.0478).
30. Z. F. Jiang, R. D. Li, S. C. Zhang and W. M. Liu, *Phys. Rev. B* **72**, 045201 2005, (arXiv:0410420).
31. Z. Wilamowski, W. Ungier, M. Havlicek and W. Jantsch, arXiv:1001.3746 (2010).
32. E. McCann and V. I. Fal'ko, *Phys. Rev. Lett.* **96**, 086805 2006, (arXiv:0510237).
33. P. R. Wallace, *Phys. Rev.* **71**, 622 (1947).
34. J. C. Slonczewski and P. R. Weiss, *Phys. Rev.* **109**, 272 (1958).

35. G. W. Semenoff, *Phys. Rev. Lett.* **53**, 2449 (1984).
36. K. S. Novoselov, A. K. Geim, S. V. Morozov, D. Jiang, Y. Zhang, S. V. Dubonos, I. V. Grigorieva and A. A. Firsov, *Science* **306** 666 (2004), (arXiv:0410631).
37. M. I. Katsnelson, *Europ. Phys. J. B* **51**, 157 (2006), (arXiv:0512337).
38. W. Zawadzki, *Phys. Rev. B* **74**, 205439 (2006), (arXiv:0510184).
39. J. M. Luttinger and W. Kohn, *Phys. Rev.* **97**, 869 (1955).
40. R. L. Kronig and W. Penney, *Proc. Roy. Soc. London* **130**, 499 (1931).
41. R. A. Smith, *Wave Mechanics of Crystalline Solids* (London: Chapman and Hall, 1961).
42. H. Feschbach and F. Villars, *Rev. Mod. Phys.* **30**, 24 (1958).
43. S. Tani, *Progr. Theor. Phys.* **6**, 267 (1951).
44. M. H. L. Pryce, *Proc. Roy. Soc.* **195 A**, 62 (1948).
45. L. L. Foldy and S. A. Wouthuysen, *Phys. Rev.* **78**, 29 (1950).
46. B. Thaller, arXiv:0409079 (2004).
47. T. M. Rusin and W. Zawadzki, *Phys. Rev. D* **82**, 125031 (2010), (arXiv:1008.1428).
48. B. Thaller, *The Dirac Equation* (Berlin: Springer-Verlag, 1992).
49. E. T. Jaynes and F.W. Cummings, *Proc IEEE* **51**, 89 (1963).
50. L. Lamata, J. Leon, T. Schatz and E. Solano, *Phys. Rev. Lett.* **98**, 253005 (2007), (arXiv:0701208).
51. D. Leibfried, R. Blatt, C. Monroe and D. Wineland, *Rev. Mod. Phys.* **75**, 281 (2003).
52. M. Johanning, A. F. Varron and C. Wunderlich, *J. Phys. B* **42**, 154009 (2009), (arXiv:0905.0118).
53. L. G. Wang, Z. G. Wang and S. Y. Zhu, *EPL* **86**, 47008 (2009).
54. S. Sasabe, *J. Mod. Phys.* **5**, 534 (2014).
55. T. M. Rusin and W. Zawadzki, *J. Phys.: Condens. Matter* **26**, 215301 (2014).
56. T. M. Rusin and W. Zawadzki, *Semicond. Sci. Technol.* **29**, 125010 (2014).
57. G. von Plessen and P. Thomas, *Phys. Rev. B* **45**, 9185 (1992).
58. S. Mukamel, *Principles of nonlinear molecular spectroscopy* (New York: Oxford University Press, 1995).
59. J. Feldmann, K. Leo, J. Shah, D. A. B. Miller, J. E. Cunningham, T. Meier, G. von Plessen, A. Schulze, P. Thomas and S. Schmitt-Rink, *Phys. Rev. B* **46**, 7252(R) (1992).
60. K. Leo, P. H. Bolivar, F. Bruggemann and R. Schwedler, *Sol. St. Comun.* **84**, 943 (1992).

61. V. G. Lyssenko, G. Valusis, F. Loser, T. Hasche, K. Leo, M. M. Dignam and K. Kohler, *Phys. Rev. Lett.* **79**, 301 (2007).
62. R. W. Boyd, *Nonlinear Optics* (San Diego: Academic Press, 2003).
63. R. Saito, M. Fujita, G. Dresselhaus and M. S. Dresselhaus, *Phys. Rev. B* **46**, 1804 (1992).
64. R. Saito, G. Dresselhaus and M. S. Dresselhaus, *Physical Properties of Carbon Nanotubes* (London: Imperial College Press, 1998).
65. A. Wirth, M. Th. Hassan, I. Grguras, J. Gagnon, A. Moulet, T. T. Luu, S. Pabst, R. Santra, Z. A. Alahmed, A. M. Azzeer, V. S. Yakovlev, V. Pervak, F. Krausz and E. Goulielmakis, *Science* **334**, 195 (2011).
66. J. S. Lauret, C. Voisin, G. Cassabois, C. Delalande, Ph. Roussignol, O. Jost and L. Capes, *Phys. Rev. Lett.* **90**, 057404 (2003).
67. T. Yajima and Y. Tara, *J. Phys. Soc. Jpn.* **47**, 1620 (1979).
68. L. Seidner, G. Stock and W. Domcke, *J. Chem. Phys.* **103**, 3998 (1995).
69. B. Wolfseder, L. Seidner, G. Stock and W. Domcke, *Chemical Physics* **217**, 275 (1997).
70. I. Stepanov, M. Ersfeld, A. V. Poshakinskiy, M. Lepsa, E. I. Ivchenko, S. A. Tarasenko and B. Beschoten, to be published.

Printed in the United States
By Bookmasters